Sustainable Development Goals Series

The **Sustainable Development Goals Series** is Springer Nature's inaugural cross-imprint book series that addresses and supports the United Nations' seventeen Sustainable Development Goals. The series fosters comprehensive research focused on these global targets and endeavours to address some of society's greatest grand challenges. The SDGs are inherently multidisciplinary, and they bring people working across different fields together and working towards a common goal. In this spirit, the Sustainable Development Goals series is the first at Springer Nature to publish books under both the Springer and Palgrave Macmillan imprints, bringing the strengths of our imprints together.

The Sustainable Development Goals Series is organized into eighteen subseries: one subseries based around each of the seventeen respective Sustainable Development Goals, and an eighteenth subseries, "Connecting the Goals," which serves as a home for volumes addressing multiple goals or studying the SDGs as a whole. Each subseries is guided by an expert Subseries Advisor with years or decades of experience studying and addressing core components of their respective Goal.

The SDG Series has a remit as broad as the SDGs themselves, and contributions are welcome from scientists, academics, policymakers, and researchers working in fields related to any of the seventeen goals. If you are interested in contributing a monograph or curated volume to the series, please contact the Publishers: Zachary Romano [Springer; zachary.romano@springer.com] and Rachael Ballard [Palgrave Macmillan; rachael.ballard@palgrave.com].

More information about this series at https://link.springer.com/bookseries/15486

Eromose E. Ebhuoma •
Llewellyn Leonard
Editors

Indigenous Knowledge and Climate Governance

A Sub-Saharan African Perspective

Springer

Editors
Eromose E. Ebhuoma
Department of Environmental Sciences
School of Ecological and Human Sustainability, College of Agriculture and Environmental Sciences
University of South Africa (UNISA)
Johannesburg, South Africa

Llewellyn Leonard
Department of Environmental Sciences
School of Ecological and Human Sustainability, College of Agriculture and Environmental Sciences
University of South Africa (UNISA)
Johannesburg, South Africa

ISSN 2523-3084 ISSN 2523-3092 (electronic)
Sustainable Development Goals Series
ISBN 978-3-030-99410-5 ISBN 978-3-030-99411-2 (eBook)
https://doi.org/10.1007/978-3-030-99411-2

This Springer imprint is published by the registered company Springer Nature Switzerland AG
The registered company address is: Gewerbestrasse 11, 6330 Cham, Switzerland

Preface

The impact of climate change has adversely compromised the livelihoods of African communities who are overwhelmingly dependent on the natural environment for food security. Indigenous Knowledge Systems is a way for African people to adapt to climate change and secure their livelihoods by uniquely applying their ancient knowledge systems. The book provides critical discussions on the linkages between indigenous knowledge, climate change, and governance. Using primarily political and participatory methods, the book explores cases within the region in order to enhance our understanding of the importance of indigenous knowledge systems to combat climate change and the factors that undermine indigenous knowledge from featuring predominately in climate change mitigation and governance. There are numerous types of land-related conflicts ranging from ownership, access, use, and management. The chapters in this volume capture the strategies put in place by Indigenous peoples in different geographical regions across Southern Africa to adapt and build their resilience to climatic risks, including how collaborations with scientific knowledge have cascaded into building people's resilience to climatic risks. This volume critically tackles the underlying issues of governance, power, and epistemic injustice that influence the exclusion of Indigenous peoples in climate change decision-making at local and national levels. The geographic spread of this book is wide and covers cases from Cameroon, Ghana, Nigeria, Zimbabwe, and Lesotho as well as regional cases from West, Central, East and Southern Africa. Key insights from this book illuminate issues that could contribute meaningfully towards the actualisation of the 13th Sustainable Development Goal (climate action) in sub-Saharan Africa. This is primarily because Indigenous peoples and climate action are inextricably linked due to their overwhelming dependence on the natural environment for their livelihood.

Johannesburg, South Africa

Eromose E. Ebhuoma
Llewellyn Leonard

Acknowledgements

The Editors of this book, Indigenous Knowledge and Climate Governance: A Sub-Saharan African Perspective, would like to thank all Contributors who agreed to contribute to this volume. The factors undermining indigenous knowledge from featuring prominently in discourses on climate governance and planning have received limited attention in the academic literature. Therefore, it is crucial to call for an open discussion whereby the role of indigenous knowledge in climate change adaptation and mitigation, as well as situating indigenous knowledge in climate governance and planning schemes, will be given a single platform to be extensively explored. We, therefore, salute the Contributors for the courage and effort they took to share their knowledge in addressing the importance of indigenous knowledge in Sub-Saharan Africa while exploring ways to improve the livelihoods of communities for climate adaptation and resilience. Furthermore, the chapters were written during the COVID-19 pandemic and under varying lockdown conditions by different African governments. A number of the Authors and their families experienced illness and bereavement during the writing process. There was also political unrest in some countries. Despite these challenges, the authors continued to preserve and work towards their final chapters. We salute all the authors for their dedication and contributions. We also acknowledge the input made by the anonymous reviewers for the initial book proposal and for their valuable and insightful comments. Similarly, we appreciate the Springer team for their guidance and support during the process. All the chapters were peer-reviewed by two reviewers, in accordance with accepted academic practices. We appreciate and thank them for their time, effort, and valuable input on improving the chapters. Finally, we thank our families and friends for their love and support. For providing the spiritual sustenance needed to pursue a more just society. Llewellyn would like to thank his wife Jessica, son Udhav, and daughter Tulsi Rasa Leonard for their support. Eromose would like to thank *Jehovah El Gibbor* for making this vision come to fruition (Zechariah 4:9). Also, Eromose would like to thank his wife Ejiro for her patience, understanding and undiluted support, especially at times when the editorial duties required my undivided attention. Finally, Eromose is indebted to family and friends for their words of encouragement.

Contents

Editors and Contributors

About the Editors

Eromose E. Ebhuoma (Ph.D.) is a postdoctoral research fellow at the Department of Environmental Sciences, College of Agriculture and Environmental Sciences, University of South Africa (UNISA). His research expertise spans climate change vulnerability and adaptation, climate services, climate change policy and governance, local and indigenous knowledge systems, political ecology, rural livelihoods, and environmental sustainability. Currently, he is a steering committee member of the South African Adaptation Network.

Llewellyn Leonard (Ph.D.) is a Professor at the University of South Africa, Department of Environmental Sciences. He received his Ph.D. in Human Geography from Kings College, University of London, School of Social Science and Public Policy. Research interests include environmental justice; human rights and ecological justice, risk communication; urban risks, climate adaptation; democracy and governance; civil society-state-industry relations; as well as industrial risks and tourism/conservation impacts. His affiliations include the Society of South African Geographers (SSAG), the International Association for Impact Assessments (IAIA), and the Rotary (New Dawn) South Africa.

Contributors

Oluwagbemiga Oluwaseun Ajayi Department of Sociology, Osun State University, Osogbo, Nigeria

Adebayo Emmanuel Akinyemi Department of Sociology, Osun State University, Osogbo, Nigeria;
Department of Sociology, University of Ibadan, Ibadan, Nigeria

Ifeoma Q. Anugwa Department of Agricultural Extension, University of Nigeria, Nsukka, Nigeria

Jane Mbolle Chah Department of Agricultural Extension, University of Nigeria, Nsukka, Nigeria

Nelson Chanza Department of Urban and Regional Planning, University of Johannesburg, Johannesburg, South Africa

Anton de Wit Department of Geosciences, Nelson Mandela University, Port Elizabeth, South Africa

Felix K. Donkor College of Agriculture and Environmental Sciences (CAES), University of South Africa (UNISA), Florida Park, Roodepoort, South Africa

Eromose E. Ebhuoma Department of Environmental Sciences, School of Ecological and Human Sustainability, College of Agriculture and Environmental Sciences, University of South Africa (UNISA), Johannesburg, South Africa

Divine Ewane Department of Animal Science, University of Buea, Buea, Cameroon

Jennifer Fitchett School of Geography, Archaeology and Environmental Studies, University of the Witwatersrand, Johannesburg, South Africa

Collen Kajongwe Department for Human Resource, School of Business Management, Manicaland State University of Applied Sciences, Mutare, Zimbabwe

Tariro Kamuti Centre for Gender and Africa Studies, University of the Free State, Bloemfontein, South Africa

Shorai Kavu Central Development Studies, School of Life Long Development Studies, Chinhoyi University of Technology, Chinhoyi, Zimbabwe

Moye Eric Kongnso Department of Geography and Environmental Planning, University of Dschang, Dschang, Cameroon

Musediq Olufemi Lawal Department of Sociology, Osun State University, Osogbo, Nigeria

Llewellyn Leonard Department of Environmental Sciences, School of Ecological and Human Sustainability, College of Agriculture and Environmental Sciences, University of South Africa (UNISA), Johannesburg, South Africa

Fulco Ludwig Water Systems and Global Change Group, Wageningen University, Wageningen, The Netherlands

Sinoxolo Magaya School of Geography, Archaeology and Environmental Studies, University of the Witwatersrand, Johannesburg, South Africa

Itumenleng Masithela Occupational Safety Department, University of Johannesburg, Johannesburg, Republic of South Africa

Kevin Mearns College of Agriculture and Environmental Sciences (CAES), University of South Africa (UNISA), Florida Park, Roodepoort, South Africa

Ntombizakhe Moyo-Nyoni Lecturer, Department of Peace and Security Studies Department, Faculty of Social Sciences, Midlands State University, Gweru, Zimbabwe

Walter Musakwa Department of Urban and Regional Planning, University of Johannesburg, Johannesburg, South Africa

Emmanuel Nyadzi Water Systems and Global Change Group, Wageningen University, Wageningen, The Netherlands

Andy B. Nyamekye Food and Agriculture Organization of the United Nations, Rome, Italy

Esdras A. R. Obossou Department of Agricultural Extension, University of Nigeria, Nsukka, Nigeria

Mulalo Rabumbulu Department of Geography, Environmental Management and Energy Studies, University of Johannesburg, Johannesburg, Republic of South Africa

Henry B. Tantoh Department of Geography and Planning, University of Bamenda–Cameroon, Bamenda, Cameroon

1 Climate Change and Indigenous Knowledge Systems in Africa: An Overview

Eromose E. Ebhuoma and Llewellyn Leonard

1.1 Introduction

Climate change has severely compromised the livelihoods of people that are overwhelmingly dependent on the natural environment, particularly those in developing countries that are usually the least able to cope with the associated social and environmental risks (Pant 2012; Evans et al. 2014; Intergovernmental Panel on Climate Change (IPCC) 2014; Poudel et al. 2017). Sub-Saharan Africa (SSA) is the region most vulnerable to climate change since warming will be greater than the global average, and agriculture, mainly rain-fed, is the primary source of subsistence for rural communities (Di Falco 2018). In 2019, the regions most severely affected by drought were in Southern Africa and many areas in the same region were affected by a protracted drought in 2014–2016. Temperatures exceeding 2 °C above the 1981–2010 average were recorded in South Africa, Namibia and parts of Angola in 2019. Large areas from the south to the north of the continent were more than 1 °C above normal levels (World Meteorological Organization 2020). Despite this, indigenous peoples in sub-Saharan Africa (SSA), through the skilful application of their knowledge systems, have been able to somewhat maintain their livelihoods, albeit not without challenges (Nyong et al. 2007; Nkomwa et al. 2014; Mafongoya and Ajayi 2017; Ebhuoma 2020). Roughly two out of three people in SSA make a living in agriculture and therefore food production is essential to improving food security and reducing poverty (Berck et al. 2018). Indigenous people possess a portfolio of valuable knowledge crucial to building climate resilience and food security (Nyong et al. 2007; Green and Raygorodetsky 2010; Hiwasaki et al. 2015; Makondo and Thomas 2018). The need for diverse knowledge to tackle climatic risks has ignited interest in understanding how indigenous knowledge systems (IKS) can assist in climate change mitigation and adaptation interventions to facilitate the achievement of the thirteenth United Nation Sustainable Development Goal (SDG 13) with a focus on taking urgent action to combat climate change and its impacts. Climate change adaptation is understood here as the process people employ to reduce the adverse effects of climate change on their livelihood and quality of life and take advantage of the opportunities provided by the changes in climatic conditions (Mugambiwa 2018).

E. E. Ebhuoma (✉) · L. Leonard
Department of Environmental Sciences, School of Ecological and Human Sustainability, College of Agriculture and Environmental Sciences, University of South Africa (UNISA), Johannesburg, South Africa
e-mail: ebhuoee@unisa.ac.za; eromose2012@gmail.com

L. Leonard
e-mail: llewel@unisa.ac.za

E. E. Ebhuoma and L. Leonard (eds.), *Indigenous Knowledge and Climate Governance*, Sustainable Development Goals Series, https://doi.org/10.1007/978-3-030-99411-2_1

SDG 13 is a clarion call for the world to take crucial steps to tackle climate change and its impacts by integrating climate change into national policies, improving climate education and building institutional capacity to scale-up climate change mitigation and adaptation strategies. This also includes a need to strengthen resilience and adaptive capacity to climate-related hazards. Additionally, Africa's Agenda 2063, concluded in 2013, takes cognisance of the fact that climate change is a fundamental challenge for the continent's development and it is already negatively affecting the ability of many African countries to achieve any of the United Nations SDGs or Agenda 2063 aspirations. Some of the reasons for these include the impacts of climate change on gross domestic product and on national budgets; climate impacts on livelihoods, communities, infrastructure and finance; and the large costs of adaptation (World Meteorological Organization 2020). Considering these challenges, there is an urgent need to explore local solutions to tackle the crisis such as IKS to assist in climate change mitigation and adaptation interventions. This is primarily because climate change and sustainable development are two fundamental issues of the twenty-first century that cannot be addressed without urgent collective action (Mthembu and Nhamo 2021). This suggests that there is a need for government officials including civil society organisations to create an enabling environment to ensure IKSs and their ways of knowing how to adapt to climate change are explicitly explored and utilised. The knowledge, sustainable land and resource management systems and practices, cultures and innovations of indigenous peoples contribute to sustainable development. They are closely linked to the fabric of United Nation's Agenda 2030—leaving no one behind—and should therefore be recognised as indispensable in contributing to the achievement of the SDGs (International Fund for Agricultural Development (IFAD) 2019). It is, therefore, not surprising to note that United Nations Educational, Scientific and Cultural Organisation (UNESCO), like other internationally reputable organisations such as the Intergovernmental Panel on Climate Change (IPCC) and UNFCCC, recognise that the involvement of indigenous peoples is fundamental to achieving the ambitious targets of the SDGs (UNESCO 2017).

This recognition stems mainly from the various scholarships that have emerged from both developed and developing countries that have showcased the usefulness of IKSs in enabling households to mitigate and adapt to climate change (Al Mamun et al. 2014; Nkomwa et al. 2014; Chinsembu et al. 2015; Nakashima et al. 2018; Ebhuoma and Simatele 2019). For instance, IKS have been successfully used to adapt to climate change in Mutoko rural district of Zimbabwe. Community-based adaptation by way of using IKS is able to reduce the vulnerability as well as improve the resilience of the local people to climatic variability and change. Subsistence farmers have used IKS and adaptive strategies over the years to assist rural community members to adapt to climate change. In this instance the community no longer grows maize in large quantities, having shifted to millet and sorghum to adapt to climate change. As such the use of IKS, through activities such as crop type change from maize to traditional millet and sorghum re-establishes the community's indigenous practices since they are made to observe the practices of previous years (Mugambiwa 2018). Indigenous people hold valuable knowledge and traditions that provide solutions to major challenges, including those related to sustainable natural resource management, climate resilience, and promoting food systems that provide healthy nutrition for all, which the United Nations SDGs focuses on, primarily because they grapple with these issues on a regular basis. For example, despite being among the most vulnerable people to climate change coupled with limited resources, indigenous people respond in creative ways to build their resilience and adaptive capacity, which is mainly hinged on the principles of social and environmental sustainability (Ebhuoma 2018; IFAD 2019). However, it must be highlighted that the western view of "knowledge" has, since its introduction in Africa, not fully taken into account the holistic nature and approach of non-western ways of indigenous knowledge production despite much propagation about the indignity of the diverse African practices. Furthermore, there has been limited effort among the various African scholars to provide their own clear narrative of knowledge based on Africa's own ideas and intellectual development (Kaya and Seleti 2018). Thus this book in part aims to provide a platform for African voices surrounding IKS and climate change adaptation and to contribute towards a more robust African knowledge system and ways of knowing.

1.1.1 Indigenous Knowledge and Climate Change Governance in Sub-Saharan Africa

In Africa, the usefulness of IKS in scaling up adaptation to climate change has been widely acknowledged. In Ngamiland district in Botswana, for example, the indigenous peoples practice flood recession farming to grow their food. This practice involves maximising the moisture left by the receding flood water and planting crops like maize, millet and sorghum that mature within a short timeframe. To maximise output, they ploughed the regions in their fields where the grasses were tall, an indication of good quality soil for optimal food production (Motsumi et al. 2012). A somewhat similar climate change adaptation strategy has been employed in farmlands that experience seasonal flooding annually in Nigeria (Ebhuoma and Simatele 2017), Ghana (Sidibé et al. 2016), Senegal (Saarnak 2003) and Mali (Traore et al. 2020) respectively. Also, Kwoyiga and Stefan (2018) revealed that indigenous knowledge helped farmers in North East Region of Ghana to locate groundwater points and determine the source and period of groundwater recharge for irrigation. To achieve this, they rely on indicators not limited to shrubs, broom plants, trees (*Ficus gnaphalocarpa*), thorns/thorn grasses, and black ants to locate groundwater availability. In a related vein, indigenes in the Dodoma city in Tanzania (Shemsanga) and Borana and Konso communities in Ethiopia (Behailu et al. 2016) have applied their unique knowledge to exploit groundwater through the identification of suitable well and pond sites. Further, Cheserek (2005) explained how village headsmen and leaders in the Kerio Valley in Kenya applied their local knowledge to manage water under harsh environmental conditions.

A remarkable feature of these ingenious climate change adaptation techniques is that they are carved out by people with low levels of formal education. Although low levels of Western education has been identified as a contributing factor to weak adaptive capacity of households to climate variability and change (Loevinsohn 2009), scholars have illustrated how the indigenous peoples in SSA skilfully maximise their meagre resources to outwit the odds of climate variability and change to grow their food (Mavhura et al. 2013; Nkomwa et al. 2014; Ebhuoma and Simatele 2017). Thus, the featuring of IKS in climate governance has been presented as essential in rural communities because of their history, values and belief systems. To appreciate the value of IKS in climate governance, it must be holistic, empowering and participatory by nature involving all major organs of the African indigenous society and their supporting entities. This involves recognising African culture and conceptions of reality (Mugambiwa 2021).

Despite its usefulness, the stereotyping of IKS as archaic (e.g. Mapara 2009) from time immemorial has created a stumbling block for indigenous people to be highly esteemed as holders of knowledge that is valuable and actionable. However, a paradigm shift regarding this framing has been observed in the last few decades among scholars whose ideologies are underpinned by Western scientific knowledge (e.g. Green and Raygorodetsky 2010). Though, there is a need for community-oriented sciences and the integration of Western scientific expertise with local community knowledge (Leonard and Lidskog 2020). Such an integration of knowledge systems can have benefits. For example, Ubisi et al. (2020) examined the role of IKS in Mpumalanga, South Africa. The study established the reliability, use and existence of indigenous knowledge in weather prediction among smallholder farmers. The local climate indicators used by the smallholders for weather and climate predictions were identified and the farmers use animal, plant and atmospheric indicators as well as human ailments to predict the weather. Farmers considered the use of IKS at a local level essential to adapting to climate variability and change and for weather predictions. Smallholders were also of the view that their local climate indicators were better than scientific weather forecasts due to the difficulty in accessing, understanding and interpreting scientific weather forecasts. However, indigenous knowledge was also in jeopardy since local climate indicators such as birds and plant species have been disappearing due to climate change. In this case the integration of IKS with modern climate science was considered useful (Ubisi et al. 2020). Yet, the infiltration of IKS into climate change governance schemes on the African continent has, to a large extent, been at a snail's pace. This is against the backdrop of the burgeoning recognition that indigenous peoples and climate action are inextricably linked due to their overwhelming dependence on the natural environment for their livelihoods (IFAD 2019). Arguably, the dearth of IKS in climate change governance initiative is partly responsible for the weak policies and adaptation strategies in localised context, which is often not tailored to adequately address the realities of local people. According to Chanza and De Wit (2016), the application of climate governance concept by the technocratic divide (policymakers and climate practitioners) to communities facing climate change impacts needs further development.

Indigenous peoples in Africa continue to face setbacks in ensuring that their knowledge and practices are enshrined in climate change governance and planning agendas aimed at mitigation and adaptation (Ratelle and Paquette 2019). These setbacks, which are fuelled by hegemonic power dynamics (Agrawal 1995) and misleading political decisions (Magni 2017), often result in the exclusion of IKS from climate change governance schemes and threaten the backbone of Agenda 2030. Failure to consider IKS when formulating climate governance schemes, we argue, could result in ill-informed and unintended policy outcomes that could jeopardise the achievement of SDG 13, whereas an accurate accounting of IKS in such schemes could inform more sustainable policy solutions. A succinct analysis of several national policies on climate change in SSA including Ghana, Zimbabwe, Kenya, Mali, South Africa and Lesotho, for example, highlight the role of IKS in boosting household resilience. Thus, in theory, while it appears that IKS seems to be gaining momentum in its acknowledgement as holders of valuable knowledge, in practice this may not necessarily be the case (Latulippe and Klenk 2020; Wheeler and Root-Bernstein 2020). For example, indigenous food production systems contribute significantly to enhancing agricultural productivity as well as guaranteeing food security and therefore governments should incorporate IKS into agricultural research and community development programmes for the conservation of the environment, promotion of sustainable agriculture and the general natural resource management systems (Muthee et al. 2019). Proponents of this vital knowledge strand may feel aggrieved by the fact that IKS may be miles away from infiltrating climate change policies and governance schemes particularly in SSA. This notion laid the foundation on which this book is based, namely to underline the reasons for its underrepresentation in climate change governance initiatives in Africa as well as its potential contribution when injected into climate change governance policies and schemes. In so doing, this volume aims to critically tackle the underlying issues of governance, power and epistemic injustices that influence the exclusion of indigenous peoples and their knowledge systems in climate change decision-making processes both at both local and national levels. The uniqueness of this book stems from the fact that to our knowledge, no volume has focused extensively on issues pertaining to factors undermining IKS from featuring prominently in climate change governance in SSA.

1.1.2 The Comprehensive Gap on Indigenous Knowledge Systems and Climate Governance

Several books have been published regarding the state of affairs of IKS in SSA. For example, Mawere (2014) and Oloruntoba et al. (2020) focused on the role of IKS in Africa's development, while Ezeanya-Esiobu (2019) presents critical philosophical and practical argument on the need for IKS to play a vital role in Africa's curricula development. In the context of climate change, some authors have either partly (Mawere 2014; Chinsembu et al. 2015; Katerere et al. 2021) or extensively (Mafongoya and Ajayi 2017; Fre 2018) explored the role of IKS in facilitating adaptation to climate change in specific geographic locations in SSA. Others have demystified the potential for IKS to facilitate development in Africa (e.g. Ngulube 2020; Ezeanya-Esiobu 2019; Oloruntoba et al. 2020). However, what is missing is a volume that collectively captures the strategies put in place by indigenous peoples in different geographical regions across SSA to adapt and build their resilience to climatic risks. Furthermore, the factors undermining IKS from featuring prominently in discourses on climate governance and planning have received limited attention in the academic literature, while no book has collectively showcased this issue. This focus is urgently needed if nations are to successfully work towards achieving SDG 13 to effectively mitigate and adapt to climate change impacts on the African continent. Thus, it is crucial to call for a frank discussion whereby the role of IKS in climate change adaptation and mitigation, as well as situating IKS in climate governance and planning schemes, will be given a single platform to be extensively explored. In so doing, this volume will critically tackle the underlying issues of governance, power and epistemic injustices that influence the exclusion of IKS in climate change decision-making both at local and national levels. This book covers regional cases (both primary and secondary data) from West, Central, East and Southern Africa (Fig. 1.1). The lead contributors to this book—all African scholars—hail from various disciplines such as environmental sociology, human and physical geography, rural development, religious studies; rural energy and peace studies; climatology; tourism geography and planning and development.

The book showcases the role of IKS in facilitating adaptation to climate variability and change, and also demystifies the challenges that prevent it from being integrated with scientific knowledge in climate governance schemes. The importance of indigenous knowledge in scaling up adaptation to climate change has also been explored. Indigenous people and their knowledge rarely feature when decision-makers prepare for future climate change. African solutions need to be part of combating and adapting to climate change. This book showcases how indigenous knowledge facilitates adaptation to climate change, including how collaborations with scientific knowledge have cascaded into building people's resilience to climatic risks. Careful attention is paid to the factors fuelling epistemic injustice towards indigenous knowledge, which prevents it from featuring in climate governance schemes across sub-Saharan Africa. Key insights from this book may illuminate issues that could contribute meaningfully towards the actualisation of the UN SDG 13 and promote mechanisms for raising capacity for effective climate change-related planning and management in sub-Saharan Africa. The valuable insights in this book will be of interest to climate change decision-makers as well as researchers and practitioners across African Studies, Geography, Rural and Urban Development.

1.1.3 Structure of the Book

The book explores three main themes. Theme 1, "The role of IKS in climate change mitigation and adaptation", comprehensively documents how indigenous peoples from different countries in sub-Saharan Africa skilfully apply their IKS to adapt to climate change while in pursuance of their livelihoods. It explores how IKS has facilitated mitigation and adaptation to climate

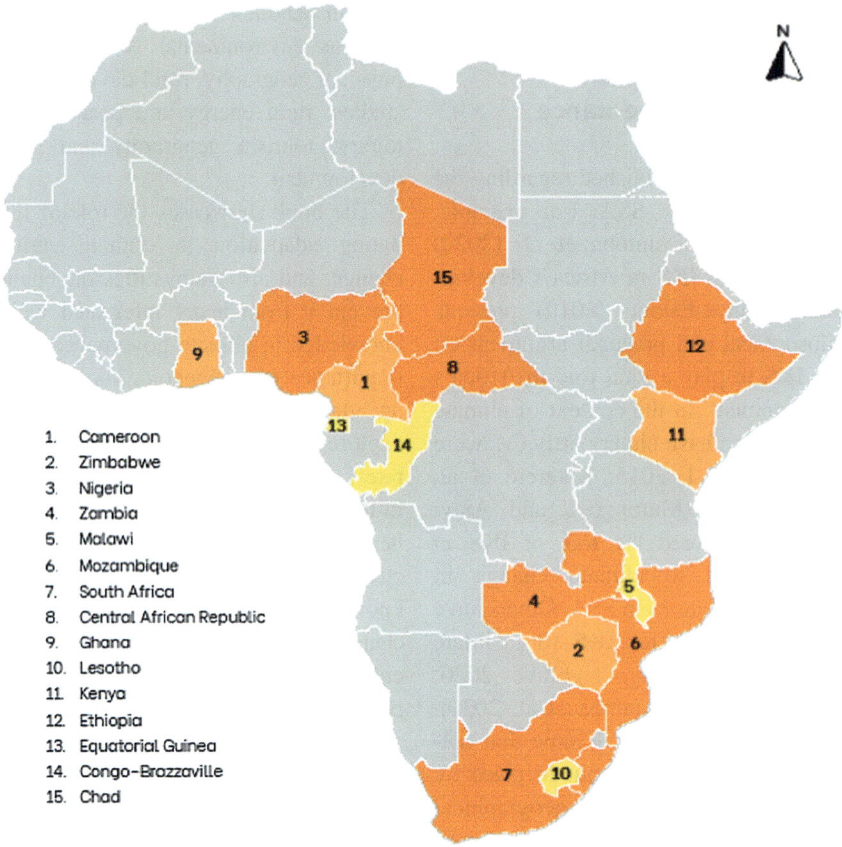

Fig. 1.1 Map of sub-Saharan Africa highlighting the locations of case studies covered in this book. *Source* Editors (2022)

change and interrogates the interventions put in place by indigenous people to ensure the preservation of the natural environment that caters to their livelihood. It provides valuable insights in ensuring that policies aimed at effective climate change governance across local scales are injected. In Chap. 2, Moye Eric Kongnso explores indigenous adaptation of pastoralists to climate variability and range land management in Cameroon. In Chap. 4, Musediq Olufemi Lawal, Oluwagbemiga Oluwaseun Ajayi and Adebayo Emmanuel Akinyemi examine the influence of indigenous spiritual beliefs in natural resources management and climate change mitigation among the Yorùbás in Nigeria. In Chap. 5, Henry Tantoh, Eromose Ebhuoma and Llewellyn Leonard investigate indigenous perspectives on gender roles, climate change and community-based adaptation in Central Africa. In Chap. 6, Shorai Kavu, Collen Kajongwe and Eromose Ebhuoma assess the role of IKS in climate change adaptation and mitigation in Southern Africa. In Chap. 3, Ntombizakhe Moyo-Nyoni examines how adopting IKS may enhance peace education programmes for climate change and adaptation in Zimbabwe. In Chap. 7, Sinoxolo Magaya and Jennifer Fitchett explore approaching positionality in research on indigenous knowledge.

Theme 2, "IKS and scientific knowledge collaboration to scale up climate resilience", acknowledges the numerous calls to integrate IKS and scientific knowledge to build people's resilience to climate change. It also recognises the limited literature documenting how such partnerships have yielded the desired fruit. This theme aims to shed light on the ways such collaborations in African communities have catalysed indigenous people's resilience to climate change, and seeks to highlight the factors that undermine the process. By showcasing the value of co-producing knowledge and theorising what different localised environments in various sub-Saharan African countries stand to benefit when IKS are integrated with scientific knowledge, this section helped to concretise the need for IKS to feature in climate change governance schemes in Africa. In Chap. 8, Emmanuel Nyadzi, Andy Nyamekye and Fulco Ludwig explore how to make climate services actionable for farmers in Northern Ghana by reflecting on the value of co-production and knowledge integration. In Chap. 9, Mulalo Rabumbulu and Itumeleng Komane explore the reconciling of indigenous and scientific knowledge to address the issue of climate change and land degradation in Lesotho. In Chap. 10, Llewellyn Leonard examines how corporate scientific expertise and privatised seeds are destroying IKS for food security and climate change mitigation in sub-Saharan Africa.

Theme 3, "Situating IKS in climate change governance and planning", concedes that IKS is widely recognised as a knowledge that can facilitate the attainment of the sustainable development goals (SDGs), yet, this knowledge, with its unique epistemological underpinning, hardly features in climate change governance and planning schemes across Africa. This theme examines the factors that impede the inclusion of IKS in climate governance and planning discourses. In so doing, this theme captures how climate governance is politicised to the detriment of indigenous people, among other issues. It highlights how the inclusivity of IKS in climate governance and planning has ramped up climate resilience in situations where such stories exist. In Chap. 14, Felix Kwabena Donkor and Kevin Mearns explore how IKS can be leveraged as a vehicle for enhanced climate change adaptation and governance. In Chap. 12, Tariro Kamuti examines the integration of IKS in the Climate Change Policy of Zimbabwe. In Chap. 13, Jane Mbolle Chah, Obossou Esdras, Eromose Ebhuoma, Ifeoma Quinette Anugwa and Divine Ewane explore the possibility of enhancing climate change adaptation through IKS and local governance in Africa. In Chap. 11, Nelson Chanza, Walter Musakwa and Anton de Wit explore the prospects for strengthening adaptation governance through IKS in SSA. In Chap. 15, Eromose Ebhuoma examined the factors fuelling the underrepresentation of indigenous knowledge in climate adaptation policies in East Africa. In Chap. 16, by drawing on key highlights of the chapters in this volume, the editors outline the implications of IKS for climate change adaptation and governance. Also, the editors argue vehemently on possible consequences for the achievement of SDG 13 if climate practitioners fail to utilise IKS and suggests future research directions.

The chapters presented in this book will be of interest to a global scholarly audience of policymakers and rural development planners, geographers, students and lecturers as well as agricultural managers. This volume can also be used as a main course textbook or a supplementary book, and includes a variety of rich cases from sub-Saharan Africa. It raises important insights into dealing with how IKS can be used to enhance governance and better adapt to climate risks and for the integration of IKS with scientific knowledge.

References

Agrawal A (1995) Dismantling the divide between indigenous and scientific knowledge. Dev Chang 26:413–439

Al Mamun MA, Shultana R, Rana MM, Mridha AJ (2014) Climate change adaptation strategies through Indigenous knowledge system: aspect on agro-crop production in the flood prone areas of Bangladesh. Asian J Agric Rural Dev 4:42–58

Behailu BM, Pietilä PE, Katko TS (2016) Indigenous practices of water management for sustainable services: case of Borana and Konso, Ethiopia. SAGE Open, 6

Berck C, Berck P, Di Falco S (2018) Agricultural adaptation to climate change in Africa. Resources of the Future Press, New York

Chanza N, De Wit A (2016) Enhancing climate governance through indigenous knowledge: case in sustainability science. S Afr J Sci 112(3–4):1–7

Cheserek G (2005) Indigenous knowledge in water and watershed management: "Marakwet" conservation strategies and techniques. FWU Water Resour Publ 3:25–33

Chinsembu K, Cheikhyoussef A, Mumbengegwi D, Kandawa-Schulz M, Kasanda CD, Kazembe L (eds) (2015) Indigenous knowledge of Namibia. University of Namibia Press, Windhoek, Namibia

Di Falco S (2018) Adapting to climate change in sub-Saharan Africa. In: Berck C, Berck P, Di Falco S (2018) Agricultural adaptation to climate change in Africa. Resources of the Future Press, New York

Ebhuoma E, Simatele D (2017) Defying the odds: climate variability, asset adaptation and food security nexus in the Delta State of Nigeria. Int J Disaster Risk Reduct 21:231–242

Ebhuoma EE (2018) Underlying the ethical principle that drives indigenous people to pursueenvironmental sustainability when adapting to climate change. https://icsd.wpengine.com/wpcontent/uploads/2018/11/Eromose-Ebhuoma.pdf. Accessed: 25 June 2021.

Ebhuoma EE, Simatele MD (2019) We know our Terrain: indigenous knowledge preferred to scientific systems of weather forecasting in the Delta State of Nigeria. Climate and Development, 11(2):112–123. https://doi.org/10.1080/17565529.2017.1374239

Ebhuoma EE (2020) A framework for integrating scientific forecasts with indigenous systems of weather forecasting in southern Nigeria. Development in Practice, 30(4):1–13.

Evans L, Milfont TL, Lawrence J (2014) Considering local adaptation increases willingness to mitigate. Glob Environ Chang 25:69–75

Ezeanya-Esiobu C (2019) Indigenous knowledge and education in Africa. In: Frontiers in African business research series. Springer Publishers, USA

Fre Z (2018) Indigenous knowledge sovereignty among African cattle herders. University College London Publishers, London, United Kingdom

Green D, Raygorodetsky G (2010) Indigenous knowledge of a changing climate. Clim Change 100:239–242

Hiwasaki L, Luna E, Syamsidik MJA (2015) Local and Indigenous knowledge on climate-related hazards of coastal and small island communities in Southeast Asia. Clim Change 128:35–56

International Fund for Agricultural Development (IFAD) (2019). Partnering with indigenous peoples for the SDGs. https://www.ifad.org/documents/38714170/41390728/policybrief_indigenous_sdg.pdf/e294b690-b26c-994c-550c-076d15190100. Accessed 21 Aug 2021

Intergovernmental Panel on Climate Change (IPCC) (2014). Africa.https://archive.ipcc.ch/pdf/assessment-report/ar5/wg2/WGIIAR5-Chap22_FINAL.pdf. Accessed 21 Aug 2021

Katerere DR, Applequist W, Aboyade OM, Togo C (eds) (2021) Traditional and indigenous knowledge for the modern era: a natural and applied science perspective. Taylor and Francis Limited

Kaya H, Seleti Y (2018) African indigenous knowledge systems and relevance of higher education in South Africa. Int Educ J 12(1):30–44

Kwoyiga L, Stefan C (2018) Groundwater development for dry season irrigation in North East Ghana: the place of local knowledge. Water 10:1724

Latulippe N, Klenk N (2020) Making room and moving over: Knowledge co-production, Indigenous knowledge sovereignty and the politics of global environmental change decision-making. Curr Opin Environ Sustain 42:7–14

Leonard L, Lidskog R (2020) Industrial scientific expertise and civil society engagement: reflexive scientisation in the South Durban Industrial Basin South Africa. J Risk Res. https://doi.org/10.1080/13669877.2020.1805638

Loevinsohn M (2009) Seasonal hunger, the 2001–03 famine and the dynamics of HIV in Malawi. In: Seasonality revisited conference 8–10 July IDS, UK

Mafongoya PL, Ajayi OC (eds) (2017) Indigenous knowledge systems and climate change management in Africa. CTA, Wageningen, The Netherlands, 316 pp

Magni G (2017) Indigenous knowledge and implications for the sustainable development agenda. Eur J Educ 52:437–447

Makondo CC, Thomas DSG (2018) Climate change adaptation: linking Indigenous knowledge with western science for effective adaptation. Environ Sci Policy 88:83–91

Mapara J (2009) Indigenous knowledge systems in Zimbabwe: Juxtaposing postcolonial theory. J Pan Afr Stud 3(1):139–155

Mavhura E, Manyena SB, Collins AE, Manatsa D (2013) Indigenous knowledge, coping strategies and resilience to floods in Muzarabani, Zimbabwe. Int J Disaster Risk Reduct 5:38–48

Mawere M (2014) Culture, indigenous knowledge and development in Africa: reviving interconnections for sustainable development. Langaa Research and Publishing Common Initiative Group, Bamenda, Cameroon

Mthembu D, Nhamo G (2021) Landing the climate SDG into South Africa's development trajectory: mitigation policies, strategies and institutional setup. Sustainability. 13:2991. https://doi.org/10.3390/su13052991

Mthembu DE, Nhamo G (2021) Domestication of the UN Sustainable Development Goals in South Africa. South African Int Affairs, 28(1):1–28. https://doi.org/10.1080/10220461.2021.1894971.

Motsumi S, Magole L, Kgathi D (2012) Indigenous knowledge and land-use policy: implications forlivelihoods of flood recession farming communities in the Okavango Delta, Botswana. Physics and Chemistry of the Earth. 50:185–195. https://doi.org/10.1016/j.pce.2012.09.013

Muthee D, Kilemba G, Masinde J (2019) The role of indigenous knowledge systems in enhancing agricultural productivity in Kenya. East Afr J Contemp Res 1 (1):22–33

Mugambiwa S (2018) Adaptation measures to sustain indigenous practices and the use of indigenous knowledge systems to adapt to climate change in Mutoko rural district of Zimbabwe. J Disaster Risk Stud 10:388

Mugambiwa S (2021) Evoking the epistemology of climate governance through indigenous knowledge systems for sustainable development in rural Zimbabwe. J Disaster Risk Stud 13(1):a1024

Nakashima D, Krupnik I, Rubis JT (eds) (2018) Indigenous knowledge for climate change assessment and adaptation. Cambridge University Press and UNESCO, Cambridge and Paris

Ngulube P (2017) Embedding indigenous knowledge in library and information science education in Anglophone eastern and southern Africa. Handbook of research on social, cultural and educational considerations for indigenous knowledge in developing countries. IGI Global, Hershey PA, pp 92–115

Ngulube P (ed) (2020) Handbook of research on theoretical perspectives on indigenous knowledge systems in developing countries. IGI Global Publishers, USA

Nkomwa EC, Joshua MK, Ngongondo C, Monjerezi M, Chipungu F (2014) Assessing indigenous knowledge systems and climate change adaptation strategies in agriculture: a case study of Chagaka Village, Chikhwawa, Southern Malawi. Phys Chem Earth 67–69:164–172

Nyong A, Adesina F, Osman Elasha B (2007) The value of indigenous knowledge in climate change mitigation and adaptation strategies in the African Sahel. Mitig Adapt Strat Glob Change 12:787–797

Oloruntoba SO, Afolayan A, Yacob-Haliso O (eds) (2020) Indigenous knowledge systems and development in Africa. Palgrave Macmillan Publishers, Switzerland

Pant KP (2012) Climate change and food security in Nepal. J Agric Environ, Futur Food: J Food, Agric Soc 5(2):9–19. https://doi.org/10.3126/aej.v13i0.7582

Poudel S, Funakawa S, Shinjo H (2017) Household perceptions about the impacts of climate change on food security in the mountainous region of Nepal. Sustainability 9:641

Ratelle M, Paquette F (2019) Climate policy needs to reflect resilience of northern Indigenous communities. https://www.climatechangenews.com/2019/11/25/climate-policy-needs-reflect-resilience-northern-Indigenous-communities/. Accessed 10 Aug 2020

Saarnak NL (2003) Flood recession agriculture in the Senegal River Valley. Geogr Tidsskr-Dan J Geogr 103 (1):99–113. https://doi.org/10.1080/00167223.2003.10649483

Shemsanga C, Muzuka ANN, Martz L, Komakech H, Mcharo E (2018) Indigenous knowledge on development and management of shallow dug wells of Dodoma Municipality in Tanzania. Appl Water Sci 8:59. https://doi.org/10.1007/s13201-018-0697-7

Sidibé Y, Williams TO, Kolavalli S (2016) Flood recession agriculture for food security in Northern Ghana. IFAD Working Paper 42

Traore K, Traore B, Synnevåg, Aune JB (2020) Intensification of sorghum production in flood recession agriculture in Yelimane, Western Mali. Agronomy 10:726.https://doi.org/10.3390/agronomy10050726

Ubisi R, Kolanisi U, Jiri O (2020) The role of indigenous knowledge systems in rural smallholder farmers' response to climate change: case study of Nkomazi local municipality, Mpumalanga, South Africa. J Asian Afr Stud 55(2):273–284. https://doi.org/10.1177/0021909619874824

United Nations Educational, Scientific and Cultural Organisation (UNESCO) (2017) Culture: at the heart of the SDGs. https://en.unesco.org/courier/april-june-2017/culture-heart-sdgs. Accessed 21 Aug 2021

Wheeler HC, Root-Bernstein M (2020) Informing decision-making with Indigenous and local knowledge and science. J Appl Ecol 57:1634–1643. https://doi.org/10.1111/1365-2664.13734

World Meteorological Organization (2020) State of the climate in Africa 2019. https://library.wmo.int/doc_num.php?explnum_id=10421. Accessed 14 Sep 2021

Eromose E. Ebhuoma (Ph.D.) is a postdoctoral research fellow at the Department of Environmental Sciences, College of Agriculture and Environmental Sciences, University of South Africa (UNISA). His research expertise spans climate change vulnerability and adaptation, climate services, climate change policy and governance, local and indigenous knowledge systems, political ecology, rural livelihoods and environmental sustainability in sub-Saharan Africa. He is a steering committee member of the South African Adaptation Network.

Llewellyn Leonard (Ph.D.) is Professor at the Department of Environmental Science, School of Ecological and Human Sustainability, University of South Africa. Before joining academia, he worked for a human rights environmental organisation working to support vulnerable communities exposed to environmental risks to shape policy and has worked to link communities in Africa, Asia, Latin America and Europe to engage in solidarity and networking. Previously he served as Vice Dean: Research at the University of Johannesburg, College of Business and Economics. His research interest include environmental justice, governance, democracy and human rights, civil-society-state-industry relations, urban risks, sustainable development; risk society and political economy/ecology to name a few. He sits on the editorial board of the Southern African Geographic Journal.

Part I

The Role of IKS in Climate Change Mitigation and Adaptation

2 Indigenous Adaptation of Pastoralists to Climate Variability and Range Land Management in the Ndop Plain, North West Region, Cameroon

Moye Eric Kongnso

2.1 Introduction

Climate variability and change have a severe impact on agro-pastoral systems, making the future uncertain for pastoralists in sub-Saharan mountains (Herrero et al. 2016). This uncertainty warrants that adaptation systems be reviewed and upgraded to lessen the negative impact of weather aberrations. Nakashima et al. (2012) and Herrero et al. (2016) determined that pastoralist communities in sub-Saharan Africa are affected by climate change in terms of rangelands, livestock and water, with extended repercussions on income and food security. Agro-pastoral systems are highly vulnerable, and adapting to change requires multiple and simultaneous responses. According to the Intergovernmental Panel on Climate Change (IPCC) (2014), adaptation implies adjustment to an actual or expected event to lessen negative impacts or exploit opportunities.

In Africa, indigenous pastoralist communities are nomadic, and mobility was a strategy to lessen the population's negative impact on the environment. According to the IPCC (2014), indigenous knowledge is an invaluable basis for responding to climate and other environmental changes, which is vital for the development of natural resource management strategies. Pastoralists have an array of traditional practices and indigenous knowledge acquired from their constant interaction with their environment. They have also developed empirical notions due to their relentless attempts to improve their quality of life (Indigenous People of Africa Coordinating Committee (IPACC) 2012; Aida et al. 2018; Salifu 2020).

Effective adaptation requires a combination of scientific and indigenous knowledge of climate policy orientations and rangeland management policies (IPACC 2012). However, pastoralists are a marginalised minority population, sensitive to the impact of climate change due to their natural resource-based livelihoods and occupation of marginal lands (Nakashima et al. 2012). Nonetheless, studies have shown that pastoralists have a long history of adapting to harsh environmental conditions such as droughts, but their efforts are undermined by inappropriate policies and strong competition for natural resources (Aida et al. 2018). Most of these traditional practices and empirical experiences have not been considered when drafting adaptation policies. This is crucial since indigenous knowledge, traditional systems of rangeland management and governance, as well as social networks, are essential to collectively respond to environmental change and build resilience among vulnerable communities.

M. E. Kongnso (✉)
Department of Geography and Environmental Planning, University of Dschang, Dschang, Cameroon
e-mail: moyeeric@yahoo.com

E. E. Ebhuoma and L. Leonard (eds.), *Indigenous Knowledge and Climate Governance*, Sustainable Development Goals Series, https://doi.org/10.1007/978-3-030-99411-2_2

The terms 'indigenous knowledge', 'traditional ecological knowledge', 'local knowledge' and 'local ecological knowledge' have been used by different authors to describe knowledge that is location-specific, acquired through long-term observation and interaction with the environment, transferred through oral traditions from generation to generation (Muga et al. 2021; Ezeala and Mbalisi 2021). In this particular work, 'indigenous knowledge' is considered as a set of 'know-how', practices, and beliefs evolving through an adaptive process in response to environmental and climate changes. Though it is time, place, and culture-specific (Speranza et al. 2009), this body of knowledge can be documented and transmitted within and between communities. Some 50 million herders and up to 200 million agro-pastoralists live in arid and semi-arid zones in Africa and rely on indigenous knowledge systems (IKS) to secure pastures for their livestock (De Haan 2016). Therefore, the integration of IKS and modern scientific knowledge is indispensable in climate change adaptation to ensure sustainable livelihoods (IPACC 2012).

The North West Region of Cameroon is actively involved in cattle rearing due to its rich natural and human resources. More than 60% of the region is viable for livestock production with 1,054,914 hectares of potential natural pastures (Ngalim 2015). Production systems vary from extensive nomadic rearing to systems of intensive ranching. Transhumance pastoralism, which is the seasonal movement of herdsmen with cattle from hill slopes and highlands to valleys/lowlands in search of pasture and water, has been embraced as a strategy to cope with recent environmental changes in the Ndop Plain. Despite the rich natural and human resources in the above-mentioned region, certain community members (or pastoralists) are often marginalised and prevented from accessing these resources due to stringent land tenure systems. Mbororo pastoralists are one such marginalised group in North West Cameroon. Due to this marginalisation and under-representation in rangeland management, termed the 'Mbororo problem' by Jibaru (2017), Mbororo pastoralists have been facing enormous challenges in ensuring they obtain a livelihood. Generally, actions to secure rangeland resources and improve rangeland management are limited. Nevertheless, a study by Azuhnwi et al. (2017) has indicated efforts are being made to improve land tenure systems in the relevant area, resource security and rangeland management. Measures to improve rangeland management include drafting the pastoral code, raising awareness, networking, as well as lobbying and litigating where a violation of statutory laws occur (Azuhnwi et al. 2017).

According to the Coalition of European Lobbies for Eastern African Pastoralism (CELEP) (2016), sub-Saharan Africa's drylands, which make up 60% of its total land area, are home to more than 110 million agro-pastoralists, and the achievement of the SDG requires tailored interventions in pastoral areas. Achieving SDG 13 (climate action) thus demands the integration of IKS into climate change adaptation strategies. However, despite efforts such as training pastoralists on the modern methods of rearing, sensitising and educating herders, and the alliance farming programme established by the Mbororo Social and Cultural Development Association (MBOSCUDA), the Mbororo communities' role in management strategies has received little attention in Cameroon's grazing zones. Against this backdrop, this work sets out to investigate the adaptation measures adopted by Mbororo pastoralists, their adaptation constraints, as well as contributions in achieving SDG 13.

This chapter proceeds in four parts. The first part provides a snapshot of public participation in environmental management in Cameroon, and briefly underlines the country's climate policy and rangeland governance in the Ndop Plain. The next part provides a succinct description of the study areas and methodology. The third part presents the findings, which are discussed focusing on how the achievement of SDG 13 can be facilitated. The final part is the conclusion, which provides policy implication guidelines based on the chapter's findings.

2.2 Public Participation in Environmental Decision-Making in Cameroon

2.2.1 Climate Change Policies and Environmental Protection in Cameroon

The Cameroon government has signed and ratified many international treaties and conventions and has implemented laws related to climate change, environmental protection, and the protection of the rights of minority groups and indigenous people. The United Nations Framework Convention on Climate Change's (UNFCCC) 16th Conference of Parties (COPs 16) in Cancun, Mexico, in 2010, encouraged developing countries to integrate climate change adaptation into national development planning systems. Inspired by this, Cameroon developed a National Adaptation Plan for Climate Change (NAPCC) aimed at reducing the vulnerability of communities and limiting climate-driven risks across all sectors.

In its implementation, the Ministry of Environment, Nature Protection and Sustainable Development (MINEPED)—in collaboration with non-governmental organisations (NGOs), such as Global Water Partnership, the German Technical Cooperation (GIZ), and the United Nations Development Programme (UNDP)—developed a road map, financial and human resources for the implementation of the NAPCC in the five agro-ecological zones of Cameroon. Among the numerous axes of NAPCC, there are projects targeted at improving local land governance in response to climate change, reducing livestock's vulnerability to climate change, and strengthening local communities' capacity to ensure their participation in the adaptation process. Equally, through the Presidential Decree No. 2009/410 of 10 December 2009, the Cameroon government created a National Climate Change Observatory that provides climate change-related information and supports local organisations' climate change adaptation and mitigation efforts.

In the domain of legislation on environmental governance, section 9 of law No. 96/12 stipulates the rights and responsibilities of everyone to protect and safeguard the environment. To do so, all parties, especially local and indigenous communities, must participate in decision-making. It equally obliges the State to provide information to the people and build their capacities through education and training. In the same light, law No. 76/166 provides a framework for rangeland management and land governance. Specifically, the management of rangeland in Cameroon is enshrined in what is called the 'agro-pastoral code' (Azuhnwi et al. 2017).

2.2.2 The Agro-Pastoral Code and Rangeland Governance in the Ndop Plain

Climate change, population pressure, and other environmental stressors have limited access to agro-pastoral resources and lead to harmful competition and conflicts between herders and crop farmers (Ngalim 2015). Actions to ensure sustainable livelihoods for minority pastoralists must include issues of rangeland governance and land tenure systems. Since the colonial era, rangeland management was regulated by the land tenure legislation; but from 2009, the Ministry of Livestock, Fisheries and Animal Industries (MINEPIA) initiated procedures to add a legal dimension on the management of agro-pastoral resources to the pastoral code. As indicated by Django et al. (2011), the land tenure legislation started in 1974 created a number of land ordinances:

- Ordinance No. 74-1 to establish rules governing land tenure in Cameroon.
- Ordinance No. 74-2 to establish rules governing State lands.
- Ordinance No. 74-3 to establish expropriation procedures for public purposes and the terms of compensation.

These ordinances were supplemented by several decrees, including:

- Decree No. 76-156 to establish the conditions for obtaining land certificates.
- Decree No. 86-167 to establish the terms and conditions for the management of private properties of the State.
- Decree No. 76/420 of 14 September 1976, amended by Decree No. 86/755 24 June 1986, regulating cattle movement and the protection of transhumance corridors.

Looking at the legal provisions for the management of grazing land, it was noted that grazing land is managed like other national lands. The management of grazing land is placed under the Land Consultative Board and the Agro-pastoral Commission; these structures have specific functions, as presented in Table 2.1.

These legal provisions and commissions have implications for the management of pastoral resources. Although the Mbororo has a special pastoral commission in MINEPIA, a critical look at the Land Consultative Board and the Agro-pastoral Commission demonstrates they are under-represented compared to crop farmers. Regarding the Ndop Plain, the land tenure systems provide for a mixed-farming zone, and, as such, transhumant pastoralism is practised on rice fields. Apart from climatic aberrations that have disrupted the pastoral calendar, other environmental stressors such as bushfires, over-cultivation, occupation of marshy zones, and the cultivation of off-season crops have led to farmers' permanent occupation of mixed-farming zones, to the detriment of pastoralists. Attempts to herd cattle to these zones have been a potential source of conflicts. Mbororos are still marginalised by the native population, and some, due to their low levels of education, are not informed of their obligations and rights as a people. The arbitrary evaluation of crop damage, killing of cattle by the native population, and disrespect for court declarations and decisions are some of the factors hindering the effective management of rangelands. Nevertheless, the MBOSCUDA has partnered with other NGOs to educate and sensitise Mbororo pastoralists on the pastoral code and other rangeland management provisions.

Table 2.1 Composition and functions of the land consultative board and the agro-pastoral commission

Agro-pastoral commission		Land consultative board	
Composition	Function	Composition	Functions
Chairperson: The divisional officer secretary: A representative of the Divisional Service for Lands Members: – A representative of the ministry of agriculture—a representative of the Ministry of Livestock – The chief of the Village concerned, and two notables – A grazer or head of the grazing community concerned	– Determines and demarcates farmland and grazing land – Defines conditions for the use of mixed-farming zones – Supervises the use of farmland and grazing land – Examines and settles conflicts between farmers and grazers (excluding criminal cases, which fall within the competence of the legal courts)	Chairperson: Divisional officer Secretary: Head of the divisional service for lands Members: – The divisional chief of Surveys – The chief of the village concerned, and two notables	– Receives applications, inspects land, and makes recommendations for the issuance of land titles or concessions – Allocates land for projects of public interest, such as schools, markets, etc. – Examines and settles land disputes

Source Author's compilation

2.3 Methodology

2.3.1 Choice and Description of the Study Area

This study was conducted in the North West Region of Cameroon, in the Ndop Plain. Ndop is a highland inter-montane plain situated on the Bamenda High Lava Plateau and falls within the Western Highland agro-ecological zone (Fig. 2.1). It has a humid tropical climate with two distinct seasons: a long rainy season (mid-March to mid-November) and a short dry season (mid-November to mid-March). The average annual rainfall is about 2 000 mm, with minimum and maximum temperatures estimated at 15.5 °C and 24.5 °C, respectively. These climatic conditions have favoured the growth of pasture.

The Ndop Plain covers a surface area of about 17,910 km^2, and altitudes range from 300 m to over 3,000 m above sea level. On a seasonal basis, Mbororo pastoralists on transhumance from the surrounding Sabga hills, Wainamah hills, and Babangki herd their cattle down the plain to feed on fresh pasture. This is because the aforementioned highland areas experience frequent dry spells, and the scorching dry-season sunshine leads to rapid degradation of pasture and water. During this period, the Ndop Plain attracts herders with its fresh pasture in swamps and fresh stocks in rice fields.

2.3.2 The Mbororo Communities in and Around the Ndop Plain

Mbororo pastoralists of the North West Region of Cameroon have a long-standing history and a recognised, well-established traditional

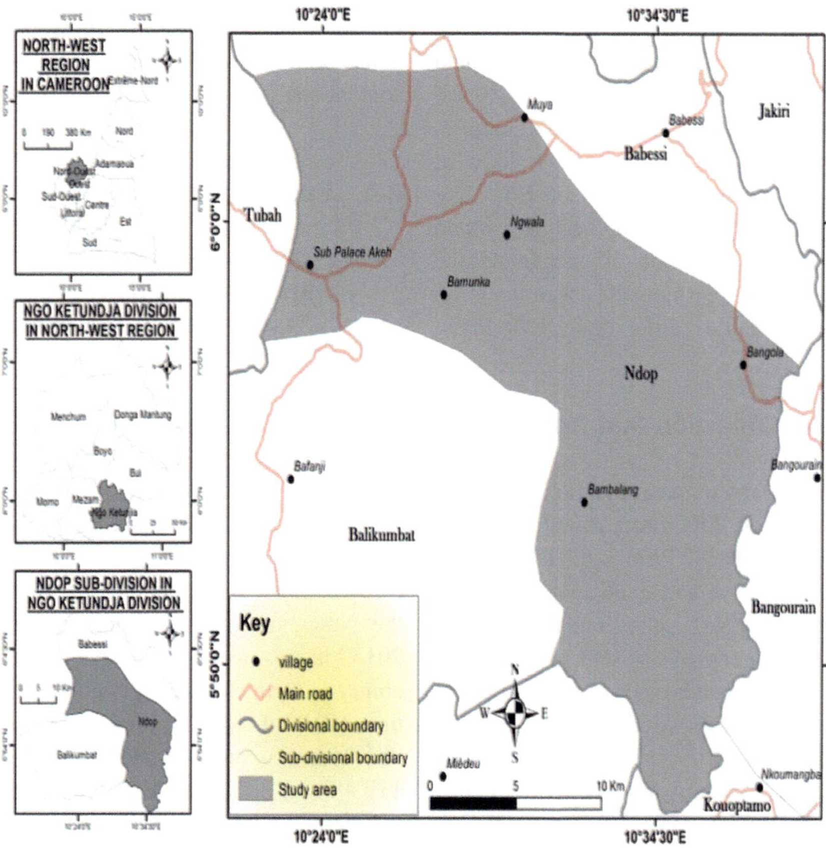

Fig. 2.1 Location of the Ndop Plain. *Source* Authors' realisation based on Images Satellite Spot (2021)

setup. Sources revealed that Mbororo pastoralists migrated into this region in the nineteenth century from Nigeria. The first group that entered the area settled in the Sabga hills that culminated in the Ndop Plain. During this time, there were still semi-nomadic communities in the area whose movement was triggered by the availability of resources and changes in environmental conditions. Interviews demonstrated that Mbororos began to settle permanently in the 1950s around the Ndop area. They settled together in isolated locations without mixing with the indigenes. Their traditional grazing unit is called the Ardorates, one of which was the *Ardo Sabga*. Their main activity was cattle rearing, and their pastoral lifestyle prevented most from getting a formal education. However, they developed local ecological skills that permitted them to handle large herds of cattle and adapt to harsh environmental conditions. This body of knowledge, although not documented, has been passed from one generation to the next.

With population increases and land scarcity, some Mbororo settlements were dispersed, but their minority status remained as they occupied mostly marginal lands, such as hilly slopes. While considered strangers by most natives, customary law permitted them the right over land, allowing some Mbororo personalities like Alhaji Danpullo to acquire many hectares of land in Sabga. The topographic advantage of the region has made transhumant pastoralism an adaptive form of cattle rearing in the Ndop Plain.

2.3.3 Data Collection and Analysis

This study adopted a purposive sampling method as participants were Mbororo pastoralists in and around the Ndop Plain. With a mixed research approach, quantitative and qualitative data were gathered. Ninety-six questionnaires were administered to purposively selected Mbororo households rearing animals for at least 10 years, while interviews were conducted with stakeholders in the agro-pastoral sector. They included two traditional Mbororo leaders, two council authorities, one authority of the Divisional Delegation of Livestock and animal husbandry, and two NGOs. A focus group discussion was also organised with seven Mbororo rearers and five crop farmers, given that Ndop is a mixed farming zone. Mbororo rearers were all male since their culture does not allow women to participate in issues regarding cattle rearing, while crop farmers were both male and female. In situ, climatic data were equally obtained from the Divisional Delegation of Agriculture and Rural Development for Ngoketunjia. Two field visits were conducted: one to the hills during the rainy season (in June 2020); and one to the plain during the dry season (in December 2020). Pastoral practices were thus observed at each phase of transhumant pastoralism.

The data obtained from these sources were analysed using quantitative and qualitative techniques. Questionnaires were coded and treated using Microsoft Excel and Sample Package for Statistical Studies (SPSS) version 20.0. The climate data were treated using coefficients of variations and the Standardised anomaly index. Interviews and focus group discussions were transcribed and analysed by using thematic and content analysis with the aid of Atlas.ti software.

2.4 Results and Discussions

2.4.1 Climate Variability and Transhumant Pastoralism in the Ndop Plain

Transhumant pastoralism in the Ndop Plain is guided by changes in seasons. Given that the onset and cessation of rains mark the beginning and end of the transhumance period, it is, thus, imperative to analyse rainfall variability trends in Ndop over the years. Computing rainfall data for the Ndop meteorological station from 1990 to 2015, it was revealed the area's rainfall variability is characterised by a coefficient of variation (CV) of 15.50% around the mean of 1] 851.96 mm and standard deviation of 119.48 mm for the study period. This CV shows that annual rainfall distribution has been varying but remains reliable for agro-pastoral activities.

Fluctuations in the number of rainy days, variations in onset dates, and cessation of rains are the main rainfall variability indicators that affect pastoralists. Pastoralists rely on rainfall for pasture growth and water, but anomalous events have been recorded in both the wet and dry seasons (Fig. 2.2a, b).

Figure 2.2 show varying episodes of positive and negative anomalies for both the wet (Fig. 2.2a) and dry (Fig. 2.2b) seasons with their overall falling trends. There are fewer years with negative anomalies than positive anomalies. No year had beyond −100 mm. Positive anomalies signify more water and possible floods, while negative anomalies signify less water and dry spells. The Standardised Seasonality Index (SSI) of 0.97 shows that rainfall is markedly seasonal with a longer, drier season. This concurs with pastoralists' perceptions of reduced rainfall and water scarcity.

Positive anomalies were seen in years that recorded rainfall above the normal for the data set. In the same light, those with negative anomalies signify rainfall below the normal and the likelihood of dry spells. Evidently, 2001, 2008, and 2012 in the wet season registered rainfall below the normal and were considered

a

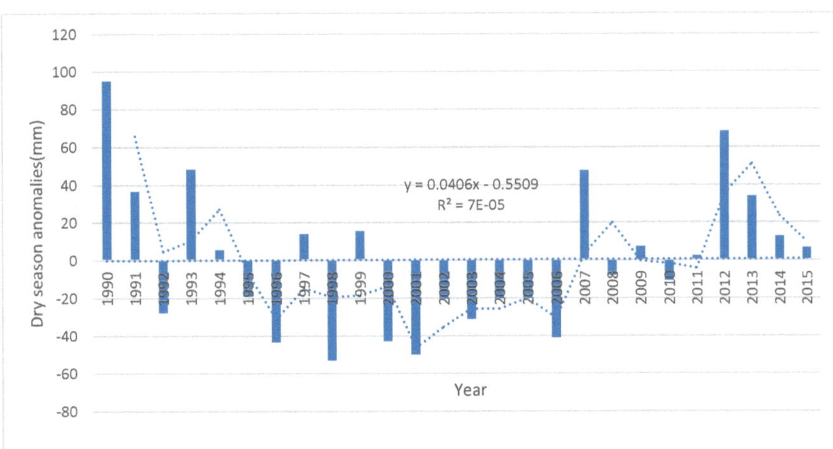

b

Fig. 2.2 **a** Wet season rainfall anomalies and trends from 1990 to 2015. *Source* Data from Ndop weather station (2020). **b** Dry season rainfall anomalies and trends from 1990 to 2015. *Source* Data from Ndop weather station (2020)

dry years. Although the dry season is characterised by little rainfall, the anomaly curve shows that the years 1992, 1998, and 2006 had exceptionally drier conditions. The highest positive anomalies above +200 mm were recorded in 1976 and 2003.

Both vertical and horizontal movements characterise transhumant pastoralism in Ndop. In the dry season (from November to February), when hills are dry with little pasture, herders move their cattle down the plain to feed on the swamp vegetation and harvested rice farms. With the arrival of rains in March, fresh grass shoots up in the hills, and the plain becomes swampy with hordes of tsetse flies. Herd men are forced to return up the hills with their cattle. Horizontally, herders equally move from surrounding lowland areas like Babungo into Bamumka that contain many swamps. Here, it is evident rainfall is an important factor in this activity. However, weather aberrations have disrupted this pastoral calendar, generating farmer-grazer conflicts due to competition over limited resources.

Rainfall unreliability and dry spells have ultimately reduced pasture availability upslope, and, thereby, have shortened the grazing duration. In such scenarios, herders tend to move downslope without respecting the timeline outlined in the agro-pastoral code. Given that Ndop Plain is a mixed-farming zone, such untimely downslope movements coincide with rice harvesting. This has reportedly been a cause of conflict between herders and rice farmers as cattle are constantly destroying crops. During focus group discussions, participants unanimously agreed that climate variability, especially dry spells, was responsible for pasture degradation, shrinking watercourses, and the outbreak of diseases that affect livestock and livelihoods. The sustainability of such a system can only be ensured if adaptation is properly mainstreamed and extended across diverse sections and actors.

2.4.2 Mbororo Perceptions of Climate Change and Indigenous Knowledge Systems

Mbororo pastoralists have a strong attachment to rangeland, and their experiences in harsh environmental conditions have enabled them to predict changes in their environment. In this regard, a herder stated:

> …we are conscious that our environment is changing and we can predict further changes. It is part of us because we have lived in the natural environment for all our lives. No one can tell us about changing environments more than we know.

This assertion was made along the slope of Sabga hill (cattle departure zone for transhumance into the Ndop Plain) as he showed us around his grazing land. It is a testimony to the fact that Mbororos are observing climate changes and can properly report them. Surveys revealed several pastoralist methods to predict changing climatic conditions and extreme events. The observation of changing wind directions, changes in the size and shapes of certain plants, the behaviour of migratory birds, dragonflies and cattle have been used to predict the arrival and departure of rains, as acknowledged by more than 85% of the population. The Mbororos' skilful herding abilities have permitted them to handle large herds along transhumance corridors in search of pasture. Transhumance provides enough time for pasture regeneration upslope, while rotational grazing allows for better manure (cow dung) distribution that acts as a source of nutrients. This nutrient distribution has boosted pasture growth in areas that suffered from overgrazing and degradation. These IKS have thus helped reduce pressure on natural pasture, provide enough time for pasture regeneration, and reduce degradation. Mbororos' ability to forecast weather changes has also been useful in fixing the period for departure on transhumance.

Mbororo communities have weather seers who read the climate, interpret to the population, and guide their actions in the field. Those who exhibit such abilities and practices are mostly the elderly with strong traditional values. Some younger generations who spend their entire lives in the fields herding cattle have also acquired such talents, which permit them to read landmark changes in the field. Still, the interpretation is performed by the old who draw conclusions based on past experiences. Asking whether there has been effective transmission of this ecological knowledge to the present generation, a Mbororo pastoralist in his 60s revealed:

> …the Mbororos of today are different from us. We lived entirely in the field and every young Mbororo was a shepherd. That is why the local people here call us "ngainako", meaning shepherd. But today, due to socio-economic changes, our youths now go to school and some are not interested in cattle rearing. Many are in Ndop town and other towns doing different jobs. This disconnection from the pastoral lifestyle has prevented them from understanding and predicting some of these changes.

This excerpt demonstrates that Mbororo traditional institutions must be strengthened, and knowledge sharing should be prioritised. The integration of IKS into local adaptation orientations will promote the sustainability of pastoral livelihoods.

2.4.3 Traditional Pastoralist Institutions and Rangeland Management Systems

2.4.3.1 MBOSCUDA: Traditional Institution for Ecological Knowledge Valorisation and Transmission

The MBOSCUDA was created in 1992 in Yaoundé from the fusion of Fulani Social Development (FUSUDA), Fulani Foundation of Cameroon (FUFOUCAM) (FUFOUCAM), and Mbororo Development Association (MDA). MBOSCUDA is represented nationwide with about 60,000 registered members. In addition to promoting the socio-economic and cultural development of the Mbororos, this association also aims to improve agro-pastoral practices to protect the environment and facilitate cooperation with people or groups who can help achieve these goals and contribute to the proper management of pastures and land.

Field observations showed that Mbororo pastoralism in the North West Region of Cameroon, in general, takes place in harsh environments and marginal lands that cannot be sustained, threatening pastoralists' livelihoods. This phenomenon has been worsened by climatic variability as frequent dry spells and erratic rainfall have exacerbated the degradation of pasture lands. In a bid to improve the Mbororo pastoralists' sustainability in the North West Region, the MBOSCUDA has been implementing a series of activities to build capacity and ensure effective transmission of indigenous knowledge from the older to the younger generation of Mbororo pastoralists.

Many pastoralists have adopted pasture improvement efforts promoted by MBOSCUDA. Moreover, the transhumant lifestyle of moving their animals down the valleys during the dry season in search of fresh pasture, crop residues and water, and going back to the plateau during the rainy season where they live with their families, has reduced the rate of environmental degradation. Moreover, pastoralists regularly give their cattle salt, and 75% of animal health problems are treated or prevented using ethnoveterinary medicines and other indigenous practices. These practices include crossbreeding local species with high-breed species that are resistant to environmental changes, pests, and diseases, and using crop residues to feed animals (MBOSCUDA 2014).

Another institutional strategy in resource management created by the MBOSCUDA is the Alliance Farming System. It entails a partnership between native crop farmers and pastoralists in sharing agro-pastoral land, water and pasture (Fig. 2.3).

In this system of farming, cattle are allowed to graze on crop residues after harvest. As the land is used for grazing, cattle dung and urine fertilise the soil, making it more productive when crop

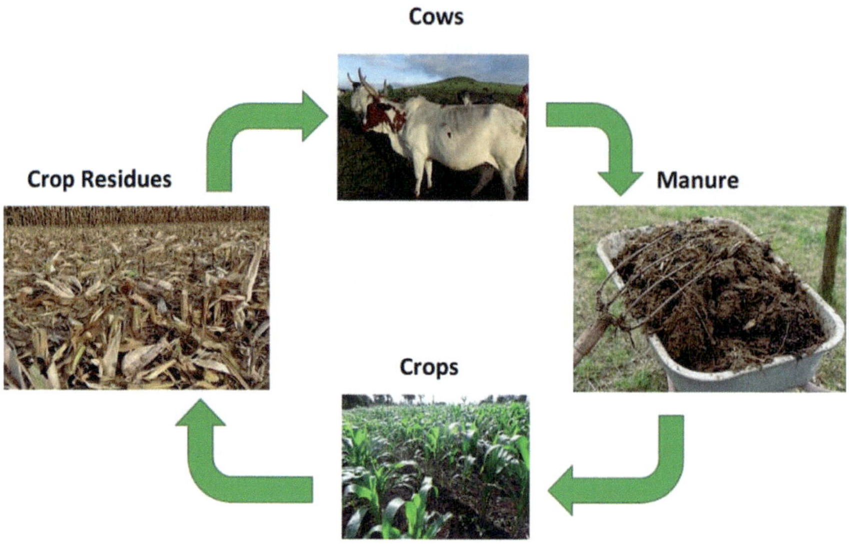

Fig. 2.3 Integrative diagram of the Alliance Farming System. *Source* MBOSCUDA (2014)

farmers return to cultivate it. This practice permits cattle to feed on rice farms during transhumance in the Ndop Plain. Moreover, in collaboration with many NGOs in the region, MBOSCUDA has developed a community-based farmer-herder dialogue platform for conflict resolution and peacebuilding. This alternative conflict management system has improved rangeland management and ensured the proper management of agro-pastoral resources. Relating to this practice, focus group participants explained that:

> …we are very thankful to MBOSCUDA, the alliance system of farming has helped us to live happily with the natives and make good use of our resources to produce more food. Cattle rearing and crop cultivation are no longer rivals as was the case before….

Based on their view of the management of grazing lands, it is evident the Mbororos' adoption of crop farming has reduced their over-dependence on animal rearing and pressure on natural pastures. The use of animal dung to fertilise pasture-land promotes pasture growth and reduces degradation. For example, a native crop farmer argued:

> …we either buy cow dung from the Mbororos or we exchange with crops from our farms. I use to agree with Mbororo rearers to allow their cattle to stay on my farm during transhumance after harvesting my rice. Crops cultivated in such areas are always different because of their fertility….

This extract shows that farmers are ready to share resources with Mbororo pastoralists. Ultimately, the grazing of cattle on rice fields has not only reduced pressure on the natural pastures but have permitted farmers and grazers to mutually benefit from the system.

2.4.3.2 The Traditional Burning of Pasture Lands to Regenerate Pasture Growth

Though regarded as environmentally unfriendly by many, pastoralists of the Ndop Plain have a culture of controlled pasture-land burning during the dry season in preparation for new growth with the arrival of the first rains in March. For this practice, ecological knowledge is used to distinguish between grass species and delimit the rangeland to avoid bush fires. For instance, based on their ecological knowledge, pastoralists know the grass species' regenerative abilities as well as species that can tolerate drought conditions. In some areas, burning is selective, and invasive and less nutritive species are burned to eliminate these from the rangeland.

2.4.3.3 Small-Scale Ranching and Planting of Improved Pasture Species

Despite the IKS' role in facilitating rangeland restoration, it is not redressing degradation as fast as it occurs in some locations on the Ndop Plain. Consequently, some Mbororo pastoralists have adopted modern rearing systems and improved the quality of pastures. Ranching is practised in the Ndawara Highlands and was introduced by the prominent Mbororo elite. While this system is still practised on a small scale, it has extended to the Sabga and Wainamah hills. Field surveys revealed that, although less than 10% of Mbororo herders are ranching, this practice is gradually gaining importance. Moreover, some herders combine this modern technique with their own IKS. Along the Sabga and Wainamah hills, some Mbororo households have demarcated small portions of land with barbed wire and planted these with improved grass species. The ranches sustain young and unhealthy animals that cannot walk long distances during the dry seasons as others are heading for transhumance. During this period, these and milking animals are kept in the ranches and fed with cultivated grass.

Another technique used in ranches during the dry seasons entails clearing and putting dried pastures on heaps. The heaps of dried pastures are mixed with salt, allowing young animals to feed while others are on transhumance. Improvements in such methods will undoubtedly reduce pressure on natural resources and ensure the sustainability of livestock production systems.

2.5 Challenges in the Integration of IKS and Realisation of SDG in Pastoral Areas

While pastoralists possess rich indigenous know-how and practices, this section shows that many challenges have limited traditional knowledge adoption, integration, and implementation in rangeland management systems. Using a questionnaire survey, a plethora of challenges were identified (Fig. 2.4). As shown in Fig. 2.4, the integration of Mbororo indigenous practices are diverse and originate from the Mbororo themselves as well as their host communities. Among these challenges, discrimination by native and administrative authorities (21%) dominates. This is because most native communities still consider Mbororos as intruding strangers in search of pasture for their animals. Institutions such as MBOSCUDA have embarked on a fight to ensure the respect of Mbororo rights as a people, but findings revealed efforts to boost the Mbororos' socio-economic development and professional integration have contributed to the erosion of traditional institutions and ecological know-how (7%).

The Mbororo themselves have also contributed to making their voices unheard. Their

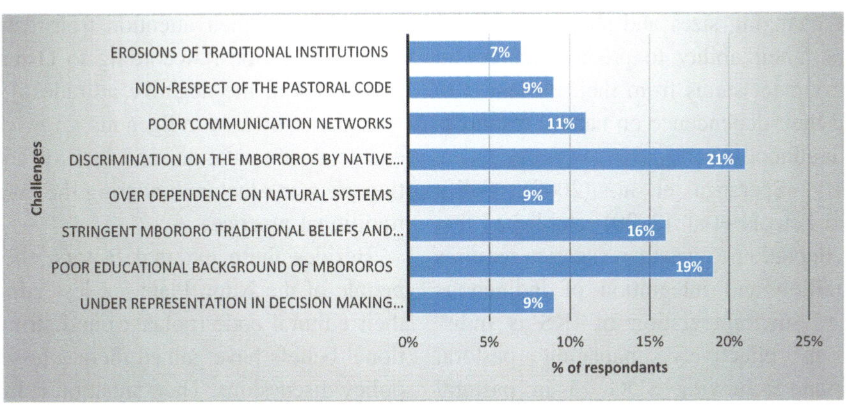

Fig. 2.4 Challenges in the implementation of Mbororo traditional knowledge systems. *Source* Fieldwork (2019)

cultural code of conduct, called "Polaako", has retarded their development, especially that of the female folk. These stringent cultural norms push the Mbororo woman to the background and prevent them from getting a formal education. These and many other factors have contributed to the under-representation of Mbororos at the scene of decision-making. Hence, achieving SDG 13 (Climate Action) in pastoral areas such as Ndop demands the integration of IKS in climate change adaptation strategies.

2.6 Discussion

2.6.1 IKS in Pastoralism and Realisation of SDG

According to Niamir-Fuller and Hubber-Sannwald (2019, p. 41), "sustainable pastoralism can synergistically link several SDG as they are critical to achieving carbon sequestration, land and water rehabilitation in pastoral areas of Africa". Traditional rearing systems, such as transhumant pastoralism practised in the Ndop Plain, are ecologically friendly since they reduce pressure on pastoral resources. The controlled burning of invasive grass species and the use of ethnoveterinary methods to treat animal diseases and pests are also environmentally friendly (MBOSCUDA 2014). Pastoralists' local practices of reading and predicting climate and environmental changes include studying changes in wind directions, the behaviour of some insects and birds, changing sizes and shapes of plants, and others. Their ability to predict changes in their environment stems from their closeness to nature and their dependence on natural resources for their livelihoods.

Similarly, Speranza et al. (2009), Salifu (2020) and Petzold et al. (2020) established that effective adaptation to weather vagaries requires the recognition and integration of indigenous knowledge. Such integration of IKS is indispensable in effectively managing pastoral resources and achieving SDG 13 in pastoral zones; community-led rangeland governance is therefore required. IKS are acquired through long-term observation and interaction with the environment and transferred through oral traditions from one generation to the next (Speranza et al. 2009; Salifu 2020). This implies that its integration into policies and management systems enhances adaptation to climate variability and mitigates future climate change impact and rangeland degradation.

2.6.2 Challenges in the Integration of Indigenous Knowledge Systems in Achieving SDGs

Mbororo pastoralists have accumulated a large body of local ecological practices and know-how that permits them to read and cope with changes in their environment. However, many challenges have prevented the adoption and implementation of their knowledge and practices in national and local adaptation programmes. Acknowledging the link between climate change and the degradation of pastoral resources, pastoralists claimed they could cope with such ecological stressors if not of external influence. Salifu (2020) has opined that indigenous people in Bona, Ethiopia, have similarly been deprived of their ability to adapt since mainstream policies failed to recognise their local knowledge. In a bid to emancipate the Mbororo community, the MBOSCUDA and other NGOs have eroded their traditional institutions and made the transmission of indigenous knowledge difficult; this is because the youths have diverted their attention from cattle rearing to other activities. According to Osman (2013), globalisation, the negative attitude of elites and foreigners towards indigenous knowledge, inadequate documentation, and lack of communication are some factors hindering the integration of traditional practices.

Besides these external factors, the Mbororo people of the Ndop Plain are less educated, and their cultural code (polaako) and stringent traditional beliefs have gained them a lesser place in policy discussions. Their situation is made worse by their minority status, the unfavourable tenure system, stigmatisation and marginalisation, which Jabiru (2017) termed 'the Mbororo problem' in the North West Region of Cameroon. The Mbororo problem has reduced their access to resources, human capacities, and networks vital in climate change adaptation. Consequently, Mbororo pastoralists' attainment of SDG 13 can only be effective if they are given access to resources (land and water) and are actively involved in rangeland management.

2.7 Conclusion and Policy Implications

This chapter examined the IKS of Mbororo pastoralists and the challenges hindering their integration into local and national climate change adaptation programmes, which could have adverse implications for achieving SDG 13. Using a mixed research method, findings demonstrated that climate change has led to a reduction in pastoral resources (land, water, and pastures) in Ndop. Mbororo pastoralists have a reservoir of traditional practices that helps them observe and predict climate changes, such as reading the behaviour of migratory birds, wind directions, and cattle behaviour. These knowledge systems have helped them adapt to the effects of climate change on pastoral resources.

The inclusion of IKS in climate change adaptation strategies is key to achieving SDG 13. Pastoralists hold valuable knowledge that provides solutions to challenges related to sustainable natural resource management and improving climate resilience. However, Mbororo pastoralists' exclusion from rangeland management institutions and their marginalisation hinders the achievement of the 2030 Agenda, which stipulates that no one is left behind (International Fund for Agricultural Development (IFAD) 2019). This is a critical challenge, as pastoralists are among the most vulnerable to the impacts of climate change as a result of their close interaction and reliance on the climate and natural systems. Pastoralists have been drawing upon their IKS to manage pastoral resources for generations, and are therefore important contributors to mitigation measures. Their involvement is required in regulating carbon levels and the climate through the sustainable management of rangeland biodiversity. Therefore, an intimate relationship exists between IKS and SDG 13 in pastoral zones of Africa.

References

Aida S, Omeny P, Pfeifer M, Olaka L, Mamo M, Rob M, Burgess N (2018) Climate change and pastoralists: perceptions and adaptation in montane Kenya. Climate Dev. https://doi.org/10.1080/17565529.2018.1454880

Azuhnwi B, Gidado M, Nsoh M, Ndamba M, Flintan F (2017)Making Rangelands more secure in Cameroon: lessons learned and recommendations for policymakers, development actors and pastoralists. Paper prepared for presentation at the "2017 World Bank Conference on Land and Poverty" The World Bank—Washington DC, March 20–24, 2017

Coalition of European Lobbies for Eastern African Pastoralism (CELEP) (2016) Pastoralism and the SDGs: how supporting pastoralism can help realise the Sustainable Development Goals

De Haan C (2016) Prospects for Livestock-based Livelihoods in Africa's drylands. World bank studies. World Bank, Washington, DC

Django S, Shei W, Duni J (2011) Legal framework for the regulation of access to, and management of pastoral resources in Cameroon. Implications for pastoral livestock production (points of interest of pastoralists of the North West Region)

Ezeala J, Mbalisi F (2021) Integrating indigenous sociocultural forms in environmental adult education for conflict resolution between pastoral nomads and host communities in Nigeria. Int J Res-GRANTHAALAYAH 9(2):345–353. https://doi.org/10.29121/granthaalayah.v9.i2.2021.3381

Herrero M, Addison J, Bedelian C, Carabine E, Havlík P, Henderson B, Van de Steeg J, Thornton P (2016) Climate change and pastoralism: impacts, consequences and adaptation. https://doi.org/10.20506/rst/35.2.2533

Indigenous People of Africa Coordinating Committee (IPACC) (2012) Influencing regional policy processes in climate change adaptation through the interaction of African pastoralist traditional knowledge and meteorological science: a contribution to the Nairobi Work programme on impacts, vulnerability and adaptation. Report on implementation of training on participatory 3D modelling in Baïbokoum, Logone Oriental, Chad. www.ipacc.org.za

International Fund for Agricultural Development (IFAD) (2019) Partnering with indigenous peoples for the SDGs Advancing solutions by working together. Via Paolo di Dono, 44 00142 Rome, Italy

Intergovernmental Panel on Climate Change (2014) Africa. https://www.ipcc.ch/pdf/assessment-report/ar5/wg2/WGIIAR5-Chap22_FINAL.pdf

Jabiru M (2017) The Mbororo problem in North West Cameroon a historical investigation. Am Sci Res J Eng, Technol, Sci (ASRJETS), ISSN (Print) 2313-4410, ISSN (Online) 2313-4402, http://asrjetsjournal.org/

MBOSCUDA (2014) Pastoralism and alliance farming: case study North West Region of Cameroon by MBOSCUDA. Annual report

Muga GO, Onyango-Ouma W, Sang R, Affognon H (2021) Indigenous knowledge of Rift Valley Fever among Somali nomadic pastoralists and its implications on public health delivery approaches in Ijara sub-County, North Eastern Kenya. PLoS Negl Trop Dis 15 (2):e0009166. https://doi.org/10.1371/journal.pntd.0009166

Nakashima DJ, Galloway-McLean K, Thulstrup HD, Ramos-Castillo A, Rubis JT (2012) Weathering uncertainty: traditional knowledge for climate change assessment and adaptation. The United Nations Educational, Scientific and Cultural Organization, Paris, France

Ngalim A (2015) Cattle rearing systems in the North West Region of Cameroon: historical trends on changing techniques and strategies. J Educ Policy Entrep Res (JEPER) ISSN: 2408-770X (Print), ISSN: 2408-6231, www.jeper.com

Niamir-Fuller M, Hubber-Sannwald E (2019) Pastoralism and achievement of the 2030 agenda for sustainable development: a missing piece of the global puzzle in book: stewardship of future drylands and climate change in the global south, challenges and opportunities for the agenda 2030.https://doi.org/10.1007/978-3-030-22464-6_3

Osman K (2013) Indigenous knowledge in Africa: challenges and opportunities. http://www.ufs.ac.za/docs/librariesprovider20/centre-for-africa-studies-documents/all-documents/osman-lecture-1788-eng.pdf?sfvrsn

Petzold J, Andrews N, Ford JD, Hedemann C, Postigo J (2020) Indigenous knowledge on climate change adaptation: a global evidence map of academic literature. Environ Res Lett. https://doi.org/10.1088/1748-9326/abb330

Salifu B (2020) Climate change impact and traditional coping mechanisms of Borana Pastoralists in Southern Ethiopia. In: Building adaptive capacity and resilience from an indigenous people's perspective. Master's thesis in indigenous studies, faculty of humanities, social sciences and education. University of Tromsø, Norway.

Speranza C, Kiteme B, Ambenje P, Wiesmann U, Makali S (2009) Indigenous knowledge related to climate variability and change: insights from droughts in semi-arid areas of former Makueni District, Kenya. Clim Change. https://doi.org/10.1007/s10584-009-9713

Moye Eric Kongnso (Ph.D.) is a lecturer in the Department of Geography and Environmental Planning of the University of Dschang, Cameroon. He holds a Ph.D. in Climatology and Development. His fields of interest are climate change adaptation, climate change communication, waste management and green economy, Indigenous knowledge systems. Epistemologically, he is interested in the concept of Political Ecology and Urban Political Ecology. Moye has worked with many local institutions and farmer groups in the domain of climate change adaptation.

3 Adopting Indigenous Knowledge Systems to Enhance Peace Education Programs for Climate Change and Adaptation in Zimbabwe

Ntombizakhe Moyo-Nyoni

3.1 Introduction

Climate change and resource scarcity are increasingly becoming multipliers of risks and threats which are aggravating existing societal problems such as poverty, social insecurity, injustice, violence, terrorism, and civil war (Froese and Schilling 2019). As Ide et al. (2021) and Dibaba (2020) argue that conflicts linked to climate change have become prevalent. Notable cases include, but is not limited to, pastoralist conflicts in Sudan (Bronkhorst 2011); land conflicts in Ethiopia (Dibaba 2020); resource-based conflicts in Democratic Republic of Congo (De Luca et al. 2012); and land distribution conflicts in Zimbabwe (Moyo 2005). In some cases, the severity of climate change related conflicts, have been interwoven with socio-economic and political problems. Scheffran et al. (2012, p. 870) asserts that "the pathways from climate change, resource access, and distribution to conflict are indirect, highly complex, and linked to other factors such as political, economic, and social conflicts, including ineffective institutions, social inequities, or low economic development". The interwoven effects of climate change, calls for transformative actions towards lasting solutions for sustainable peace.

The establishment of the global SDGs by the United Nations in 2015 became a proactive indication of the global need for sustainable peace. SDG 13 was of paramount importance in addressing the climate change problem. SDG 13 called to take urgent action to combat climate change and its impact (UNCT and Government of Zimbabwe 2014). Three of the five targets were defined as guidelines for action against climate change under SDG 13, these include: integrate climate change measures into policies and planning; build knowledge and capacities to meet climate change; strengthening of resilience and adaptive capacity to climate-related disasters. The remaining two targets are defining actions to be taken to reduce climate change. In this chapter, SDG 13 is of great significance in focusing on the resilience and adaptive capacities, as adopted by the people of Lusulu, in transforming conflicts as a strategy to combat climate change. Additionally, SDG 13 influenced Zimbabwe's Climate Policy (2016), which proposes that the challenge of climate change requires Zimbabwe to access and develop technologies relevant for the implementation of appropriate mitigation and adaptation projects and actions. Zimbabwe's climate change policy became reflective of the desires of the people of Zimbabwe, as outlined in the Constitution of Zimbabwe (2013), which sets out the environmental rights of citizens, i.e.,

N. Moyo-Nyoni (✉)
Lecturer, Department of Peace and Security Studies Department, Faculty of Social Sciences, Midlands State University, Gweru, Zimbabwe
e-mail: moyo-nyonin@staff.msu.ac.zw

E. E. Ebhuoma and L. Leonard (eds.), *Indigenous Knowledge and Climate Governance*, Sustainable Development Goals Series, https://doi.org/10.1007/978-3-030-99411-2_3

where citizens can enjoy an environment which is not harmful to their health and well-being.

In spite of global sustainable development agenda environmental pollution and ecological degradation remains common, affecting the overall wellbeing of individuals and communities (Svodziwa 2015). Resource based conflicts have become prevalent in Africa (Ide et al. 2021; Dibaba 2020). In some cases, violence emerges as an attempt to resolve differences; societies seem to opt for common and quick solutions in resolving conflicts (Froese and Schilling 2019; Moyo 2005). Evidence of recurrent conflicts regarding resources, in various parts of the world (Froese and Schilling 2019; Dibaba 2020; Moyo 2005), implies that violence cannot be used to adapt and combat climate change, nor its impact. Brock-Utne (2001) proposes that wars persisted longer due to the fact that IKS conflict resolutions were not considered. War violates people's fundamental human rights and strains relationships. Once relationships are strained, the chances of resolving conflicts become slim. Lederach (2003, p. 13) observes that conflict transformation "is to envision and respond to the ebb and flow of social conflict as life–giving opportunities for creating constructive change processes that reduce violence, increase justice and respond to life problems in human relationships". This might imply that conflict transformation, in the context of climate change, can be effective if the underlying factors of these conflicts are resolved. Conflict transformation seems to fit well in preventing the escalation of climate change related conflicts. Constructive transformation of conflicts, which are linked to climate change, might mean restoration of community relationships, which, in turn, might foster cooperation in that members may prioritize meeting the community obligations created by climate change.

Zimbabwe's climate change policy (2016) proposes that an effective response to climate change should incorporate IKS, culture and science. IKS can be a vital source of information in supplementing modern practices in combating climate change. IKS refers to the collective memory of traditions which are still in place (Fisher et al. 2017). IKS is the information, skills and practices acquired by indigenous people through past experiences, observations, informal experiments and an intimate understanding of their environment (Marango 2017). IKS presupposes that each community has its own specific knowledge which is related to its history; its cultural practices; and its traditional ways of dealing with problems (*ibid.*). Both knowledge and wisdom have been demonstrated by some communities in Africa when dealing with climate change related conflicts and in reducing the likelihood of the recurrence of such conflicts (Ide et al. 2021; Dibaba 2020; Bronkhorst 2011). Moreover, it is evident that conflicts which are left unresolved, or which are resolved unsatisfactorily, become a threat to the social fabrics and social institutions which hold communities together (Botcharova 2001; Scheffran et al. 2012). Thus, unresolved, or poorly resolved, conflicts might become a major hindrance in the fulfilment of key national policies on climate change, as well as the SDG 13.

There has been extensive research on climate change and its devastating effects on Zimbabwe. Chirimuta and Mapolisa (2011) write on climate change and food security; Dube and Phiri (2011) wrote on the impact of climate change on the household; and Chiwome et al. (2002) wrote on traditional leaders taking initiatives in reinforcing the IKS in climate change. There seems to be less research on the use of IKS in resolving conflicts which are linked to climate change within Zimbabwe. However, failure to resolve such conflicts constructively can prevent sustainable peace. Thus, this chapter seeks to assess the impact created by the use of IKS in transforming climate change related conflicts, in order to facilitate the adaptation of peace education programs for climate change within Zimbabwe.

Peace education is very difficult to define (John 2018), but it revolves around the creation of global consciousness and the value of transforming situations into something desirable (Bar-Tal 2002). According to Reardon (1988), the overall goal of peace education is to foster the growth of genuine environmental consciousness. This will enable people to operate as global citizens, wherein the current human state can be changed by altering the social structures and thought patterns which led to it. Bar-Tal and Reardon seem to imply that even climate change, as a current global problem, might be transformed through peace education. John (2018), states that the contents of peace education should not be in any way given universally or eternally, because it has to be related to peace problems at any time and in any place. Karimi and Ahmadipor (2016) and Naoufal (2014), propose that environmental education and awareness are legitimate forms of peace education. IKS of various societies can be researched to enhance peace education programs for climate change and adaptation. The objective of the study is to explore possibilities of adopting IKS to enhance peace education programs for climate change and adaptation. This chapter proceeds in five parts: Firstly, the theoretical framework; secondly, a description of the study area; thirdly, the results of the study; fourthly, the findings of the results; and lastly, the conclusion and recommendations.

3.2 Theoretical Framework

Environmental peacebuilding refers to activities adopted for lasting peace with the environment. According to Dibaba (2020), the process of peacebuilding involves fixing root causes of conflicts and backing the local population's capacity for peaceful conflict management and resolution. On a similar note, Ide et al (2021), posits that environmental peacebuilding comprises of multiple approaches and pathways, where management of environmental issues are integrated in and can support conflict prevention, mitigation, resolution and recovery. SDG 13 is an essential instrument in enhancing environmental peacebuilding at global level, as it proposes taking action in addressing the root causes of climate change. Environmental peacebuilding encourages community members to cooperate and work together in identifying the root causes of climate change, and to seek ways to transform them. Dibaba (2020) argues that peacebuilding processes can be twofold, firstly, in working to address destructive conflicts; and, secondly, in advancing existing peaceful practices. In the Lusulu context, and elsewhere, peace education can be used to enhance environmental peacebuilding through transformation of values, attitudes and behaviour.

3.3 Research Methodology

3.3.1 Study Area

The study area is the Lupane district, which is located in the Matabeleland North Province of Zimbabwe (Fig. 3.1). Zimbabwe is located in a semi-arid belt of Southern Africa, with 80% of its population dependent on rain fed agriculture (Zimbabwe's climate change policy 2016; Muromo et al. 2021; Chitongo 2019). Zimbabwe is classified under five Agro-ecological regions (Muromo et al. 2021). The regions are classified according to rainfall patterns, with region one being the wettest and region five being the driest. Chrimuta and Mapolisa (2011) observed that Zimbabwe has an abundance of rich natural resources, such as large tracks of land, and yet is on the peripheries of global food security. Zimbabwe has been experiencing economic, political and social instabilities for over two decades (United States Institute for Peace 2003). The situation exposed civilians to immense poverty, with wage conflicts being linked to food, economic and communal insecurity (Moyo 2005). It is acknowledged that poor and developing nations are more vulnerable to climate change, when considering the limited technology and resources at their disposal to mitigate the effects of climate change (Mutsvangwa 2010).

Lupane is in region IV, which is characterized by its poor soil and erratic rainfall (Muromo et al. 2021). According to Zimbabwe Statistics (2021), Lupane has a population of 198,600 (Zim Stat 2021). The main source of livelihood is cattle rearing and rain fed agriculture (Svodziwa 2015). Lupane has latitude of 18°55′48.05″S and a longitude of 27°45′33.58″E (www.distancesto.com). The coordinates of Lupane are near the Zambia and Botswana borders; 184.4 km SSE of Maamba, Southern, Zambia and 196.5 km N of

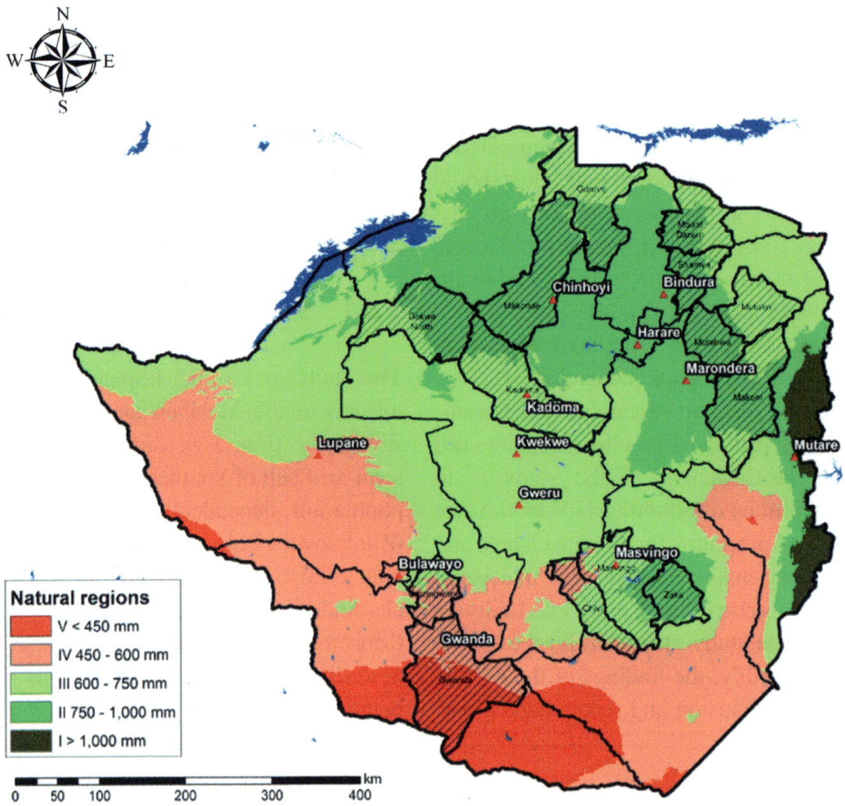

Fig. 3.1 Map of Zimbabwe. *Source* Kassie et al. (2017)

Masunga, North-East, Botswana. The elevation for Lupanen is: 983.96 m (ibid). The environmental degradation and climate change in Lupane, like most of rural Zimbabwe, is due to the attributes of both humans, and nature itself (ibid.). Drought has been prevalent for almost two decades, characterized by the reduction of rainfall; the increase in atmosphere temperatures; land degradation; poor resources management; deforestation; scarcity of grazing lands; and water shortages; amongst others (Chitongo 2019; Svodziwa 2015). Lupane town, which is considered as the centre of the district, is located 170 kms North of the city of Bulawayo, the second largest city in Zimbabwe and the major city of the Matabeleland Region. Lusulu ward is one of the wards in the Lupane rural area, located 60 kms from Lupane centre, which is at the border of Binga and Lupane. By virtue of being at the border of Lupane and Binga, Lusulu ward has a mixed population of Ndebele and Tonga people, who are both geographically and socially marginalized (Muromo et al. 2019; Mutsvangwa 2010).

Lusulu, like most parts of Zimbabwe, survives from rain fed agriculture and cattle rearing which has, unfortunately, been affected by climate change. As mentioned above, the area is characterized by erratic and low rainfall (less than 450 mm/year), as well as poor types of soil, predominantly the Kalahari sandy soil. Unfortunately, these conditions fail to provide security when crops are planted, which leads to food insecurity in the area. The area is "remote and very poor, with a poverty prevalence of 88.3% in 2012, which is high even for Zimbabwe, where the national average was 72% in the same year. This concentration of people living in poverty has been attributed to high levels of deprivation, illiteracy and recurrent droughts" (Muromo et al.

2021). Local climatologists predict that there will be reduced productivity of crop-livestock systems in the country's marginal rural areas, as a result of climate change (Chitongo 2019), and Lusulu is already falling under this category. Since there is not much written about Lusulu, a study of this community will be beneficial. The map below shows the location of the Lupane district.

3.3.2 Data Collection Methodology

A qualitative research design was adopted, given the importance of acquiring in-depth knowledge involving human attitudes and behaviour. Data was collected through focus group discussions (FGDs), as well as individual interviews. Forty participants were purposively selected to include various community stakeholders: traditional leaders, community leaders, church leaders, women, and youth. Four FGDs were held with ten people per group, two of these were composed solely of females and the other two, solely of males. Ages of participants were from 21 to 74, with the majority of the participants being in their late forties and early fifties. FGDs were supplemented with individual interviews; ten participants were drawn from the FGDs; six of these were males and four were females. The average age of the individual interview participants was 45. Individual interviews created a platform for pertinent issues which were then raised during FGDs. Responses from both the FGDs and individual interviews are crucial towards unearthing strategies which households can utilize to build resilience and the capacity to adapt to climate-related disaster; and it can be used to build knowledge and the capacity to address climate change, all of which are essential in facilitating the attainment of SDG. The gist of the discussions and interviews were to solicit views on possible strategies which can be used to transform conflicts into actions to combat climate change; and to attain insight of the Lusulu ward's IKS on climate change. Data was analyzed thematically.

3.4 Findings

3.4.1 Knowledge About Climate Change

Discussions in all of the four focused groups revealed that participants are aware of the effects of climate change. Effects included changes in rainfall patterns; reduction in the amount of rainfall; drought and floods; changes in temperature; early winters; and short rain seasons, among other aspects. Participants believe that the reduction in rainfall, which seems to be the main concern, is a spiritual issue linked to ill behaviour which angers the spiritual worlds and thus rainfall was reduced as punishment. The sentiments were encored in all groups, but FGD-1 was very vocal on this. One participant said that:

> There is no way rain can fall normal, there is a lot of ill practices conducted by people. People are evil… and thus the heavens are angry.

On another note, ninety percent of the participants fully believe that their suffering was largely linked to the current economic, political and social instabilities, and not so much to climate change. The FGDs had to be redirected several times, due to the participants shifting the discussion from climate change, and focusing on economic and political dynamics. The community did not consider climate change as a contributor to their problems. For example, participants blamed the government for failing to construct dams and drilling boreholes.

3.4.2 Effects of Climate Change

3.4.2.1 Unequal Distribution of Resources

Lusulu ward has been plagued by the adverse effects of climate change for decades. The resource scarcity was subsequently crystallized and resulted in unequal distribution of resources, which in turn, resulted in severe food shortages. Seventy percent of participants indicated that community members were living in precarious conditions; community members lacked the economic means to break the cycle of poverty in which they have become so deeply entrenched. Lusulu ward, being dependent on rain fed agriculture, experienced a change in rainfall patterns, which affected the production levels of crops. The community is still holding onto old systems and agricultural practices. Forty percent of the participants revealed that individuals or groups intentionally exclude other individuals or groups, from benefiting from available resources. A very common crisis referred to, was with regards to the distribution of governmental food aid and farming contributions, which then commonly turned into conflicts. The social fabric of the community was affected by these conflicts, and hindered community members in working together to take action to combat climate change, as proposed in the national climate change policy (Zimbabwe's climate change policy 2016).

3.4.2.2 Water Shortages

Discussions revealed that due to limited rainfall, water sources like rivers, dams and wells dry up which exposes the community to water shortages. Unavailability of water is said to be a main driver of interpersonal, communal and human-animal conflicts. Sixty percent of the participants revealed that there were years in which the situation became so severe that community members prevented each other's domestic animals from drinking water at their usually drinking sites, in fear that the animals will contaminate or finish the limited water. One participant explained:

> People have been made to be cruel to each other… villagers drive cows of their enemies to pits so that they can go unnoticed and die there because of wanting to reduce cows drinking from the well….

3.4.2.3 Human-Animal Conflicts

Human-animal conflicts were raised as a major threat to human survival, and that people are being denied their constitutional right to a safe environment. Due to water and food shortages, animals are forced to seek water from the same sources as the humans. It was eluded to that there have been cases in which crocodiles had attacked livestock and humans, and that, in turn; this resulted in the crocodiles being killed by the humans. This practice may, of course, lead to crocodiles becoming an endangered species. On the other hand, all participants agreed that that no, or limited rainfall often results in livestock being moved from grazing lands to areas around farming lands. This results in livestock encroaching on farming fields, which in turn, results in interpersonal conflicts. Bronkhorst (2011) expresses a similar concern in Sudan, where conflicts typically happen near, or along, pastoral migration routes, where pastoralists' livestock encroach on farm land and vice versa. Discussions revealed that in some cases, conflicts resulted in community members harming each other and the animals which were involved. When explaining this situation, a participant revealed that:

> We have witnessed people axing each other right in this community, fighting for grazing land.

However, the community had the strength to address some of these conflicts before they escalated into violence.

3.4.3 Conflict Intervention Strategies Adopted by the People of Lusulu

The FGDs and individual interviews revealed that most interpersonal and communal conflicts were linked to resource scarcity, linked to climate change. As indicated by Froese and Schilling (2019), climate change and resource scarcity are multipliers of social problems, such as, poverty, social insecurity, injustice, and violence, amongst others. The situation in Lusulu ward is not any different. Lusulu ward experiences political, economic and social conflicts, the roots of which are linked to resource scarcity caused by climate change. However, the community did not see the escalation of conflicts, rather the implementation of the IKS of conflict resolution. IKS focuses on addressing conflicts in their earliest stages to promote the social fabric of the community and the building of community relations. This is fundamental in building resilience and taking action against climate change.

3.4.4 Conflict Transformation Processes Led by Traditional Leaders

Two sources of Zimbabwe's law are common and customary laws. The common law (Roman-Dutch and English) as introduced by colonization; and secondly, there is indigenous/customary law, which is the laws which guided relations before Zimbabwe was colonized (Legal Resources Foundation (LRF) (1984). The people of Lusulu ward indicated that they are still guided by the customary laws in dealing with non-criminal incidents. Customary laws are used to transform conflicts, where the conflict transformation processes are guided by traditional leaders during a meeting known as *Inkudla* or *idale*, a Ndebele term which literally translates to peacemaking gatherings. The LRF (1984) refers to these gatherings as a court ("traditional court"), and indicates that the traditional leaders which led the traditional courts, were composed of chiefs (overseeing a number of wards), Headmen (leader of a number of villages), Kraal head (leader of a number of village heads) and Village heads (leader of a number of households). Participants indicated that the traditional courts are conducted hierarchically, using a bottom-up approach to conflict resolution. Thus, should a matter of conflict arise, it is first handled at an interpersonal/family level through negotiation and mediation processes. If the conflict remained unresolved, the matter was reported to the village head's court; if still unresolved, the matter would be taken up to the Kraal head's court; if still unresolved, the matter would be taken to the headman's court; and, if still unresolved, the matter would be forwarded to the chief's court (the traditional court).

Discussions with participants indicated that traditional courts involve the individuals or groups, with a stake in the conflict, as well as the entire community. The process is guided by the traditional leaders and those designated to help them according to the traditional leader's act, which governs traditional leaders. The conflict resolution process was said to be a participatory approach in nature, and aims at establishing community associations, relationships and collectiveness. The people of Lusulu ward appreciated the role played by traditional leaders in leading the conflict transformation processes. The traditional courts are said to have been guided by the concept of natural justice, which was linked to fact-finding missions aimed at verifying situations around conflicts. In this regard, a participant during individual interview explained that:

> Courts were held in a way where the natural sense of the right and wrong were used in handling conflicts.

3.4.5 Use of Restorative Versus Retributive Measures

Restorative justice system is different from criminal justice law which focuses more on retribution, whereas, this system focuses on the restoration of relationships; meeting obligations created by a conflict; and meeting obligations which were created by the conflict (Zehr 2019). Discussions revealed that conflict transformation approaches, as used in the Lusulu ward, were aimed at restoration. Restorative justice is aimed at establishing justice through restoration of needs; and honouring of obligations which resulted due to a conflict (Zehr 2019).The restorative justice process allowed each party to a conflict the opportunity to state their case; this was to promoted fairness and created a platform where parties could come together for the transformation of a situation (ibid.). The FGDs indicated that the views of the entire community were considered through a participatory engagement process, so as to attain the views of community members who were indirectly affected by a conflict. An interview with the headman indicated that participatory community engagement has the possibility to promote ownership and acceptance of the outcomes of the processes.

Restorative justice is the opposite of retributive justice, where, in the words of Christie (1977), retributive justice takes away a conflict from its owners, who will never recover from the effects of that conflict. Retributive measures have the possibility of tearing a community apart, in a way which widens the scope for conflict. Furthermore, resentment may be harboured between community members, which can possibly hinder efforts to mitigate and adapt to climate change. Thus, indirectly, restorative approaches can be an essential ingredient in dealing with issues of climate change.

Participants revealed that the conflict resolution approach involves acknowledgment from the wrongdoer, who then truthfully responds to the questions asked by the victim. The process was said to be essential as it has the potential of enhancing forgiveness, reconciliation and the healing of emotional wounds created by conflict. It was stated that if forgiveness, reconciliation and healing was not attained, it might result in the recurrence of a conflict, or the escalation thereof to violence. An example was given where two villagers had a conflict over grazing land. The conflict was not resolved satisfactorily, and as a result, the villagers held deep-rooted grudges towards each other. Subsequently, the one axed the other during a football match, and the axed villager later died due injuries. It was indicated that, to date, the villagers are still pained and bitter about the incident.

A common conflict which was raised was that of cows encroaching on someone else's fields. In this instance, the traditional leader in charge, and those assigned to assist him, will evaluate the value of the damage caused to the property; and the possible value of the outcome which could be produced by the cow owner. The results will be presented to the peace-making circle, which will, together with the offender and the claimant, discuss the terms and conditions of the compensation. Payment was derived from a place where both the offender and the claimant would benefit, usually from the profits which proceeded from the produce of the cow owner. Participants did, however, indicate that, sometimes, the community would receive limited bags of maize seeds from the government, which was meant for the entire community. As a proactive measure to curtail conflict linked to unequal distribution of resources, the traditional leaders, together with the people, resolved to open the maize seeds bags and measure them equally for the benefits of the entire community.

3.4.6 Indigenous Practices Adopted in Building Peace with the Environment

Domesticated animals are believed to exist for the welfare of people and thus, people should take care of them so that they would benefit from the wellbeing of their animals. Three FGDs revealed that there are traditional norms observed to protecting both domesticated and wild animals, for example, if animals, such as donkeys and cows were used in farming the fields, these animals would be released by mid-day, at the latest, for rest. Furthermore, animals should not work under high temperatures and should not be over worked. In cases where someone is using a cow or donkey cart, the load carried should be reasonable, and the driver has to avoid over speeding of the cart, to prevent the exploitation of the animals. Save for the practice of killing a crocodile that would have killed a person the community seemed to be very concerned about the conservation of wild animals. Discussions revealed that the custom of the land is that immature animals should not be killed for food; and no animal, bird or fish should be killed for anything other than food.

The two FGDs revealed that there are traditional beliefs among the people of Lusulu about the conservation of trees for the benefit of people presently, and in future; for example, certain species of trees are prohibited to be used as fire wood; people are discouraged to cut down live trees, for any purpose (although some did); traditional healers who make medicine from trees are trained not to destroy them, but to use only parts of the trees such as the leaves or bulks or parts of the roots. A traditional healer revealed that:

> Our training oriented us well…we should collect medicine from trees in a way that will enable us to collect medicine again….

The implication of this is that the community is taking action against climate change with regards to the conservation of trees. Furthermore, it was indicated that they are bound by traditional beliefs, which advocated that the forest should be protected because the ancestors and other spirits reside there. The belief is that if people protect the forest, the forest will protect them through giving them food and fire wood, among other things. All these seem to be in line with the Constitution of Zimbabwe with regards to environmental rights. Nyahunda and Tirivangasi (2019) observed that the envisaged determinants of adaptive capacity are still far from being a reality in Zimbabwe's rural communities; where poverty is at centre stage and people's livelihoods are aggravated by climate change. One group revealed that the people of Lusulu has a norm of rotating crops. Six out of ten participants indicated that they were still holding on to the practice. The common reasons cited for not holding on to the practice was that, due to poverty, some community members cannot afford to rotate their crops as they are dependent on the government to donate maize seeds. Thus, they do not have the luxury of choosing from a variety on seeds. The community indicated that traditionally, crop rotation was a strategy believed to rejuvenate the soil, since different nutrients were deposited into the land by the different crops which were planted. The act was said to promote natural ways of replenishing the soil, which is in line with maintaining peace with land where pollution and ecological degradation has thus far been prevented. Reference was also made to a traditional practice in which a field was set aside by the traditional leaders to be worked by the community. The produce from this field was distributed among the less fortunate. This act was believed to result in the blessing of the community. Blessings were associated with abundance of rainfall at the right time, which resulted in a good harvest.

3.5 Discussion

The findings of the study revealed that climate change is a major problem experienced by the people of the Lusulu ward, as in most parts of rural Zimbabwe (Muromo et al. 2021; Svodzwa 2015; Dube and Phiri 2011). Climate change has contributed to other social, economic and political problems, which are likely to contribute, in turn again, greatly to the escalation of the problem of climate change. Dibaba (2020) claimed, in the Ethiopian context, that climate change exposes latent conflicts. The Lusulu ward experiences poverty, marginalization, resource scarcity, human-animal conflicts, interpersonal conflicts, and communal conflicts, amongst others. Muromo et al. (2021) and Mutsvangwa (2010) observed that climate change in developing countries has increased the level of poverty. Mutsvangwa (2010), in the context of Lupane, predicted that climate change will likely be severe and will increase with time because of poverty.

Lack of knowledge on climate change is one problem that can aggravate the problem. As observed in the Lusulu ward, members act out of ignorance by acting and making decisions which violates the environment. Human actions will continue to hinder progress made towards the actualization of SDG 13. Peace education is one approach which can be adjusted to transform values, behaviours and practices to reduce changes in the climate. The Zimbabwe climate change policy (2016) proposed educating people on climate change, however, based on information from the Lusulu ward, these educational programs were unknown to them. The government should, perhaps, take active initiatives to educate even remote communities on climate change.

The community was aware of climate change, but they were not fully educated on it. They uphold the tradition of IKS on conflict transformation, which enabled them to identify conflicts in their earliest stages and manage them. Other than leaving conflicts to escalate to violence, was raised by Brock-Utne (2001), who believes that traditional methodologies of transforming resource-based conflicts will always seek constructive solutions, other that destructive outcomes linked to war. The Lusulu community was still holding onto some IKS which promoted conflict transformation. Such IKS can be combined with modern knowledge for a collective strategy against climate change, and towards the fulfilment of the targets as set out in SDG 13. Additionally, cooperation and community peacebuilding seem to be one aspect valued by the people of Lusulu in transforming problems of climate change. What can be drawn from this is that community cooperation is essential in building community resilience against climate change.

3.6 Conclusion and Recommendations

This study explored climate change related problems experienced by the people of Lusulu. Climate change has impacted the people of Lusulu in various ways. The people of Lusulu have experienced interpersonal and communal conflicts linked to climate change and they have maintained IKS ways in transforming them. The conflict transformation skills used in Lusulu, promotes community unity, which can strengthen the community in resisting the pressure produced by climate change. IKS can be adopted to influence peace education programs, so as to empower both young and old with constructive conflict transformation processes and ways of maintaining peace with the environment. Findings from this study can be instrumental in facilitating the achievement of SDG 13, in that it can be used as a guideline to develop peace education programs. Henceforth, research and systematic observations are needed to influence strategies towards resilience against climate change in the Lusulu ward and beyond. Finally, the government should invest in providing information and tools to communities to enable them to effectively adapt to climate change.

In light of the findings of this chapter, transformative action is needed in developing strategies to cope with climate change; and as an approach to prevent the escalation of the problems (Dube and Phiri 2011). Peace education can be presented to people of all ages by using both formal and informal ways. Peace education is one approach that can be adopted by communities to instil a culture of maintaining environmental peacebuilding, as it exhibits the potential of transformation of values, attitudes and human behaviour (Bar-Tal 2002; Harris 1999). Peace educators can draw lessons from the IKS valued by local people and combine it with scientific initiatives to develop a curriculum which can be used in communities and schools to educate people on climate change.

Additionally, peace educators can draw learning and teaching points from SDG 13; the national climate change policies; and other similar documents for transformation of attitudes and actions of learners. Moreover, learners can be taught skills of mitigating climate change related conflicts. IKS for conflict transformation can be adopted as an ongoing strategy to mitigate and adapt to conflicts brought on by climate change and as a way of reducing wars linked to resources scarcity. Based on the study conducted by Dibaba (2020) in Ethiopia, it become evident that climate change related conflicts can escalate to wars when left unresolved. Peace education programs should be influenced by systematic research and observation of Zimbabwe's climate change policy (2016).

References

Bar-Tal D (2002) The elusive nature of peace education. In: Salomon G, Nevo B (eds) Peace education: The concept, principles and practices around the world. http://construct.haifa.ac.il/~cerpe/

Botcharova O (2001) Implementation of track two diplomacy: developing a model for forgiveness. In: Helmick R, Petersen R (eds) Forgiveness and reconciliation: religion, public policy and conflict transformation. Templeton Foundation Press, Radnor, PA

Brock-Utne B (2001) Indigenous conflict resolution in Africa. In: A draft presented to the week-end seminar on indigenous solutions to conflicts held at the University of Oslo. Institute for Educational Research 23–24 of February 2001

Bronkhorst S (2011) Climate change and conflict: lessons from conflict resolutions from Southern Sahel of Sudan. ACCORD, Durban

Chitongo L (2019) Rural livelihood resilience strategies in the face of harsh climatic conditions. The case of ward 11 Gwanda, South, Zimbabwe. Cogent Soc Sci 5:1617090

Chiwome E, Furusa M, Gambahaya Z (2002) Indigenous knowledge and technology in African and Diasporan communities: multi-disciplinary approaches. Southern African Association for Culture and Development Studies, Harare

Chrimuta C, Mapolisa T (2011) Centering the Peripherised systems: Zimbabwe indigenous knowledge systems for food security. Zimb Int J Open Distance Learn 1(2)

Christie N (1977) Conflict as property. Br J Criminol 17:1–19

Constitution of Zimbabwe (2013) Government printers, Harare

De Luca G, Maystadt J, Ulimwengu J, Folledo R (2012) Mineral resources and conflict in the Democratic Republic of the Congo: a case of ecologically fallacy, International food policy research institute, discussion paper

Dibaba MA (2020) Environmental justice and peacebuilding practices in Ethiopia. Int J Ecosyst 10(1):23–29. https://doi.org/10.5923/j.ije.20201001.02

Dube T, Phiri K (2011) Rural Livelihoods under stress: the impact of climate change on Livelihoods in South Western Zimbabwe. Am Int J Contemp Res 3(5). www.aijcrnet.com

Fisher J, Westermann J, The Chikukwa community (2017) Journey to healing: choosing your own path using the three circles of knowledge. Printworks, Harare

Muromo F, Madhanzi T, Manjeru P, Isaac I, Matunhu J (2019) Building resilience to climate change through the adoption of grain and vegetable amaranth in Bings district of Matabeleland North, Zimbabwe. World J Agric Res 9(1):9–14. https://doi.org/10.12691/wjar-9-1-2

Froese R, Schilling J (2019) The nexus of climate change, land use, and conflicts. Curr Clim Chang Rep 5:24–35. https://doi.org/10.1007/s40641-019-00122-1

Harris IM (1999) Types of peace education. In: Raviv A, Oppenheimer L, Bar-Tal D (eds) How children understand war and peace. Jossey-Bass, San Francisco, pp 299–310

https://www.distancesto.com/coordinates/zw/lupane-latitude-longitude/history/80117.html

Ide T, Bruch C, Carius A, Conca K, Dabelko GD, Matthew R, Weinthal E (2021) The past and future(s) of environmental peacebuilding. Int Aff 97:1, 1–16. https://doi.org/10.1093/ia/iiaa177

John VM (2018) Peace education in Po-apartheid South Africa: needs, responses, and constraints. Asian J Peacebuilding 6(1):55–74

Karimi M, Ahmadipor Z (2016) Peace education and environmental sustainability. LAP Lambert academic publishing, Sunnyvale

Kassie GT, Aabdulai A, Greene WH, Shiferaw B, Abate T, Tarekegne A, Sutcliffe C (2017) World Dev 94:465–477. https://doi.org/10.1016/j.worlddev.2017.02.008

Lederach JP (2003) The little book of conflict transformation. Good Books, Intercourse

Legal resources foundation (1984). http://www.lrf.co.zw. Accessed 20 Apr 2021

Marango T (2017) Potential strategies for harnessing indigenous rainmaking practices to combat the negative effects of climate change in Chimanimani District of Zimbabwe, Ph.D. thesis, Institute for Rural Development, University of Venda, Thohoyandou, South Africa

Moyo S (2005) Land and natural resource redistribution in Zimbabwe. Afr Asian Stud 4(1–2):188–223. www.brill.nl

Muromo F, Madanzi T, Manjeru P, Isaac I, Matunhu J (2021) Building resilience to climate change through the adoption of grain and vegetable Amaranth in Binga North of Matebeleland North Zimbabwe. World J Agric Res 9(1):9–14

Mutsvangwa EP (2010) Climate change and vulnerability to food insecurity among smallholder farmers: a case study of Gweru and Lupane districts in Zimbabwe. M. A. Thesis, University of Free State

Naoufal N (2014) Peace and environmental education for climate change: challenges and practices in Lebanon. J Peace Educ 11(3). https://doi.org/10.1080/17400201.2014.954359

Nyahunda L, Tirivangasi HM (2019) Challenges faced by rural people in mitigating the effects of climate change in the Mazungunye communal lands, Zimbabwe. Jamba: J Disaster Risk Stud 11(1):a596. https://doi.org/10.4102/jamba.v11i1.596

Reardon BA (1988) Comprehensive peace education: educating for global responsibility. Teachers College Press, New York

Scheffran J, Brzoska M, Kominek J, Link PM, Schilling J (2012) Climate change and violent conflict. Science 336(6083):869–871. https://doi.org/10.1126/science.1221339

Svodziwa M (2015) The feasibility of small grains as an adoptive strategy to climate change. Rjoas 5(41):40–55

Svubure O, Gumbo T, Soropa G, Rusere F, Ndeketeya A, Moyo D (2011) Evaluation of the Sand Abstraction Systems for Rural Water Supply: the case of Lupane District, Zimbabwe. Int J Eng Sci Technol 3(1)

Tanyanyiwa VI, Chikwanha M (2011) The role of indigenous knowledge systems in the management of forest resources in Mugabe area, Masvingo, Zimbabwe. J Sustain Dev Afr 3:132–149

UNCT and Government of Zimbabwe (GoZ) (2014). Zimbabwe Country Analysis

United States Institute for Peace (2003) Zimbabwe and the prospects for non-violent political change, Special Report 103. United States Institute for Peace, Washington DC

Zehr H (2019) Restorative justice beyond crime: a vision to guide and sustain our lives. UPPADO, p 21

Zimbabwe Climate Policy (2016) http://newfour.ncuwash.org/wp-content/uploads/2017/08/Zimbabwe-Climate-Policy-2016.pdf. Accessed 25 Feb 2021

Zimbabwe Statistics (2021) Population size by province. https://www.zimstat.co.zw/. Accessed 17 Sep 2021

Ntombizakhe Moyo-Nyoni (Ph.D.), Mcom, Ba.h.Theo Ba. Theo), is a lecturer in the department of Peace and Security Studies. She is a peace building practitioner, specialising violent prevention, trauma healing, peace education Zimbabwe. Ntombizakhe Moyo-Nyoni is a member of Alternative to Violence Project Zimbabwe, serving with Alternative to violence project international as a member of the Board of Trustee.

4 Influence of Indigenous Spiritual Beliefs in Natural Resources Management and Climate Change Mitigation and Among the Yorùbás in Nigeria

Musediq Olufemi Lawal, Oluwagbemiga Oluwaseun Ajayi, and Adebayo Emmanuel Akinyemi

4.1 Introduction

Societies all over the world are inextricably linked to their natural environment, mainly because it is essential both for their survival and for satisfying human wants (Tobias 2015). Human relations and social activities in most geographic regions are significantly shaped by the availability of natural resources, weather patterns and a variety of other characteristics, such as the ability to withstand competition and uncertainties (Smith et al. 2019; WHO 2016). Natural resources provide wealth and subsistence for both the public (e.g., local government and forest authorities) and private (e.g., communities and private investors) actors engaged in local governance. Many people residing in developing countries depend on natural resources for subsistence, which also serve as a source of employment and income generation. However, the effect of human activities on the global environment has increased greatly in recent centuries (Temitope and Ogunbode 2019; Butler 2019), with severe consequences for natural resources. Unsustainable agriculture, unregulated extractive industries, rapid urbanisation, rampant coastal development, illegal and unregulated hunting trade in wildlife products and illegal logging are causing fragmentation and degradation of natural resources (Nwadike et al. 2020; Arora et al. 2018; Ogbanga et al. 2018; Claire 2017).

For indigenous people that depend on healthy and productive ecosystems to meet their daily needs, the degradation of the natural environment may put their survival in jeopardy. There are between 370 and 500 million indigenous people worldwide in over 90 countries. This constitutes just 5% of the global population and accounts for about 15% of the extremely poor. More than other segments of the world population, they rely on the resources endowed by nature for at least a significant part of their livelihoods. Studies (UNEP and IRP 2020; Mittal and Gupta 2015) revealed that of the Earth's original six billion hectares of forest cover, 30% of the Earth's land surface area (3.9 billion hectares) have been depleted. Different studies have shown that Africa's forest cover has decreased by net forest area in recent years (FAO 2015, 2016b; UNEP-WCMC 2016; Keenan et al. 2015). More than 4 million hectares of forest are lost in Africa every year, which is more than double the global average deforestation rate (Doyle 2018).

M. O. Lawal (✉) · O. O. Ajayi · A. E. Akinyemi
Department of Sociology, Osun State University, Osogbo, Nigeria
e-mail: musediq.lawal@uniosun.edu.ng

A. E. Akinyemi
Department of Sociology, University of Ibadan, Ibadan, Nigeria

E. E. Ebhuoma and L. Leonard (eds.), *Indigenous Knowledge and Climate Governance*, Sustainable Development Goals Series, https://doi.org/10.1007/978-3-030-99411-2_4

Between 2000 and 2010 Africa lost 34 million hectares of forest (Marx 2018). The forested area of Sub-Saharan Africa decreased by 30.6%, 27.7%, and 27.1%, respectively, in 1990, 2010, and 2015.

The loss of resources endowed by nature poses several threats to communities that depend directly on them to meet their daily needs. Such threats include danger to food supplies (Ogundipe et al. 2020; Poore and Nemecek 2018; FAO 2016a), reduced opportunities for recreation and tourism (Stefanica et al. 2021; Lee and Chen, 2021; Bandoi et al. 2020), reduced sources of fuelwood and medicinal herbs (Ulla et al. 2021; Suleiman et al. 2017), and interferes with natural ecological functions such as greenhouse gas absorption (Bologna and Aquino 2020; Koch et al.2019) and reducing the ability of ecosystems to deal with natural or human-induced disasters such as floods (Echendu 2020). Climate change, as several studies show (Norwegian Red Cross, 2019; Thomas 2017), has contributed significantly to the degradation of natural resources in Africa. Climate change presents the single biggest threat to sustainable development globally and its widespread, unprecedented impacts disproportionately affects the poorest and most vulnerable.

In recognition of climate change threats to sustainable development, the 13th Sustainable Development Goals (SDG 13), as specified in the 2030 Agenda for Sustainable Development, is a clarion call to comprehensively embrace climate action to drastically reduce its adverse impacts globally (Miyamoto 2020; UN 2017). For the indigenous people of Africa who are the most vulnerable to climate change, a meaningful lifestyle can only be guaranteed through such dedicated climate action. Indigenous Africans have fought against existential threats arising from capitalist exploitation (FAO 2020; Curtis et al. 2018). These exploitations included the construction of mega-dams across rivers and unsustainable logging in the forests (Berazneva and Byker, 2017). The epistemological philosophy that informs their actions is underpinned by values, myths, taboos, beliefs and practices (Abugiche et al. 2017; Asante et al. 2017; Diawuo and Issifu 2017).

The traditions and belief systems of the indigenous people often mean that they regard nature with deep respect, as well as having a strong sense of place and belonging. Indigenous people are custodians of the landscapes and ecosystems that are central to efforts to mitigate climate change and adapt to its effects. As Ayaa and Waswa (2016) highlight, natural resource conservation is not foreign to the indigenous people. While making it a part of their tradition, the management and utilisation of natural resource conservation are also expressed variously in their beliefs and practices (Olowu 2015). For indigenous people, the evolving nature-society interactions gave rise to sociocultural practices of identifying certain parts of the landscape as sacred forests and groves, sacred corridors and a variety of ethno-forestry practices (Eneji et al. 2012). These sociocultural belief systems become the institutional framework within the body of indigenous knowledge systems (IKS) for natural resource management practices (Olowu 2015). Through this lens, the attitude of respect towards the earth as 'mother' became prevalent among indigenous communities, specifically in the developing countries of Africa, Asia and Latin America (Ayaa and Waswa 2016).

Against this background, this chapter contributes to existing scholarly literature by analytically exploring how indigenous spiritual beliefs and practices among two communities, each in both Osun (Iba and Ore) and Oyo (Kishi and Sepeteri) States in Nigeria, scale up natural resource management and climate change mitigation. The indigenous language spoken in Osun and Oyo States is Yorùbá. The focus on Nigeria is largely driven by the fact that the country is listed as one of the developing countries with an exceedingly high rate of deforestation (Oyetunji et al. 2020; Ogundele and Adebisi 2016), and because it has been identified as one of the most vulnerable countries in Africa to climate change (Ndeye et al. 2019; Ibrahim et al. 2015). This chapter proceeds in four parts. The first part succinctly highlights how indigenous belief systems among the Yorùbá facilitate the management of natural resources. The second part describes the study areas and methodology applied to obtain primary data. The third part features the results and discussion, while the final part—the conclusion—underscores how the findings of the study could scale-up the actualisation of SDG 13 in not only the Osun and Oyo States, but also in other indigenous communities in developing countries where spiritual beliefs influence the management of natural resources.

4.2 Indigenous Belief Systems Among the Yorùbá's

The Yorùbá always tie the importance of food security to the prudent management of the environment from time immemorial. This informs the elevation of environmental resources to a spiritual level with reference to the role of the gods in its sustenance (Stanton 2020; Wood 2017). As earlier indicated, the African traditional religion acknowledges the supremacy of the gods over the universe. Till now, the belief in and respect for the gods as the owners of the entire universe persists among the Yorùbás. Before the advent of modern religion in Nigeria, each individual, family or state had its own deity which was worshipped and pacified. The worship of the different deities did not constitute denominations and there were no complex organisations or centrally planned worships (Agai 2021; Ojo 2017). Besides the common belief that the deities were but agents of the supreme god, there was mutual respect among the different devotees of the different deities (Ajang 2016; Okunola and Ojo 2016).

As it relates particularly to physical environment and its sustenance/conservation, the Yorùbá value system conduces to a cosmic and pragmatic interdependence among humans, nature and deities (Bewaji 2018; Aderinto 2017; Bádéjo 1996). This cosmic interdependence necessitates a reciprocal interaction of ethical obligations among the three categories of being. This interdependence and reciprocity promote conservation. Aderinto (2017) illustrated that trees, forests, hills and mountains are regarded by the Yorùbá to be abodes of goddesses and gods. Not only does this belief make people mindful of the cosmic power of nature, but it also demands of them a practice of veneration and worship (Van der Meer 2017; Usman and Falola 2019a), which protects and conserves natural spaces and environmental resources. Probably the very first nature item to ever be venerated, rivers are among the most natural features deified in many societies (Ray 2020). Diedre Bádéjo's (1996) seminar work on the Osun goddess and river in southwestern Nigeria supports Rays' assertion. It equally exemplifies how sacralisation and deification of natural resources promote nature conservation among the Yorùbá people. These sacred practices of pro-environmentalist/conservationist values and beliefs are associated with other divinities among the Yorùbá (Olu-Osayomi 2017).

Africans, including the Yorùbás, saw the forest and the resources endowed by nature therein as a life-giving force. The plants in the forest were revered and regarded as the most mystical and enigmatic of all living things. Consequently, the Yorùbás do not meddle with nature in a casual manner because of this. Every element of nature, according to the Yorùbás, has a specific function, and one of those is to produce a healthy, sustainable and attractive home for humans. Osun groove in Osogbo, whose woodland and river are banned for hunting and fishing, is an example of the deep desire for care of the environment (Owoseni 2017). In addition, certain of these natural features are believed to function as dwellings for certain deities. The Yorùbá believe that the spirits of deities such as Ọbàtálá, Ọbalúayé, Orisa oko, etc., inhabit virtually all objects, animate and inanimate (Oyebola 2019). It is, therefore, no surprise that many different spirits are linked with hills and mountains, as well as rivers, cliffs, cave dwellings, trees, streams, lakes and thick woods.

Some natural features, such as Ibadan Hill (Oke-Ibadan) and Olumo Rock in Abeokuta, provided continual protection to citizens during Yorùbá civil wars (Dada et al. 2020; Usman and Falola 2019b; Okunade 2019; Salami 2019; Bennett 2010). It was believed that these natural beings were continually showering benefits on the inhabitants of the villages, no matter where they may be found. Each of these spirits oversee the natural and the environmental aspects of Yorùbá life. The Yorùbás do not usually tamper with nature carelessly. They are expected to treat all of the creatures with respect and reverence. In a typical Yorùbá setting one cannot just dig the ground for any purpose without due permission from the soil or the earth goddess (such as Orisaoko and Yemoja). This is anchored in the belief that the soil possesses a life of its own (Owoseni 2017).

The Yorùbá belief system has always been environmentally respectful in spite of the indiscriminate destruction of forests and unsustainable harvesting of their products. The traditional cultural system still showed evidence of in-situ conservation. The adherents of the Yorùbá religion are often involved in regular reforestation and preservation schemes. This manifests through the creation of shrines that serve as protectors and for nature preservation (such as Osun groove). The establishment of shrines and groves where log felling is prohibited greatly contributed to conservation and biodiversity in numerous Yorùbá communities (Oladeji et al. 2021). This is maintained through the knowledge of existing taboos and myths about the use of environmental resources. The deities' capabilities to strike and kill humans, destroy structures or incapacitate transgressors into the forbidden forests sustained the management of natural resources (Daniel et al. 2015; Daye and Healey 2015). This shows that Yorùbá life is a vital intersection where the natural meets the supernatural, and where ancestors and the divine meet the human race in a unique way (Owoseni 2017).

4.3 Materials and Methods

4.3.1 Description of Study Areas

Kisi and Sepeteri are located in the Oke Ogun area of Oyo State. Kisi is the headquarters of the Irepo Local Government area, while Sepeteri is the largest town in the Saki East Local Government area. The Sepeteri community plays host to the field command of Old Oyo National Park, which is one of Nigeria's largest conserved areas. The climate is equatorial, with dry and wet seasons and with a notably relative high humidity. The average daily temperature ranges between 25 °C (77.0 °F) and 35 °C (95.0 °F) almost throughout the year (Owolabi 2017; Oloukoi 2014). The vegetation pattern of this area is that of Guinea savannah, which gives way to grassland interspersed with trees. This climate favours the cultivation of crops such as maize, yam, cassava, millet, rice, plantain, cocoa tree, palm tree and cashew.

Iba is an ancient town in the Ifelodun Local Government Area of Osun State in the northeast of Ọṣun State. Ore is a town within the Inisa/Ore/Agbeye district in Odo Otin Local Government area of Osun State. The vegetation of these communities could be classified as lowland forest zone and semi-deciduous moist forest. There are two prominent seasons in the areas, namely the rainy and dry seasons. The relative humidity in the early morning is generally high, usually over 90% throughout the year. Petty trading and farming are the predominant occupations in these communities (Borisade et al. 2021). Arable crops (such as tomato, sorghum, cocoyam, yam, cassava, potato and maize) and cash crops (such as cocoa, palm oil, kolanut, coffee and plantain) are prevalent. The environments of Iba, Ore, Kisi and Sepeteri are relatively safeguarded against wanton destruction from day-to-day human activities. Reserved forests in the four communities were selected for the study (Fig. 4.1). The four forests perfectly fitted into what is conventionally known as protected areas (Ali et al. 2020; Ayivor et al. 2020).

4.3.2 Methodology

A qualitative method was employed in this study. The unique features of the environmental conditions of Iba, Ore, Kisi and Sepeteri communities informed their purposive selection for this research, made possible by their geographical locations and the residents' penchant for adherence to traditional rites. The study areas have become the home to a diverse fauna and flora.

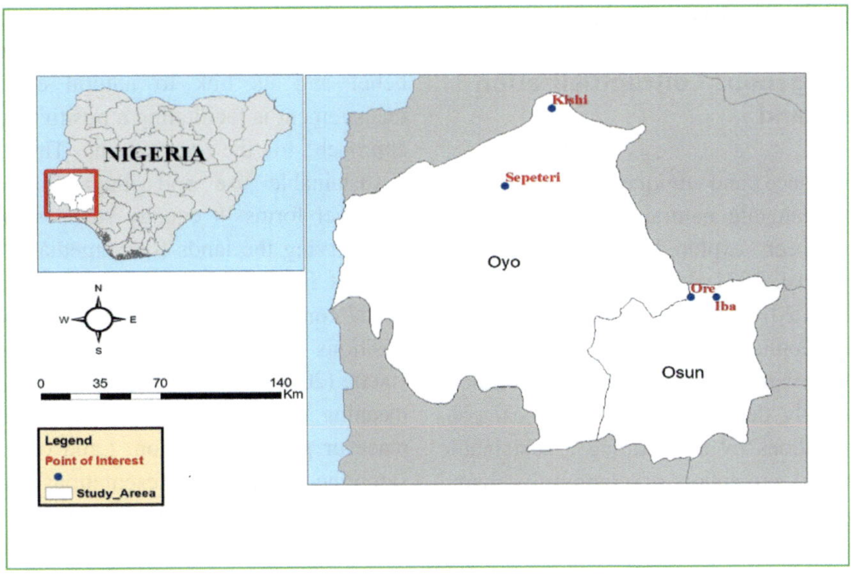

Fig. 4.1 Map of Osun and Oyo States, Nigeria showing the study locations. *Source* Authors (2021)

For better understanding of the research problem, structured interviews were used to obtain primary data collection. In line with this, the target (the study population) was classified into primary, second and tertiary stakeholders. The primary stakeholders were the community leaders. The secondary stakeholders were the clerics of traditional religious bodies in the study areas, while the tertiary stakeholders comprised the members of traditional ruling councils. From each of these communities a total of 12 participants were sampled. The need for precise information on practice and process of natural resource management, climate change mitigation and ecological conservation informed the choice of these respondents. The selection of participants for this study was made easier through information about their constant involvement in community development and environmental management activities. Key informants who had resided in each community for a minimum of 20 years, as well as snowball sampling procedures, were used to recruit eligible participants. However, circumstantial meeting of other qualified persons necessitated accidental sampling methods. At the end of the exercise 48 interview sessions were conducted among respondents in all four communities (Table 4.1). Interviews were conducted in Yorùbá language and later translated into English. Thematic analysis that involved identification of patterns in meaning across the in-depth interviews was adopted for data analysis.

Table 4.1 Breakdown of participants in interview sessions

Participants	Communities				Total
	Iba	Ore	Kisi	Sepeteri	
Primary stakeholders (male and female community leaders)	4	4	4	4	16
Secondary stakeholders (clerics of traditional religious bodies within the study areas)	4	4	4	4	16
Tertiary stakeholders (members of traditional ruling councils)	4	4	4	4	16
Total	12	12	12	12	**48**

4.4 Results and Discussion

4.4.1 Indigenous Conceptualisation of Land

The significance and uniqueness of IKS in enabling sustainable natural resource management have been explored by scholars (e.g., Mabvurira et al. 2021; Maru et al. 2020; Desta and Smithson 2010). From all indications, IKS is traditionally applied in harmony with the natural and spiritual world. These traditional practices are ingeniously designed to address local ecological limitations by maintaining a sustainable utilisation and protection of commonly shared natural resources. This manifested in how the people in the study areas perceive the land, its uses and the production outcomes. For example, a participant in her late 60s who resides in Iba community stated:

> *Land is a sign of God's bene*volence on human race. Land, as nature, has a determining influence in moulding the life, occupations and standard of living of a people. This is the rationale behind varieties in occupational inclination of several human communities. For example, Iba community has comparative advantage in agricultural products like yam, cocoa, and plantains… We know the implications of mismanagement of our land; hence, our resolve to reasonable practices that will ensure we don't run out of our good soil nutrients.

The peoples' interaction with the land is also underpinned by spirituality. A community leader in Iba, in his late 70s, stated:

> Our people have a deep spiritual relationship with the land; they feel at one with their ancestral territory. We feel responsible for the healthy maintenance of the land and its associated resources such as waters and soils, plants and animals. This is not limited to the present time of our existence but for future generations. To ensure that land is not traded unnecessary for the highest bidders, the practice of burying our people on family lands was common. Such lands where the ancestors are buried are not often trespassed upon; they are avoided for purchase and usually turned to sacred places and revered.

The sustainable utilisation of natural resources was relatively effective through the application of indigenous belief systems. For these people, deceased individuals and communal spirits have transformed into deities. The deployment of this belief and its link to natural environmental management is facilitating a positive behavioural approach to the ecosystem. This prevented unsustainable use and illegal trespassing in whatever forms. It equally sustains the practice of reserving the lands for immediate family use only. It is noteworthy to mention that respondents' conceptualisation of land corroborated the positions of Lines et al. (2019), Rowhani and Hatala (2017) and Durkalec et al. (2015) that its meaning transcends access to a literal material place or physical location. Land often involves symbolic or sacred representations and spiritual relationships with broader, more universal notions of Mother Earth (see also Korff 2021).

In many mythologies, what individuals will be and how well such a person is going to live depends on their relationship with land. The economic prosperity of a community is closely linked to the richness of the endowed natural resources and how they are handled while making use of them. This was stressed by a participant in Iba community, in his 60s, when he revealed that:

> The realisation of the sacredness of land informs the prominent position individual societies usually accord it. For instance, in Iba community, the belief is that how the land is treated will determine the type of climatic condition we are likely to live with. In view of this, individuals within the community have been inducted into the value system of treating the land with care.

4.4.2 The Utilisation of Natural Resources and Emerging Environmental Challenges

Participants demonstrated how essential the land is for the development of vegetation cover to improve food production through soil enrichment. From their perspectives, the relevance of nutrients in the land denotes the substance that is needed to keep the plants growing. In this regard, a farmer and community leader, in his 70s, from Kisi community commented:

> From years of experience on the job, the people appreciated the importance of loamy soil with good humus content, good drainage and aeration, as well as adequate nutrient content. Awareness of inappropriate usage of natural resources and misapplication of methods for their usage can lead to significant environmental degradation, especially in ecologically sensitive areas. Clearing of natural vegetation usually exposes the soil to the direct effect of rain and sun intensity and causes erosion. The loss of organic matter (humus) through bush burning, high temperature while clearing of natural vegetation exposes the soil to the direct effect of rain and sun intensity through the removal of topsoil containing organic matter, thereby leaving the soil barren.

Participants also attributed the causes of the often-pronounced environmental damage to destructive land use patterns, which involved the removal of existing vegetation to make way for the development of agriculture and the need to provide grazing lands for livestock. Large areas of the protective forest cover were destroyed through illegal tree felling, bush burning, overgrazing by cattle, sheep and goats, peeling of the tree bark for the preparation of indigenous medicines, and extensive deforestation due to increased demand for fuel and timber. At that point in time, the community started to experience unfavourable weather conditions. These anthropogenic activities resulted in a further loss of fuel wood and fodder, intensification of drought, soil erosion and loss of crops and livestock. Virtually all the people within the communities are aware of the consequences of the failures at putting this trend under control. Typical in the aftermath of this, according to a woman leader, in her late 50 s, from Sepeteri community are:

> ... increases in temperature, variable rainfall, flooding, drought, land degradation, more frequent extreme weather events and destruction of habitats in the wild, and subsequent loss of plant and animal life therein.

To curb the over-exploitation of natural resources, some of the sites were partly protected with physical barriers from daily human encroachment. Cases in point included Igbo Ojupo and Yemoja grove in Sepeteri Community, Aluku forest (Kisi town), Awuru (Ore town) and Igbo'ta (Iba town). A collective resolution was also adopted for the empowerment of traditional institutions to issue proclamations on standard practice expected within the communities that were studied. This resolve may be underpinned by the fact that, for indigenous peoples the health of land is central to their identity and culture. Land is perceived as their mother; it is steeped in their culture, thereby making caring for the land the responsibility of every community member. It is obvious that the people are aware of the inherent importance in the rational use of natural resources and the deployment of various indigenous methods for their protection. Efforts at conserving natural resources are viewed as a product of understanding the interactions between nature and humanity. The rational use of environmental resources was underlined as being able to assist the populace by minimising the risk associated with exhaustive usage of the natural environment, and in so doing, mitigate the occurrence of climate change.

4.4.3 Indigenous Beliefs and the Management of Natural Resources

Speaking about the Yorùbá belief system, Awolalu and Dopamu (1979) theorised that the forest and other natural habitats serve as the abode for ancestors and divinities. Such natural habitats include hills, mountains, rivers, rocks, caves, trees, brooks and thick forests that are found across the Yorùbá land. It is, therefore, not surprising that prudent management of the forest and the resources therein is given paramount attention to ensure the daily survival of households through sustainable use of environmental resources in Iba, Ore, Kisi and Sepeteri communities, as well as the designation of some parcels of land as 'protected areas'.

Some rules were made to prevent the unsustainable usage of resources within such areas. These rules usually mandate time to venture into protected areas for the usage of resources in the forest, specify and limit such resources to items such as firewood and the hunting of animals for domestic use, and specify an age restriction for those qualified to enter the forests. Aside the foregoing, some forests are reserved for the burial of passing kings with elaborate burial rites. The general belief is that the kings who were buried transform into deities. The knowledge of negative repercussions from the deities that reside in the lands and the forests always enhance full compliance with the rules. While these rules provided an avenue to halt unsustainable usage of the resources, it equally provided an opportunity for the environment to replenish itself. This eventually gave rise to the emergence of protected areas and subsequently assisted the process of uninterrupted biotic succession. For instance, a community leader in Ore community, in his 60s, explained that:

> The creation of protected areas came out of the abilities of the forefathers to see into the future and the divine guidance of the oracles. Most of the time, oracles prescribed what should be done and exact spots that should be classified as protected forests. These forests are usually given names in line with the peculiarities of the people of the areas.

A participant in Kisi community revealed that in their community they have Adasobo hill and forest as well as Aluku forests. In a related vein, participants in Iba, Ore and Sepeteri communities revealed that they have such protected areas. In Iba, both Ayee and Igbo'ta forests are designated protected areas. In Ore, the protected forest is called Awuru while in Sepeteri they are known as Ojupo and Yemoja forests. The generally known mechanism for controlling transgression in these protected areas is the application of taboos due to their emotional attachment to the traditionally protected areas, which has assisted in protecting these resources from vandalisation. In highlighting the potency of traditional beliefs underpinned by taboos in the protection of the natural forest environment in the town, a community leader in Sepeteri, in his 70s, stated:

> Taboos play considerable roles in the sustenance of the entire forests/groves found within the community. For instance, it is a taboo for the non-initiates to enter the classified heritage sites. It is a taboo to fetch fuel wood from the forests/groves, clear or cultivate land in the grove, fell trees for lumbering, collect/gather fruits, including wild nut, clear or cultivate land in the groves, palm kernel and so on and set fire in the forest. There are specific sanctions against the violation of these taboos. The repercussions range from banishment from the community, payment of atonement and infliction of the violators with diseases.

Regulating entry and harvesting of environmental resources in sacred forests help to mitigate climate change through carbon sequestration (Maru et al. 2020; Kihonge 2017). The limitation of this chapter is the inability to quantify the amount of carbon sequestered from the atmosphere by the sacred forests in the study areas. Nonetheless, the importance of untouched forests in sequestering carbon from the atmosphere is widely acknowledged and this is partly responsible for the tree planting campaign by various climate scientists globally (Daniel et al. 2015; Daye and Healey 2015).

Some participants emphasised that community alertness of movement around the protected areas was a factor in ensuring compliance against vandalisation. Furthermore, the widespread practice of Ifa divinities enhanced the regular consultation of deities for guidance in almost everything regarding the people's life, specifically among the indigenous people of the study areas. According to a participant who doubled as high priest in one of the study areas:

> There are laid down procedures or instructions that were given as parts of conditions for sustainable livelihood that we must adhere to. Notwithstanding the incursion of Christianity, Islam and Western culture, we consult Ifa on regular basis. The mutual and robust relationship between the community and our ancestors is made possible through the regular consultation of Ifa as intermediary between other deities. This reflects in how we manage our environmental resources. (Male IDI/Community Leader/Iba, Osun State, Aged 72 years).

A part of IKS relevant to enforcing total regulation of how environmental resources are utilised, is the reservation of certain areas of the environment for traditional rites. Members of the communities automatically avoid areas set aside, thereby facilitating their transformation to the present status as 'thick forests'. Typical examples of such cited in Oyo State included Igbo Ojupo that was set aside as the burial grove for the first set of kings that ruled Sepeteri town. Another is Yemoja grove, which is a forested grove in Sepeteri that contains shrines of different deities. Aluku forest in Kisi is a forested stretch of land that serves as an abode for shrines. Adasobo is a forested hill where a reasonable amount of biodiversity is found. People are only allowed into this forest annually during the community festival named after the hill. During this period able-bodied members of the community are allowed to harvest animals within this natural enclave. The harvest is regulated to prevent wastage.

Ayee is a hilly natural feature that is located in Iba community in Osun State. The sacred nature of this hilly structure and the protection it enjoys against vandalisation equally extend to the adjoining ecosystems. Within this community is Igbo'ta, which is a large expanse of land that has been conserved as the burial site for every king that ruled the community until the present time. People are forbidden to enter Awuru forest (Ore town, Osun State), Igbo'ta (Iba town, Osun State), Aluku forest (Kisi town, Oyo State) and Igbo Ojupo (Sepeteri town, Oyo State). Only initiates are allowed into these forests. There are sets of indigenous methods that are spiritually inclined and used in managing these forested and preserved heritage sites.

Some participants also further stressed that community members appreciated the importance of natural resources within their domains. This informed their common desire to ensure that these resources are protected. As a result, some community members who believed in traditional institutions constantly saw to the upkeep of these resources through periodic public gatherings to appraise the effectiveness of protective measures put in place, and ensured the enforcement of penalties on those who violated such rules. Because communal lands are governed under community-based tenure systems, the communities have under their relative control the rules, authorities, institutions, rights and norms that govern access to and control over land and related resources. In spite of the havoc of former colonial land policies that overlaid established patterns of land distribution, a community land tenure system still exists. Similar to the exposition of IFAD (2018) and the Rights and Resource Initiative (RRI 2015), the right to own or manage terrestrial natural resources is held at the community level. The right is exercised through established local institutions and practices that have, over time, assisted in sustaining fragile ecosystems.

4.5 Conclusion

The conventional protected areas explored in this study are forests located in Osun and Oyo States. With other natural features such as hills, shrines and burial sites found in them, practices associated with the sites revealed that they embody indigenous principles of conservation and resource management. The application of physical barriers for protection of some sites in Kisi, Sepeteri, Ore and Iba communities against daily human encroachment was equally adopted by the people. From the outcome of the study it is obvious that portions of land set aside as ancestral lands are jealously regarded as very distinct and the source of life for the people in each of the sampled communities. The beginning of the communities and their sustenance is defined in relation to these forests set aside for protection. Because community members are acutely aware of the possible consequences that could adversely affect them due to harm inflicted on their forest, they jealously guard it from unsustainable harvesting. Therefore, a vital lesson to be drawn from this to facilitate the actualisation of SDG13, is to raise awareness regarding how climate change would affect households in SSA, and globally if sustainable and environmentally friendly behavioural changes are not made.

Since the practice of deviance cannot be ruled out, taboos exist which guide the conduct of the people within the environment. including forests. The taboos as well as sanctions that go along with violations are fully ratified by community members due to their emotional attachment to the traditionally protected areas. Therefore, another vital lesson useful to actualise SDG 13 may be to draw on issues people are emotionally attached to and state how climate change could affect them to foster the adoption of pro-environmental behaviour. It is important to state that indigenous spiritual belief systems influence the sustainability of all the protected areas in this study. This finding suggests that it may be worthwhile to ascertain how place-based indigenous spiritual beliefs influence climate change mitigation and to develop policies towards reinforcing such practices to facilitate the actualisation of SDG 13. However, for SDG 13 to be achieved in Osun and Oyo States, complementary efforts must come from government and non-governmental agencies. These complementary efforts relate to harnessing the identified traditional and spiritual mechanisms for not just mitigating and adapting to the impacts of climate change, but also for enhancing resilience to them.

References

Abugiche AS, Egute TO, Cybelle A (2017) The role of traditional taboos and custom as complementary tools in wildlife conservation within Mount Cameroon National Park Buea. Int J Nat Resource Ecol Manag 2(3):60–68. https://doi.org/10.11648/j.ijnrem.20170203.13

Aderinto S (2017) Introduction: uncovering Africa's past. In: Aderinto S (ed) African kingdoms: an encyclopedia of empires and civilizations. ABC-CLIO, Santa Barbara, pp xiii–xxvi

Agai JM (2021) Reflection on the theory of the Arab origin of the Yorùbá people. Theologia Viatorum 45 (1):a77. https://doi.org/10.4102/tv.v45i1.77

Ajang, J.A. (2016) Our deities, our strength, the example of Nfam and Oolim in Ogoja Upper Cross River Region of Nigeria. *Lwati: A journal of Contemporary Research,* 13(3):31–40.

Ali N, Goettsch B, Shi Y, Hardcastle J, Oldfield S (2020) Study on the linkages between protected areas and the conservation of biodiversity for food and agriculture–thematic study for the state of the world's biodiversity for food and agriculture. FAO, Rome

Arora NK, Fatima T, Mishra I, Verma M, Mishra J, Mishra V (2018) Environmental sustainability: challenges and viable solutions. Environ Sustain 1:309–340. https://doi.org/10.1007/s42398-018-00038-w

Asante EA, Ababio S, Boadu KB (2017) The use of indigenous cultural practices by the Ashantis for the conservation of forests in Ghana. SAGE Open 7(1). https://doi.org/10.1177/2158244016687611

Awolalu JO, Dopamu PA (1979) West African traditional religion, 1st edn. Onibonoje Press and Book Industries, Ibadan, p 310

Ayaa DD, Waswa F (2016) Role of indigenous knowledge systems in the conservation of the bio-physical environment among the Teso community in Busia County-Kenya. Afr J Environ Sci Technol 10 (12):467–475

Ayivor JS, Nyametso JK, Ayivor S (2020) Protected area governance and its influence on local perceptions, attitudes and collaboration. Land 9(9):310. https://doi.org/10.3390/land9090310

Bádéjo DL (1996) Osun Séègésí: the elegant deity of wealth, power, and femininity. Africa World Press, Trenton

Bandoi A, Jianu E, Enescu M, Axinte G, Tudor S, Firoiu D (2020) The relationship between development of tourism, quality of life and sustainable performance in EU countries. Sustainability 12 (4):1628

Bennett E (2010) Encyclopedia of Africa. Oxford University Press, Oxford

Berazneva J, Byker TS (2017) Does forest loss increase human disease? evidence from Nigeria. Am Econ Rev 107(5):516–521. https://doi.org/10.1257/aer.p20171132

Bewaji JAI (2018) Yorùbá values and the environment. Yorùbá Stud Rev 3(1):229–249

Bologna M, Aquino G (2020) Deforestation and world population sustainability: a quantitative analysis. Sci Rep 10:7631. https://doi.org/10.1038/s41598-020-63657-6

Borisade TV, Odiwe AI, Akinwumiju AS, Uwalaka NO, Orimoogunje OI (2021) Assessing the impacts of land use on riparian vegetation dynamics in Osun State, Nigeria. Trees Forest People 5:100099. https://doi.org/10.1016/j.tfp.2021.100099

Butler RA (2019) Consequences of deforestation. https://rainforests.mongabay.com/09-consequences-of-deforestation.html

Claire K (2017) The environmental effects of non-eco-friendly packaging. Green J 12(3):45–56

Curtis PG, Slay CM, Harris NL, Tyukavina A, Hansen MC (2018) Classifying drivers of global forest loss. Science 361(6407):1108–1111. https://doi.org/10.1126/science.aau3445

Dada OT, Agbabiaka HI, Badiora AI, Odufuwa BO, Ojo DB (2020) Patronage variability in Olumo Rock and its impacts on a typical Nigerian community. Int J Tourism Cities 7(1):158–176. https://doi.org/10.1108/IJTC-02-2020-0023

Daniel SK, Jacob DE, Udeagha AU (2015) Tree species composition in selected sacred forests in Nigeria. Int J Mol Ecol Conserv 5(7):1–10. https://doi.org/10.5376/ijmec.2015.05.0007

Daye DD, Healey JR (2015) Impacts of land-use change on sacred forests at the landscape scale. Global Ecol Conserv 3:349–358

Desta A, Smithson S (2010) Indigenous knowledge in the context of natural resources management: an information systems perspective. In: UK academy for information systems conference proceedings, 16. http://aisel.aisnet.org/ukais2010/16/

Diawuo F, Issifu AK (2017) Exploring the African traditional belief systems in natural resource conservation and management in Ghana. J Pan Af Stud 8(9):119–126

Doyle A (2018) Africa's deforestation twice world rate, says atlas. www.reuters.com/article/us-africa-environment/africas-deforestation-twice-world-rate-says-atlas-idUSL1064180420080610

Durkalec A, Furgal C, Skinner MW, Sheldon T (2015) Climate change influences on environment as a determinant of indigenous health: relationships to place, sea ice, and health in an Inuit community. Soc Sci Med 136:17–26

Echendu AJ (2020) The impact of flooding on Nigeria's sustainable development goals (SDGs). Ecosyst Health Sustain 6:1. https://doi.org/10.1080/20964129.2020.1791735

Eneji CV, Ntamu GU, Ajor JO, Ben CB, Bassey JE, Williams JJ (2012) Ethical basis of African traditional religion and sociocultural practices in natural resources conservation and management in Cross River State, Nigeria. J Res Peace Gender Develop 2 (2):034–043

Food and Agriculture Organisation (FAO) (2015) Climate change and food security: risks and Responses. FAO

Food and Agriculture Organisation (FAO) (2016a) Global Forest Resources Assessment 2015: how are the world's forests changing? 2nd edn. FAO, Rome. http://ww.fao.org/3/a-i4793e.pdf

Food and Agriculture Organisation (FAO) (2016b) State of the world's forests 2016b. forests and agriculture: land-use challenges and opportunities. FAO, Rome. http://www.fao.org/3/a-i5588e.pdf

Food and Agriculture Organisation (FAO) (2020) Transforming agriculture and food systems: halting deforestation and promoting sustainable production and consumption of forest products. In: Twenty-fifth session of the committee on forestry, 5–9 Oct 2020. http://www.fao.org/3/nd565en/nd565en.pdf

Ibrahim A, Iheanacho AC, Bila Y (2015) Economic analysis of causes and impact of deforestation in Nigeria. J Agric Econ Soc Sci 1(1):142–150

International Fund for Agricultural Development (IFAD) (2018) Indigenous peoples' collective rights to lands, territories and natural resources: lessons from IFAD-supported projects. https://www.ifad.org/documents/38714170/40272519/IPs_Land.pdf/ea85011b-7f67-4b02-9399-aaea99c414ba

Keenan RJ, Reams GA, Achard F, de Freitas JV, Grainger A, Lindquist E (2015) Dynamics of global forest area: results from the FAO global forest resources assessment. For Ecol Manag 352:9–20. https://doi.org/10.1016/j.foreco.2015.06.014

Kihonge EW (2017) Assessing contribution of sacred natural sites to climate change effects on dryland ecosystem: case study: the Gabbra community, Marsabit County. M.Sc. thesis, University of Nairobi. http://erepository.uonbi.ac.ke/bitstream/handle/11295/101333/Kihonge_Assessing%20Contribution%20of%20Sacred%20Natural%20Sites%20to%20Climate%20Change%20Effects%20on%20Dryland%20Ecosystemcase%20Study-%20the%20Gabbra%20Community%2C%20Marsabit%20County.pdf?sequence=1&isAllowed=y

Koch A, Brierley C, Maslin MM, Lewis SL (2019) Earth system impacts of the European arrival and great dying in the Americas after 1492. Q Sci Rev 207 (1):13–36. https://doi.org/10.1016/j.quascirev.2018.12.004

Korff J (2021) Meaning of land to Aboriginal people. www.creativespirits.info/aboriginalculture/land/meaning-of-land-to-aboriginal-people

Lee CC, Chen MP (2021) Ecological footprint, tourism development, and country risk: international evidence. J Clean Product 279:123671. https://doi.org/10.1016/j.jclepro.2020.123671

Lines LA, Division YDFNW, Jardine CG (2019) Connection to the land as a youth-identified social determinant of Indigenous Peoples' health. BMC Public Health 19:176

Mabvurira V, Muchinako GA, Smit EI (2021) Shona traditional religion and sustainable environmental management: An Afrocentric perspective. Afr J Soc Work 11(3):111–118

Marx E (2018) Fightback against deforestation in Africa focuses on small farmers. https://www.reutersevents.com/sustainability/fightback-against-deforestation-africa-focuses-small-farmers

Maru Y, Gebrekirstos A, Haile G (2020) Indigenous ways of environmental protection in Gedeo community, Southern Ethiopia: a socio-ecological perspective. Cogent Food Agric 6:1. https://doi.org/10.1080/23311932.2020.1766732

Mittal I, Gupta RK (2015) Natural resources depletion and economic growth in present era. SOCH-Mastnath J Sci Technol (BMU, Rohtak) 10(3). https://ssrn.com/abstract=2920080

Miyamoto M (2020) Poverty reduction saves forests sustainably: lessons for deforestation policies. World Develop 127:104746. https://doi.org/10.1016/j.worlddev.2019.104746

Ndeye SD, Issa O, Zougmoré RB, Ouedraogo M, Partey ST, Gumucio T (2019) Factors influencing gendered access to climate information services for farming in Senegal. Gender Technol Develop 23 (2):93–110. https://doi.org/10.1080/09718524.2019.1649790

Norwegian Red Cross (2019) Overlapping vulnerabilities: the impacts of climate change on humanitarian needs. Norwegian Red Cross, Oslo, pp 19–22

Nwadike GU, Idenedo WO, Okoro RC (2020) Unsustainable marketing practices and environmental degradation in Niger delta region of Nigeria. Int J Manag Sci 7(5):58–75

Ogbanga MM, Amadi LA, George-Anokwuru C (2018) Environmental degradation as constraint to sustainable rural entrepreneurship in the Niger Delta, Nigeria. Ann Ecol Environ Sci 2(2):52–67

Ogundele A, Adebisi O (2016) Deforestation in Nigeria: the needs for urgent mitigating measures. IIARD Int J Geogr Environ Manag 2(1):16–18

Ogundipe AA, Obi S, Ogundipe OM (2020) Environmental degradation and food security in Nigeria. Int J Energy Econ Policy 10(1):316–324

Ojo O (2017) The slave ship Manuelita and the story of a Yorùbá community, 1833–1834. Tempo 23(2). https://doi.org/10.1590/TEM-1980-542X2017v230209

Okunade J (2019) #Egbaland: the famed Olumo rock, Abeokuta. https://mywovenwords.com/2019/09/egbaland-the-famed-olumo-rock-abeokuta-ogun-state.html

Okunola R, Ojo MOD (2016) Re-assessing the relevance and efficacy of Yorùbá gods as agents of punishment: a study of Sango and Ogun. Issues Ethnol Anthropol 7 (4):1057–1076

Oladeji SO, Osanyinleye GD, Lawal A (2021) Assessment of the conservation values of Osun Osogbo Sacred Grove, Osun State, Nigeria. Tanzania J for Nature Conserv 90(2):97–114

Oloukoi G (2014) Factors influencing women's access to water in Oke Ogun Region. Ethiop J Environ Stud Manage 7(6):654–663

Olowu D (2015) Indigenous knowledge systems and the African agenda for development in the age of globalisation. Maiden distinguished public lecture, Centre for Black Culture and International Understanding (UNESCO-Affiliated Institution), Osogbo, Nigeria, 30 June 2015

Olu-Osayomi O (2017) Dramatic aspect of Ese ifa in Yorùbáland. Int J Stud English Lang Lit (IJSELL) 5 (10):12–18

Owolabi JA (2017) Geo-historical backgrounds and religious traditions of Oke-Ogun People. J Arts Contemp Soc 9(1):4–5

Owoseni AO (2017) Water in Yorùbá belief and imperative for environmental sustainability. J Philos Culture Relig 28:13–19

Oyebola DO (2019) African traditional medicine: practices among the Yoruba of Nigeria. Christian Faith Publishing, Meadville

Oyetunji PO, Ibitoye OS, Akinyemi GO, Fadele OA, Oyediji OT (2020) The effects of population growth on deforestation in Nigeria: 1991–2016. J Appl Sci Environ Manag 24(8):1329–1334

Poore J, Nemecek T (2018) Reducing food's environmental impacts through producers and consumers. Science 360(6392):987–992. https://doi.org/10.1126/science.aaq0216

Ray C (2020) Holy wells and sacred springs. In: Ray C (ed) Sacred waters: a cross-cultural compendium of hallowed springs and Holy Wells. Routledge, London, pp 1–32

Rights and Resource Initiative (RRI) (2015) Who owns the world's land? a global baseline of formally recognized indigenous and community land rights. RRI, Washington, D.C.

Rowhani M, Hatala AR (2017) A systematic review of resilience research among indigenous youth in contemporary Canadian contexts. Int J Health Wellness Soc 7(4):45–58

Salami OB (2019) The role of slaves in Ibadan wars and diplomacy in the pre-colonial South Western Nigeria, 1835–1893. KIU J Soc Sci 5(3):335–343

Smith L, Panter J, Ogilvie D (2019) Characteristics of the environment and physical activity in midlife: findings from UK Biobank. Prevent Med 118:150–158. https://doi.org/10.1016/j.ypmed.2018.10.024.

Stanton J (2020) Orúmilà ní òò wàá bá ní: Examining the quantum measurement problem through the Yorùbá philosophy of Ifa divination. B.Sc. thesis, Brown University, USA

Stefanica M, Sandu CB, Butnaru GI, Haller AP (2021) The nexus between tourism activities and environmental degradation: romanian tourists' opinions. Sustainability 13:9210. https://doi.org/10.3390/su13169210

Suleiman MS, Wasonga OV, Mbau JS, Elhadi YE (2017) Spatial and temporal analysis of forest cover change in Falgore game reserve in Kano, Nigeria. Ecol Process 6:11. https://doi.org/10.1186/s13717-017-0078-4

Temitope AJ, Ogunbode TO (2019) Environmental impact of women exploitation of forest resources in some forest reserves in Osun State, Nigeria. MOJ Ecol Environ Sci 4(4):160–164. https://doi.org/10.15406/mojes.2019.04.00148

Thomas V (2017) Climate change and natural disasters: transforming economies and policies for a sustainable future. Routledge, New York. https://doi.org/10.4324/9781315081045

Tobias JK (2015) 'We are the Land': Researching environmental repossession with Anishinaabe elders. Ph.D. thesis, Electronic Thesis and Dissertation Repository, 2784. The University of Western Ontario. https://ir.lib.uwo.ca/etd/2784

UNEP and IRP (2020) Sustainable trade in resources: global material flows, circularity and trade. United Nations Environment Programme, Nairobi, Kenya. UNESCO

UNEP-WCMC (2016) The state of biodiversity in Africa: a mid-term review of progress towards the Aichi biodiversity targets. UNEP-WCMC, Cambridge

United Nations (UN) (2017) Sustainable development goals report 2017. https://unstats.un.org/sdgs/files/report/2017/TheSustainableDevelopmentGoalsReport2017.pdf

Usman A, Falola T (2019a) Religion and world view. The Yoruba from prehistory to the present. Cambridge University Press, Cambridge, pp 277–280

Usman A, Falola T (2019b) The nineteenth century: wars and transformations. The Yoruba from prehistory to the present. Cambridge University Press, Cambridge, pp 159–240

Van Der Meer T (2017) Spiritual journeys: a study of Ifá/Òrìṣà practitioners in the United States initiated in Nigeria. https://aura.antioch.edu/etds/337

Wood FE (2017) Objects and immortals: the Life of Obi in Ifa-Orisa religion. Ph.D. thesis, Harvard University. https://dash.harvard.edu/bitstream/handle/1/41140237/WOOD-DISSERTATION-2017.pdf?sequence=1&isAllowed=y

World Health Organization (WHO) (2016) Global health observatory data repository: prevalence of insufficient physical activity among adults. http://apps.who.int/gho/data/node.main.A893?lang=en

Musediq Olufemi Lawal (PhD) is currently a member of the faculty and senior lecturer in the Department of Sociology, Osun State University, Nigeria. His teaching and research areas include health, cultural anthropology, environment, population and development.

Oluwagbemiga Oluwaseun Ajayi (PhD) is a lecturer, faculty member and currently the head of Department of Sociology, Osun State University, Osogbo, Nigeria. His research interests are in the areas of child and youth development, religion and social problems. He has contributed to books and published in several local and international academic Journals.

Adebayo Emmanuel Akinyemi is a doctoral student in the Department of Sociology, University of Ibadan, Nigeria. He is equally a member of the faculty and lecturer in the Department of Sociology, Osun State University, Osogbo, Nigeria. His area of research interest includes religion, environment, social theory, social problems and development. He has participated in several conferences and workshops at both local and international levels. He is a fellow, SSRC NextGen 2018.

5 Indigenous Women's Vulnerability to Climate Change and Adaptation Strategies in Central Africa: A Systematic Review

Henry B. Tantoh, Eromose E. Ebhuoma, and Llewellyn Leonard

5.1 Introduction

Despite climate change being a global phenomenon, its impacts on various demographic groups are felt differentially (Ebhuoma et al. 2020; Nyantakyi-Frimpong 2020). Thus, the generalised notion of environmental threats distorts responsibilities and vulnerabilities that are distributed disproportionately between developed and developing countries, urban and rural communities, and between men and women. For instance, in the developing world, communities are dependent on natural resources and ecosystems; therefore, they rely on the natural environment for their livelihoods and wellbeing (Tantoh and Simatele 2017; Tantoh and Mckay 2021). To this end, these communities share a complex cultural relationship with their natural surroundings. Unfortunately, this renders them vulnerable to the effects of climate change (International Union for Conversation of Nature (IUCN) 2008; International Labour Organization (ILO) 2019a, b). This is worse for women who are disproportionately affected by climate change in most rural communities as they are subjected to marginalisation based on gender (International Fund for Agricultural Development (IFAD) 2004; Food and Agriculture Organization of the United Nations (FAO) 2010; Pentlow 2020; Tantoh and McKay 2020), despite being responsible for about 80 per cent of the food production in sub-Saharan Africa (SSA) (Africa Development Bank (AfDB) 2011). Gender is a hierarchical concept with a range of features relating to, and differentiating between, femininity and masculinity. These may include, for example, gender identity, biological sex, and sex-based social structures such as gender roles based on the context (Haig 2004). In the context of this chapter, gender inequality and discrimination refers to the bias and barriers faced by women and girls such as the management of natural resources, decision-making processes and access to credits, among others.

It is argued that women in SSA are vulnerable to climate aberrations not only because of their

H. B. Tantoh (✉)
Department of Geography and Planning, University of Bamenda–Cameroon, Bamenda, Cameroon
e-mail: bikwibilith@gmail.com

E. E. Ebhuoma · L. Leonard
Department of Environmental Sciences, School of Ecological and Human Sustainability, College of Agriculture and Environmental Sciences, University of South Africa (UNISA), Johannesburg, South Africa
e-mail: ebhuoee@unisa.ac.za

L. Leonard
e-mail: llewel@unisa.ac.za

E. E. Ebhuoma and L. Leonard (eds.), *Indigenous Knowledge and Climate Governance*, Sustainable Development Goals Series, https://doi.org/10.1007/978-3-030-99411-2_5

gender, but also because of the socially constructed roles as primary caregivers (United Nation Development Programme (UNDP) 2010; Ebhuoma et al. 2020). Furthermore, women bear most of the brunt of the impacts of climate change because they are among the poorest of the poor and are often marginalised in decision-making processes (Carvajal-Escobar et al. 2008; ILO 2017; Tantoh and Mckay 2020). In addition, women have restricted rights, limited access to resources, limited mobility and limited voice in the community and household decision-making, which are directly related to cultural barriers (Tantoh and Mckay 2020). This makes them much more vulnerable than men to the effects of climate change (Ebhuoma et al. 2020). Consequently, dangers associated with climate change have the potential to crystallise gender imbalances and even erode progress that has been made towards gender equality in many developing countries. This unfair treatment can result in unfortunate consequences for all as women play a unique role in the stewardship of natural resources and support to households and communities. However, despite being exposed to different kinds of environmental transformations, they have developed coping strategies to face these dire situations (UNDP 2010). For example, the Chipko movement in Uttarakhand, India, which was inspired by women the 1970s and which vehemently resisted government's proposal for commercial forestry on their land, became famous for the tactic of embracing trees as their engagement with subsistence agriculture and their role in nurturing and protecting communities, as they were acutely aware of the ecological and social threats posed by commercial forestry (Brown 2014). Therefore, women may be holders of valuable knowledge that can be useful to scaling up mitigation and adaptation to climate change (IUCN 2008).

Arguably, a vital piece of this expertise is hinged on their indigenous knowledge of natural resource management, agricultural practices, predicting weather patterns and upholding customary institutions for their sustenance for centuries (Borrini-Feyerabend et al. 2004; ILO 2019a, b). This knowledge is strongly linked to traditional and religious beliefs, customs, folklore, land-use practices and community-level decision-making processes, and have historically been dynamic, responding to changing environmental, socio-economic and political conditions to ensure the sustainability of natural resources and the provision of tangible and intangible benefits for posterity (Parrotta et al. 2009). Generally, indigenous people continue to face social, economic and environmental vulnerabilities. Nevertheless, they have devised and continue to devise ways to cope with and confront the changes occurring in their surrounding environment (UNDP 2010; ILO 2019a, b). This reinforces the viewpoint of Adeyeye, Hagerman and Pelai (2019) that indigenous knowledge is a 'living thing', which will continue to evolve. Consequently, new knowledge will consistently be produced based on adaptation and innovation because indigenous knowledge is living knowledge. It is, therefore, not surprising that indigenous people are considered as "the vanguard of running a modern economic model based on the principles of a sustainable green economy" (ILO 2017:1). This is reinforced by the overwhelming value and role of traditional knowledge in the management of natural resources. This distinction is further exemplified by the commitment of the international community to ensure inclusivity towards sustainability through the 2030 Agenda for Sustainable Development and the 2015 Paris Agreement on Climate Change.

In the context of SSA, indigenous knowledge and practices have sustained the livelihoods, cultures and agricultural practices of local communities for decades (Borrini-Feyerabend et al. 2004; Ebhuoma et al. 2019). It is recognised that subsistence farmers in SSA depend on indigenous knowledge to determine the time, quantity and types of food crops to produce in a particular farming season (Nkomwa et al. 2014; Ebhuoma et al. 2020). However, the effects of climate change represent one of the most challenging threats to agriculture in particular, and yet, it is a source of livelihoods to about 70% of SSA

(Hellmuth et al. 2007; Tantoh et al. 2021). It is also argued that women represent about 40% of the agricultural workforce in 46 of Africa's 54 countries (AfDB 2011). In this context, the human threats inherent in climate change are crucial and may be more serious for women in SSA and, particularly, in Central Africa where approximately 60% of its population live in extreme poverty—by far the highest regional poverty rate in Africa (Institute for Security Studies (ISS) 2018). While several studies have examined the nexus between gender roles, agriculture and climate change adaptation in Central Africa (Carvajal-Escobar et al. 2008; UNDP 2010; Ahearn and Tempelman 2010), gender and water (Tantoh and Mckay 2020), gender inequalities and adaptation to climate change (Balehey et al. 2018) and indigenous people, climate change and gender perspectives (ILO 2017; Pentlow 2020; Ebhuoma et al. 2020), the dynamics that lead some indigenous women to become more vulnerable than their male counterparts in pursuance of climate change adaptation have not received tremendous scholarly attention.

Against this background, this study examined the vulnerability of women in Central Africa to climate change and how they adapt to the effects of climate change. Doing so could enable interested groups, including development practitioners, to easily pinpoint issues that need to be addressed to redress indigenous women's vulnerability to climate change, which in turn, facilitates the actualisation of the thirteenth sustainable development goal (SDG 13)—climate action. This chapter proceeds as follows: first, it describes the linkages between indigenous knowledge, gender and climate change in SSA followed by a detailed description of the systematic review process of the relevant literature. The next section discusses the themes that emerged from the systematic review in relation to what they could mean for the actualisation of SDG 13, while the final section is the conclusion and makes recommendations that could dismantle the gender divide that makes women more vulnerable to climate change than their male counterparts.

5.2 Gender Roles, Climate Change and Inequalities Among Indigenous Communities

Mindful of the gendered nature of the subject of climate change, it is widely acknowledged that effective adaptation to climate change will require a by-line of the gender dimension of climate change (Gurung and Mwanundu 2006). A gendered approach to climate change requires more than a set of disaggregated data showing that climate change has a differential impact on women and men; an understanding of existing inequalities between men and women and how climate change can exacerbate these inequalities is also required (Alyson et al. 2008). A gendered approach also requires taking into consideration such imbalances in developing climate change adaptation and mitigation policies (UNDP 2010; 2012), identifying the gendered inequality in vulnerability to climate (Carvajal-Escobar et al. 2008), and gender mainstreaming of any strategy or plan of action implemented to improve adaptation to climate change. The concept of gender in development recognises that men and women both hold different positions, responsibilities and decision-making authorities in the household and community. They also play different roles in society, have dissimilar control over and use of resources, and often have different views and needs (UNDP 2010). While rural women and men may play complementary roles in farming activities, agriculturalists observed that women tend to play a greater role in natural resource management and ensuring nutrition and the provision of potable water in the household (AfDB 2011).

In Africa, women often grow, process, manage, and market food and other natural resources. They are responsible for raising small livestock, managing vegetable gardens and, as already noted, provisioning fuelwood and water. For example, rural women in SSA provide about 80% of food production as opposed to men who are more engaged in cash crops for export (AfDB 2011). However, there are cases where men and women both grow commercial crops. This is common in countries like Burkina Faso,

Table 5.1 Possible climate changes and their impacts on women and indigenous people

Climate change impacts	Possible threats	Examples	Possible effects on women and Indigenous people
Direct effects	Increased temperature in water bodies (lakes, streams, rivers, oceans)	An increasing rate of coral bleaching due to thermal stress	Loss of coral reefs can affect the tourism industry, a sector in which women comprise 46% of the workforce
	Increased drought and water shortage	-This leads to crop and animal loses - Damage to crop and animal quality	-Women and girls in developing countries are often the primary collectors, users and managers of water -Water scarcity will endanger particularly rural livelihoods, increase their workloads, affect performance in school and reduce the time for income-generating activities
	Limited access to resources (land, irrigation technology, credit facilities)	- Reduction in yields -Drop in productivity of crop land and the propagation of insect infection/plant diseases	-Poor rural women are the most vulnerable to the effects of climate change due to their poor adaptation capacities -Low-income earners
	Increased extreme weather events	Frequent and intense natural disasters (heat waves, cyclones, hurricanes)	-Natural disasters affect women and children most as they are always left behind after an episode to clear the damage while the men look for employment elsewhere
Indirect effects	Proliferation of diseases	Climate variability plays a critical role in malaria epidemics in Central Africa (Cameroon, Chad, Central Africa Republic, Gabon, Equatorial Guinea, Congo-Brazzaville)	- Women have less access to medical services than men - They care for the sick, which increases their workloads - Rural households with HIV/AIDS-infected patients have fewer resources to adapt to the effects of climate change - Differential vulnerability between female-headed households
	Loss of species	Climate variability and change result in species extinction	- Climate change affects food security, stressing women more because of their traditional roles (sustenance of household needs-food, water, fuelwood, etc.)
	Decreased crop production	Reduction of crop and livestock yields due to extreme weather events	-Rural women, in particular, are responsible for half of the world's food production and produce between 60–80% of the food in most developing countries

Adapted from Lambrou and Piana (2006), Gurung and Mwanundu (2006), UNDP (2010)

Tanzania and Zambia. In Kenya, South Africa, Uganda and Zimbabwe, where production is dominated by females, they also engaged in non-traditional agricultural exports (FAO 2010; Ebhuoma et al. 2020). In this context, the human threats inherent in climate change are crucial and they may be more serious for women in such regions (Table 5.1).

Among the marginalised, women and children, particularly handicapped women, are disproportionately marginalised and vulnerable to the effects of climate change (AfDB 2009, 2011). It is argued that climate change has specific effects on women and men because of the different roles they play in society and their differentiated access to social, economic and physical resources (AfDB 2011). Women may further suffer from cultural exclusion and marginalisation in decision-making processes, for example, because of their sex and traditional gender roles exacerbating the inequalities they experienced in climate change (see Table 5.1). Therefore, the roles of women in the family as well as their social positions change due to the effects of climate change—they are aggravated by the effects of climate change on essential aspects such as food, water and energy supply (AfDB 2009).

In terms of household tasks, the roles of women and girls are likely to be more severely impacted by changes in the climate. For example, rural women, and girls to some extent, frequently provide households with water and fuelwood for heating and cooking (see Table 5.1). The time needed for their work of gathering water and fuelwood will likely increase due to water shortages and potential deforestation (Tantoh and Mckay 2018). The reduction of available opportunities for women for food production and preparation, as well as their participation in income-generating activities, will likely affect household food security and nutritional wellbeing. This is because of women's heavy reliance on subsistence agriculture and their contribution to the production of economic crops, agriculture for their livelihoods (AfDB 2011), which further expose them to the effects of climate change. In this regard, indigenous people, particularly women and children, are uniquely at risk because they are often placed at the forefront of the direct effects of both climate change and climate-related mitigation and adaptation actions, despite being among those who have contributed the least to climate change (IPCC 2018). While the risks faced are manifold, indigenous people and women are increasingly recognised as agents of change in achieving strong and meaningful climate mitigation and adaptation action (Moyo 2014; Pentlow 2020). Therefore, several positive developments have focused on indigenous people's issues and traditional knowledge more specifically, creating many avenues for taking forward multiple approaches to sustainable development and climate action.

Furthermore, benefactors and international organisations have become more gender-sensitive to climate change responses. One example of this is the United Nations Convention which advocates for the Elimination of All Forms of Discrimination against Women (CEDAW) in its response to climate change. This entails understanding existing inequalities between women and men and how climate change can exacerbate these inequalities, and understanding how these inequalities can intensify the impacts of climate change on all individuals and communities (UNDP 2010).

5.3 Methodology

This paper is based on a systematic review of existing academic literature on climate change, indigenous perspectives, gender roles and adaptation strategies in SSA. Systematic reviews are objective with a consistent technique to acquire results (Ahn and Kang 2018). They usually have high levels of evidence-based studies that are facilitated by the research questions. SSA is home to some of the oldest indigenous populations like the pygmies in the South-Eastern part of Cameroon. Furthermore, about 70% of the population of this region derives their source of livelihood and wellbeing from environmental resources which are, unfortunately, exposed to the effects of climate change. In addition, gender inequality, for example, is a major characteristic with the prevalence of patriarchal tendencies. The Central African sub-region encompasses the equator and contains the Congo Basin, with the second-largest rainforest in the world, after the Amazon Basin in South America. The region comprises Cameroon, Chad, the Central African Republic, Equatorial Guinea, Gabon and the Congo-Brazzaville. However, three countries were explored (Cameroon, Chad, Equatorial Guinea and Congo-Brazzaville) because they are in the Gulf of Guinea (with the exception of Chad) with similar characteristics, such as the equatorial tropical climate, vegetation, and indigenous communities (pygmies), among others. Furthermore, the region is home to some of the most susceptible indigenous populations of the world because of the sensitivity and fragility of its natural environment, and its high dependence on environment-based livelihoods such as agriculture, an issue that is equally at play in Chad (Edmond 2019). Gender mainstreaming constitutes significant support for development planning and an entry point for the successful management of natural resources in the region (Pentlow 2020; Tantoh et al. 2021).

An internet search was done on databases such as Google Scholar, ScienceDirect, Scopus of indexed journals and other peer-reviewed publications using keywords such as climate change, indigenous knowledge, gender roles and adaptation strategies. These search engines—ScienceDirect, Scopus and Google Scholar—are among the most effective for literature review search. An estimated total of 475 peer-reviewed journal articles and reports were generated from 2004–2021 and methodically reviewed using a four-stage process (Fig. 5.1). This was done to reduce the literature and select only those articles and reports that were relevant to the study. This process resulted in the selection of 56 different articles and reports which formed the basis for the current chapter.

Thirty-two out of the 56 pieces of literature examined in this study was obtained from SSA, while 10 was from elsewhere for comparative reasons. The topics of the literature examined related to indigenous perspectives on climate change, gender roles and adaptation strategies in SSA, which were later used to structure the results and discussion. The abstracts of these academic literature generated were originally reviewed and those directly related to the topic were retained and further perused. The references of the retained academic literature were also exploited to trace other important studies, which were further examined. This guided the structure and formed the basis of our study. Some of the academic literature consulted beyond 2010 was used because of their pertinence to the topic under discussion.

5.4 Results and Discussion

The results for this chapter were obtained from academic literature from different regions and countries in SSA. The major themes that were extracted from the findings were harmonised and this guided the structure and analysis of the results. However, other academic literatures from South America that were directly related to the subject under discussion were used for comparison (Table 5.2).

5.4.1 Vulnerability of Indigenous Women in Central Africa to Climate Change

5.4.1.1 Differentiated Gender Roles at Household Level

Climate change is a global reality and indigenous communities and poor rural women in the developing world are the most affected group. Women are mostly affected because they do most of the work to adapt, as they are historically responsible for food supplies and rearranging the household after disasters and the men go out to look for jobs (UNDP 2012; ILO 2019a, b; Pentlow 2020). This disparity in Central Africa, for example, is inherent in culture and the social position of women in the family and is exacerbated by the effects of climate change, particularly on their means of sustenance (AfDB

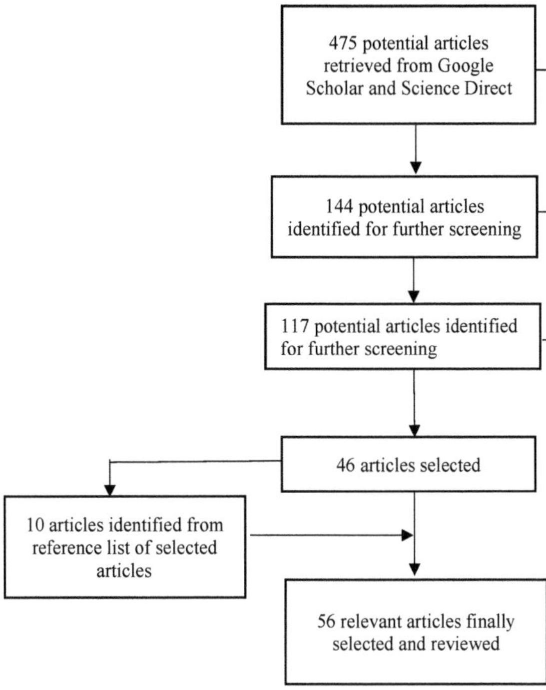

Fig. 5.1 Flow chart of article screening and selection process

2011; UNDP 2012; ILO 2019a, b). Furthermore, the effects of climate change modify the tasks of men and women in society. Rural women, for example, are responsible for cooking and providing water and fuelwood for heating and cooking. These activities are time consuming and are expected to increase as the adverse effects of climate compromise the availability of natural resources (water, forests).

The division of roles, in which women are 'restrained' from actively engaging in activities within the domestic space while affording their male counterparts the liberty to make informed decisions regarding the sort of tasks to engage in to obtain their livelihoods, means that threats emanating from climatic risks can, in the worst situation, compel men in the agrarian sector to seek alternative sources of livelihoods. Women are restrained from exercising this level of freedom because of their deep-rooted attachments to the domestic sphere, including, in some cases, the added tasks of acting as caregivers, and this stifles their ability to pursue occupational change (Tantoh and Mckay 2018; Ebhuoma et al. 2020; Pentlow 2020). In this light, the human threats integral in climate change are crucial and are more severe for indigenous communities, particularly women involved in agriculture and other natural resource exploitation activities. However, it is noteworthy to mention that while discrimination against women is more common, men also often face discrimination if they wish to perform commercial and family roles that are typically seen as 'feminine' (UNDP 2010). Although gender inequality has restricted the contribution of women to better manage natural resources and hydro-meteorological hazards in constructing processes of adaptation to climate change, it is essential to acknowledge the productive, reproductive and community roles that both men and women play in society (Carvajal-Escobar et al. 2008).

Table 5.2 The regions, countries and major themes of academic literature used in the study

Region	Country	Number	Themes
Central Africa	Cameroon	2	Gender and climate change
		1	Gender roles & natural resource management
		2	Gender equality & women empowerment
	Chad	1	Gender issues
	Equatorial Guinea	1	Gender issues
	Central Africa Republic	1	Community-based adaptation strategies
West Africa	Nigeria	1	Asset vulnerability and climate change
	Ghana	1	Gender issues
		1	Smallholder farmers & climate extremes
East Africa	Ethiopia	1	Gender and climate change
		1	Gender, inequality & climate change adaptation
	Kenya, Uganda & Zimbabwe	2	Gender issues
Southern Africa	South Africa	1	Subsistence farming and vulnerability
		1	Women, poverty alleviation & development
		1	Indigenous people and protected areas
	Zambia, Democratic Republic of Congo	2	Gender issues
Sub-Saharan Africa	The entire SSA, excluding the aforementioned	3	Climate risk management
		2	Indigenous people & nature conservation
		3	Vulnerability and adaptive capacity
		13	Climate related issues, vulnerability and adaptation
South America	Mexico and Uruguay	1	Women's role & climate change adaptation
		2	Gender and climate change
		1	Environmental governance
		1	Traditional forest related knowledge
World		10	Climate change and community-based adaptation strategies

Source Authors (2021)

5.4.1.2 Patriarchal Influence on Access to Resources and Institutional Practices

Existing research such as studies done by Ahearn and Tempelman (2010), AfDB (2011), Balehey et al. (2018), ILO (2019a, b), Pentlow (2020), among others, recognise the critical and differentiated roles of men and women in society and their differentiated access to land, capital and other natural resources. At the household level, the ability to adapt to climatic changes depends on access and control over land, money, credit and technology (Lambrou and Piana 2006). As such, women are often less able to adapt to climate change than men since they represent the majority of low-income earners and they are often denied rights to property and land, which make it difficult for them to access credit and agricultural extension services (Namubiru-Mwaura 2014). In communities where women are disenfranchised from active participation in economic or public functions that their male counterparts take part in, they become more vulnerable to climate change due to the greater exposure to the risks they are exposed to. According to Onwutuebe (2019, p. 4), "men whose privileged power status (in terms of financial, economic, social, and political strength) depends largely on concessions provided by patriarchy invariably possess stronger adaptive capability to cope with adverse impacts of climate change". This is implicitly and explicitly crystallised in studies conducted in Cameroon (Fonjong and Gyapong 2021), the Democratic Republic of Congo (Peterman et al. 2011) and Chad.

Patriarchy embodies normative values and judgements, which, on the basis of biological differences between males and females, provide for the division of labour, privileges and opportunities between men and women. Patriarchy reinforces male supremacy, as it does not recognise any form of equality between men and women (Onwutuebe 2019, p. 2). While studies in Central Africa advocated for gender equality in access to resources in rural areas to dismantle patriarchy (Peterman et al. 2011; Masson et al. 2019; Fonjong and Gyapong 2021), the campaign for women empowerment is faced with climate-related problems that are heavily underpinned by patriarchy. Meza (2010), for example, asserted that because of social contexts that fuel uneven power relations, men and women are unequally equipped to adapt to disasters, which implies that the adverse impact of climate change is not uniform across genders.

5.4.1.3 Underrepresentation of Women in Rural Institutions

In the developed world, there is a growing recognition of rural women's roles in and contribution to agricultural production and the achievement of the Sustainable Development Goals (SDGs), in particular in ensuring food security, improved nutrition and promoting sustainable agriculture (SDG 2) and in drawing on their wealth of experience to build their resilience and adapt to climate change (SDG 13) by virtue of being custodians of indigenous knowledge (ILO 2019a, b). However, persistent discriminatory social institutions hinder rural women's empowerment. These institutions include discriminatory practices, which undermine women's rights to own, control or use land and non-land assets; discriminatory practices that restrict women's access to financial services; and social norms imposing that women's assets be mediated only by men. Discriminatory social institutions refer to both formal and informal laws, social norms and practices that restrict women's and girls' rights, access to empowerment opportunities and resources (Organisation for Economic Co-operation and Development (OECD) 2014). Women and girls are especially vulnerable to food insecurity, partly because they are often denied fundamental human rights such as the right to own and control the land they cultivate and to have access to productive resources. This matter has been highlighted in Cameroon (Tchoundjeu et al. 2010) and Equatorial Guinea (Castillo-Rodriguez and Valenciano-Mañé 2020).

A catalyst for maintaining the aforementioned status quo is partly that women are underrepresented in rural institutions, an issue not just prevalent in Central Africa but across the African continent (Okeke et al. 2017). This has implications for distribution of resources to improve the welfare of farmers. For example, gender biases in some institutions often reproduce assumptions that it is men who are the farmers (Gurung and Mwanundu 2006). While this reference is from more than a decade ago, it can be argued that this sort of bias may be prevalent in some rural communities in both Central Africa and the entire SSA region for the selfish gains of men, especially when government officials plan to distribute resources to farmers. This may be one way in which rural men assert their masculinity, as they feel threatened by the widespread advocacy for gender equality (Peterman et al. 2011; Masson et al. 2019; Fonjong and Gyapong 2021).

5.4.2 Adaptation Strategies of Indigenous Women in Central Africa to Climate Change

Even though women and indigenous groups are victims of climate change, they also contribute considerably to adaptation. The dependence of women and indigenous people on the natural environment has put them in good stance to appreciate and transform their livelihoods and wellbeing adaptation strategies. For example, rural women and indigenous people are the first to learn about deteriorating water quality and quantity. It is argued that successful adaptation to climate change requires the acknowledgement of the gender dimension of climate change (Gurung and Mwanundu 2006; Carvajal-Escobar et al. 2008). In addition, it is acknowledged that women are often in a better position to note certain environmental hazards (Moyo 2014; Pentlow 2020) and are knowledgeable about what is needed and which native trees should be protected to ensure the sustainability of their livelihood activities. The following section sheds light on the techniques employed by indigenous women in Central Arica to reduce their vulnerability to climate change.

5.4.2.1 Social Capital

Findings from the literature highlighted that indigenous women establish networks with other women, which increases their social capital (Epo 2012). At local levels one finds the women who have greater clarity about what diverse social groups lose in the short term after hazards such as inundation, cyclones, earthquakes and drought in the community (Carvajal-Escobar et al. 2008). Membership, especially in informal saving groups, is an indicator of the cognitive aspects of social capital such as trust, solidarity and sharing. In Cameroon, for example, a study discovered that belonging to a religious faction and not being married contributed to building social capital (Epo 2012). Social capital in a community can facilitate social learning, which has been documented to facilitate the sharing of knowledge needed to enhance adaptation to climate change (Srang-iam 2013; Ensor and Harvey 2015).

5.4.2.2 Community-Based Adaptation Strategy

The effects of climate change specifically affected the world's poorest and most marginalised countries and communities in SSA Africa. These effects are mostly felt in water resources through extreme weather events like droughts, floods and violent storms (Nguimalet 2018). Droughts and floods, for example, are hydro-climatic extremes and have historically upset the livelihoods of communities in the watersheds in Central African Republic (CAR) (Nguimalet 2018). However, international development organisations such as the United Nations Framework Convention on Climate Change (UNFCCC) and the Intergovernmental Panel on Climate Change are making efforts to upscale and empower developing countries, in particular, to adapt to the effects of climate change (Reid et al. 2007). While such efforts are not always aimed specifically at community-based adaptation (CBA) (Reid et al. 2007), some of them, such as the Adaptation Fund, target the most vulnerable communities.[1] In addition, however, most adaptation efforts have been top-down in nature with little attention directed to communities' experiences of climate change and their efforts to cope with their changing environments. Therefore, rural communities initiated strategies to mitigate and adapt to the effects of climate change. Among the various strategies practiced by rural communities in Central Africa are migration against environmental hazards (floods, draughts, etc.), incorporating climate change information (local and scientific knowledge will help individuals and communities to make informed decisions), participatory disaster risk reduction methods (community mapping, seasonal calendars, community mapping and modelling, stakeholder analysis, etc.) and changing activities (Reid et al. 2007; Nguimalet 2018; Donkor et al. 2020, 2021). In the case of watersheds in the Central African

[1] https://www.adaptation-undp.org/projects/upscaling-community-based-adaptation-ethiopia#tab-links.

Republic, for example, most rural communities advocate for raising awareness and giving training in climate change, increasing the number of boreholes to certify regular water supply, installing traditional wells, protecting water and forests against bushfires, and revising agricultural calendars as pathways to advance communities' adaptation approaches (Nguimalet 2018). Despite the fact that these adaptation strategies generally sought to improve the livelihoods and wellbeing of rural dwellers in Central Africa, different approaches and frameworks work differently for different communities.

5.5 Conclusion

Climate change is a global problem that affects different communities, individuals and ecosystems differently due to the inherent disparities in their vulnerability. Furthermore, climate change and the potential risk of disasters are developmental problems that constrain the sustainability of indigenous communities, while the effect is often worse for women, partly due to gender inequalities at grassroots level. This study, therefore, examined why indigenous communities and particularly women in Central Africa are vulnerable to climate change and how they adapt to the effects of climate change. The findings of this systematic review indicated that women and indigenous communities are both victims of climate change due to their dependence on subsistence agriculture and natural resources with which they share a complex relationship for their livelihoods and wellbeing. Rural women are the poorest of the poor, particularly handicapped women, and are disproportionately vulnerable to the effects and potential disasters of climate change. Indigenous women in Central Africa are marginalised when it comes to ownership of land, management of natural resources and access to credit facilities—exposing them more to the brunt of climate change. This is attributed to the differentiated gender roles at household level, patriarchal influence on access to resources and institutional practices and underrepresentation of women in rural institutions. This reinforces pre-existing traditional inequalities with men.

Despite the aforementioned challenges that worsens indigenous women's vulnerability to climate change than, they draw on their social capital and utilise community-based adaptation to minimise their vulnerability and build resilience. In this regard, faced with the manifestations of climate change, their knowledge and experiences in striving to build their resilience and adaptive capacity should be taken into account when planning adaptation processes for vulnerable communities. This would likely facilitate the attainment of SDG 13. However, it should be acknowledged that men play a decisive role in these processes. Thus, inclusive and consultative adaptation strategies that give participation to both men and women exposed to risks are essential for gender equity and sustainable outcomes. Community-based adaptation solutions have the potential and traditionally appropriate strategies to counteract the burdens of climate change, promote adaptation and contribute to sustainable human development at the grassroots level in Central Africa. Adaptation is also an avenue for questioning and changing traditional gender relations in society.

References

Adeyeye Y, Hagerman S, Pelai R (2019) Seeking procedural equity in global environmental governance: indigenous participation and knowledge politics in forest and landscape restoration debates at the 2016 World Conservation Congress. Forest Policy Econ 102006:1–14

AfDB (2011) Climate change, gender and development in Africa 1(1). https://www.afdb.org/sites/default/files/documents/publications/climate_change_gender_and_development_in_africa.pdf. Accessed 25 Mar 2021

AfDB (2009) Gender equality and women's empowerment: an updated gender plan of action (UGPOA) 2009–2011. ADB/BD/WP/2009/84; ADF/BD/WP/2009/60. African Development Bank Group, Tunis

Ahearn MC, Tempelman D (2010) Gender issues in agricultural and rural household well-being. Paper presented at the Third global conference on agricultural and rural household statistics, 24–25 May 2010, Washington DC

Ahn E, Kang H (2018) Introduction to systematic review and meta-analysis. Korean J Anesthesiol 71(2):103–112

Alyson B, Justina D, Emily E (2008) Gender and climate change: mapping the linkages: a coping study on knowledge and gaps. Mekelle, Tigray, Ethiopia. http://www.bridge.ids.ac.uk/sites/bridge.ids.ac.uk/files/reports/Climate_Change_DFID.pdf. Accessed 10 Mar 2021

Balehey S, Tesfay G, Balehegn M (2018) Traditional gender inequalities limit pastoral women's opportunities for adaptation to climate change: evidence from the afar pastoralists of Ethiopia. Pastoralism: Res Policy Practice 8:23. https://doi.org/10.1186/s13570-018-0129-1

Borrini-Feyerabend G, Kothari A, Oviedo G (2004) Indigenous and local communities and protected areas: towards equity and enhanced conservation. Switzerland and Cambridge, IUCN, Gland

Brown T (2014) Chipko Legacies - sustaining an ecological ethic in the context of agrarian change. Asian Stud Rev 38(4):639–657

Carvajal-Escobar Y, Garcia-Vargas M, Quintero-Angel M (2008) Women's role in adapting to climate change and variability. Adv Geosci 14:277–280

Castillo-Rodriguez S, Valenciano-Mañé A (2020) Women in equatorial guinea. Oxford Res Encyclopedia Afr History:1–34. https://doi.org/10.1093/acrefore/9780190277734.013.657

Donkor FK, Kevin M, Enokenwa O, Tantoh HB et al (2020) Attitudinal change towards agriculture through the generational lens and impact on engagement in related activities: case study from a mountainous area. In: Squires VR, Gaur MK (eds) Food security and land use change under conditions of climatic variability. https://doi.org/10.1007/978-3-030-36762-6_10

Donkor FK, Kevin M, Enokenwa O, Tantoh HB et al (2021) Mainstreaming education into disaster management to facilitate disaster resilience. In: Eslamian S, Eslamian F (eds) Handbook of disaster risk reduction for resilience. Cham, Springer. https://doi.org/10.1007/978-3-030-61278-8_10

Ebhuoma E, Simatele MD, Tantoh HB, Donkor FK (2019) Asset vulnerability analytical framework and systems thinking as a twin methodology for highlighting factors that undermine efficient food production, Jàmbá: J Disaster Risk Stud 11(1):a597. https://doi.org/10.4102/jamba.v11i1.597

Ebhuoma E, Donkor FK, Ebhuoma OO, Leonard L, Tantoh H (2020) Subsistence farmers' differential vulnerability to drought in Mpumalanga Province, South Africa: Under the political ecology spotlight. Cogent Social Sciences 6(1):1792155

Edmond C (2019) These indigenous women are paying the price of climate change. https://www.weforum.org/agenda/2019/12/lake-chad-africa-climate-change-water-indigeous-women/. Accessed 8 Feb 2022

Ensor J, Harvey B (2015) Social learning and climate change adaptation: evidence for international development practice. Wires Clim Change 6:509–522. https://doi.org/10.1002/wcc.348

Epo BN (2012) Implications of access to microcredit and social capital for female entrepreneurship in Cameroon. https://media.africaportal.org/documents/Implications_of_Access_to_Microcredit_.pdf Accessed 15 Nov 2021

Food and Agriculture Organization (2010) Gender dimensions of agricultural and rural employment: differentiated pathways out of poverty. Status, Trends and Gaps. FAO, Rome

Fonjong L, Gypong A (2021) Plantations, women, and food security in Africa: interrogating the investment pathway towards zero hunger in Cameroon and Ghana. World Dev 138:105293. https://doi.org/10.1016/j.worlddev.2020.105293

Gurung JD, Mwanundu S (2006) Gender and desertification: expanding roles for women to restore drylands. Rome. http://genderandenvironment.org/resource/gender-desertification-expanding-roles-women-restore-dryland-areas/. Accessed 10 Mar 2021

Haig D (2004) The inexorable rise of gender and the decline of sex: social change in academic titles, 1945–2001. Archives Sexual Behav 33(2):87–96

Hellmuth ME, Moorhead A, Thomson MC, Williams J (eds) (2007) Climate risk management in Africa: learning from practice. New York

International Union for Conversation of Nature (IUCN) (2008) Indigenous and traditional peoples and climate change. Issues Paper. https://www2.ohchr.org/english/issues/climatechange/docs/IUCN.pdf. Accessed 25 Mar 2008

International Fund for Agricultural Development (IFAD) (2004) Gender issues in IFAD projects in Western and Central Africa: a stocktaking exercise. International Fund for Agricultural Development, Rome

Intergovernmental Panel on Climate Change (IPCC) (2014) Synthesis report. In: Pachauri RK, Meyer LA (eds) Contribution of working groups I, II and III to the Fifth assessment report of the intergovernmental panel on climate change. IPCC, Geneva

International Labour Office (ILO) (2017) Indigenous peoples and climate change: from victims to change agents through decent work/International Labour Office. Gender, Equality and Diversity Branch. ILO, Geneva. http://www2.ohchr.org/english/issues/climatechange/docs/IUCN.pdf. Accessed 29 July 2021

ILO (2019a) Indigenous peoples and climate change: emerging research on traditional knowledge and livelihoods. Gender, Equality and Diversity & ILOAIDS Branch, Geneva, Switzerland

ILO (2019b) Empowering women in the rural economy: policy guidance notes. https://www.ilo.org/wcmsp5/groups/public/—ed_dialogue/sector/documents/publication/wcms_601071.pdf. Accessed 10 Nov 2021

Institute for Security Studies (ISS) Africa and climate change: projecting vulnerability and adaptive capacity. https://media.africaportal.org/documents/Africa_and_climate_change.pdf. Accessed 6 May 2021

Intergovernmental Panel on Climate Change (IPCC) (2018) Summary for policymakers. In: Masson-Delmotte V, Zhai P, Portner HO et al (eds) Global Warming of 1.5 °C, An IPCC Special Report on the impacts of global warming of 1.5 °C above pre-industrial levels and related global greenhouse gas emission pathways, in the context of strengthening the global response to the threat of climate change, sustainable development, and efforts to eradicate poverty (Geneva, Switzerland, World Meteorological Organization)

Lambrou Y, Piana G (2006) Gender: the missing component in the response to climate change. Food and Agriculture Organization of the United Nations, Rome, 46 p. http://www.fao.org/sd/dimpe1/docs/pe1.051001d1b.pdf

Meza LER (2010) Climate change, poverty and migration processes in Chiapas, Mexico. Int J Labor Res 2:188–210

Moyo CS (2014) 'Active participation of rural women in developmental issues: poverty alleviation lessons for South Africa. Gender Behaviour 12(1):5994–6001

Nguimalet C (2018) Comparison of community-based adaptation strategies for droughts and floods in Kenya and the Central African Republic. Water Int 43(2):183–204. https://doi.org/10.1080/02508060.2017.1393713

Nkomwa EC, Joshua MK, Ngongondo C, Monjerezi M, Chipungu F (2014) Assessing indigenous knowledge systems and climate change adaptation strategies in agriculture: a case study of Chagaka Village, Chikhwawa, Southern Malawi. Phys Chem Earth 67–69:164–172. https://doi.org/10.1016/j.pce.2013.10.002

Namubiru-Mwaura E (2014) Land tenure and gender: approaches and challenges for strengthening rural women's land rights. Women's voice, agency, & participation research series 6, World Bank, Washington DC. http://documents.worldbank.org/curated/en/309091468153867580/pdf/927600NWP0Wome00Box385358B00PUBLIC0.pdf. Accessed 17 Nov 2021

Nyantakyi-Frimpong H (2020) Unmasking difference: intersectionality and smallholder farmers' vulnerability to climate extremes in Northern Ghana. Gend Place Cult 27(11):1536–1554. https://doi.org/10.1080/0966369X.2019.1693344

Okeke IN, Babalola CP, Byarugaba DK, Djimde A, Osoniyi OR (2017) Broadening participation in the sciences within and from Africa: purpose, challenges, and prospects. CBE Life Sci Educ:16(2). https://doi.org/10.1187/cbe.15-12-0265

Onwutuebe CJ (2019) Patriarchy and women vulnerability to adverse climate change in Nigeria. SAGE Open:1–7

Organisation for Economic Co-operation and Development (OECD) (2014) Social institutions and gender index 2014 synthesis report. OECD, Paris. https://www.empowerwomen.org/en/resources/documents/2014/11/social-institutions-and-gender-index-2014-synthesis-report. Accessed 10 Nov 2021

Parrotta JA, Oteng-Yeboah A, Cobbinah J, IUFRO (2009) Traditional forest-related knowledge and sustainable forest management in Africa. IUFRO World Series, 23. Vienna, Austria

Pentlow S (2020) Indigenous perspectives on gender, power and climate-related displacement. Gender Equality Consultant, Cuso International https://cusointernational.org

Peterman A, Palemo T, Bredenkamp C (2011) Estimates and determinants of sexual violence against women in the Democratic Republic of Congo. Am J Public Health 101(6):1060–1067

Reid H, Alam M, Berger R, Cannon T, Huq S, Milligan A (2007) Community based adaptation to climate change: an overview. https://pubs.iied.org/sites/default/files/pdfs/migrate/G02608.pdf? Accessed 14 sNov 2021

Srang-iam W (2013) Social learning and climate change adaptation in Thailand. In: ISSC/UNESCO, World Social Science Report 2013: Changing Global Environments. OECD Publishing and Unesco Publishing. https://www.researchgate.net/profile/Antonio-Ioris/publication/315038696_The_Politics_of_Climate_Change_and_Grassroots_Demands/links/58d52915aca2727e5ea2d3f6/The-Politics-of-Climate-Change-and-Grassroots-Demands.pdf#page=30. Accessed 15 Nov 2021

Tantoh HB, Simatele D (2017) Community-based water resource management in North-west Cameroon: The role of potable water supply in community development. S Afr Geogr J 99(2):166–183

Tantoh HB, McKay TJM (2018) Rural self-empowerment: the case of small water supply management inNorthwest, Cameroon. Geo J. https://doi.org/10.1007/s10708-018-9952-6

Tantoh HB, McKay TJM (2020) Investigating community constructed rural water systems in Northwest Cameroon: leadership, gender and exclusion. Int Develop Plan Rev 42(4):455–478. https://doi.org/10.3828/idpr.2020.4

Tantoh HB, McKay TJM (2021) Assessing community-based water management and governance systems in North-West Cameroon using a cultural theory and systems approach. J Clean Product 125804. ISSN 0959-6526. https://doi.org/10.1016/j.jclepro.2021.125804

Tantoh HB, McKay TJM, Donkor FE, Simatele MD (2021) Gender roles, implications for water, land and food security in a changing climate: a systematic review. Front Sustain Food Syst 5:259

Tchoundjeu Z, Degrande A, Leakey RRB, Nimino G, Kemajou E, Asaah E, Facheux C, Mbile P, Mbosso C, Sado T, Tsobeng A (2010) Impacts of participatory tree domestication on farmer livelihoods in West and Central Africa. Forests Trees Livelihoods 19(3):217–234. https://doi.org/10.1080/14728028.2010.9752668

UNDP (2010) Gender, climate change and community-based adaptation: a guidebook for designing and implementing gender-sensitive community-based adaptation programmes and projects. United Nations

Development Programme (UNDP), New York. www.undp.org
UNDP (2012) Gender and climate change. Capacity development series, Africa. Training Module 2. United Nations Development Programme (UNDP), New York. https://www.uncclearn.org/wp-content/uploads/library/undp06.pdf. Accessed 13 Apr 2021

Henry B. Tantoh (Ph.D.) is a lecturer in the Department of Geography and Planning, University of Bamenda–Cameroon. His research explores the social, economic and political aspects of natural resource management and environmental governance and their relationships to wider societal change. Much of his recent work has been concerned with the politics of natural resource management in sub–Sahara Africa, with a particular focus on co–management of water resources, systems that promote stakeholder participation, climate change adaptation and sustainability.

Eromose E. Ebhuoma (Ph.D.) is a postdoctoral research fellow at the Department of Environmental Sciences, College of Agriculture and Environmental Sciences, University of South Africa (UNISA). His research expertise spans climate change vulnerability and adaptation, climate services, climate change policy and governance, local and indigenous knowledge systems, political ecology, rural livelihoods and environmental sustainability in sub-Saharan Africa. He is a steering committee member of the South African Adaptation Network.

Llewellyn Leonard (Ph.D.) is Professor at the Department of Environmental Science, School of Ecological and Human Sustainability, University of South Africa. Before joining academia, he worked for a human rights environmental organisation working to support vulnerable communities exposed to environmental risks to shape policy and has worked to link communities in Africa, Asia, Latin America and Europe to engage in solidarity and networking. Previously he served as Vice Dean: Research at the University of Johannesburg, College of Business and Economics. His research interest include environmental justice, governance, democracy and human rights, civil-society-state-industry relations, urban risks, sustainable development; risk society and political economy/ecology to name a few. He sits on the editorial board of the Southern African Geographic Journal.

6 Climate Change Experts' Perspective on the Role of Indigenous Knowledge Systems in Climate Change Adaptation in Southern Africa

Shorai Kavu, Collen Kajongwe, and Eromose E. Ebhuoma

6.1 Introduction

Studies have revealed that the economic gains that developing countries have accumulated during the previous decades are at risk of being reversed by the adverse effects of climate change (Ajani 2013; Stuart 2017). Southern Africa is by no way spared from this. The literature suggests that the phenomenon of climate change portends to further impoverish or afflict populations with additional hardships, especially those in poor societies with weak infrastructure and economic well-being in Southern Africa (Simatele and Simatele 2015; Tume et al. 2019; Mavhura 2020). The region is prone to frequent and intense El Niño-Southern Oscillation (ENSO) events, leading to widespread droughts in some areas and erratic rainfall and increased flooding in others. Climate models generally show decreasing rainfall for Southern Africa (Chanza 2014; Sietz et al. 2008), which is likely to result in a decline in available surface water and decrease in river runoff (Conceição Junior and United Nations Human Settlements Programme 2010; Sietz et al. 2008). With one in three households in Southern Africa living below the global poverty line of less than $1.90 US dollars daily (Hamel et al. 2019), such extreme weather conditions will put the livelihood of the poor who are overwhelmingly dependent on land-based activities in a precarious position. This is partly because of limited capability and access to resources to cope with the adverse effects of climate change (Simatele and Simatele 2015). Notwithstanding, like in poor rural households in other parts of sub-Saharan Africa (SSA), indigenous knowledge systems (IKS) play a pivotal role in cushioning and ensuring rural households in Southern Africa can continuously produce their food (Jiri et al. 2015; Mavhura et al. 2013). Thus, the importance of the perceptions, experiences, and knowledge of indigenous peoples has gained prominence in the discourse revolving around climate change mitigation and adaptation in SSA among numerous international development organisations (Nkomwa et al. 2014) and scholarly literature (Ebhuoma 2020; ILO 2019; Petzold et al. 2020; Tume et al. 2019).

The Inter-Governmental Panel on Climate Change (IPCC), for example, acknowledged the

S. Kavu
Central Development Studies, School of Life Long Development Studies, Chinhoyi University of Technology, Chinhoyi, Zimbabwe
e-mail: kavkajongwe@gmail.com; skavu@energy.gov.zw

C. Kajongwe (✉)
Department for Human Resource, School of Business Management, Manicaland State University of Applied Sciences, Mutare, Zimbabwe
e-mail: codzakajongwe@gmail.com; collen.kajongwe@staff.msuas.ac.zw

E. E. Ebhuoma
Department of Environmental Sciences, School of Ecological and Human Sustainability, College of Agriculture and Environmental Sciences, University of South Africa (UNISA), Johannesburg, South Africa
e-mail: ebhuoee@unisa.ac.za

E. E. Ebhuoma and L. Leonard (eds.), *Indigenous Knowledge and Climate Governance*, Sustainable Development Goals Series, https://doi.org/10.1007/978-3-030-99411-2_6

need to consider IKS in the international assessments during the fifth assessment (AR5) and agreed to ensure consideration of IKS in the sixth assessment report (AR6) and in future reports (Petzold et al. 2020). Also, a study conducted by Petzold et al. (2020) sought to establish the level of research publications on the subject of IKS in different geographical locations globally. It was found that sub-Saharan Africa (SSA) was among the regions on which a substantial literature gap on the role of IKS regarding climate change mitigation and adaptation still exists (Petzold et al. 2020). Furthermore, it is argued that the use of IKS in climate and other developmental policies was minimal in the same regions (Ajani 2013; Habibu et al. 2017). In SSA, efforts to evaluate the role of IKS in climate change adaptation have focused mainly on the perception and agricultural practices of local people. Minimal attention has been given to climate change experts' perception of IKS and its potential to contribute effectively to climate adaptation interventions. In this chapter, climate change experts refer to individuals with a proven record of published research work and/or traceable work experience in the field of climate change (Ajani 2013; Habibu et al. 2017).

In an effort to bridge this gap, this chapter aimed at understanding climate change experts' knowledge regarding the role of IKS in facilitating rural households' adaptation to climate change with a particular focus in Zambia, Malawi and Mozambique, countries situated in Southern Africa. El Niño-induced cyclones have occurred more frequently over the last decade in both Malawi (Dube et al. 2021) and Mozambique (Lupele 2020) than over any other decade in history. Also, cyclonic activities in the Indian Ocean are expected to increase in Mozambique (Conceição Junior and United Nations Human Settlements Programme 2010; Petzold et al. 2020; Sietz et al. 2008) and Malawi (World Bank 2021), partly due to increased sea surface temperatures. This will likely result in widespread flooding in both countries. While Zambia has been adversely affected by climate change (Simatele and Simatele 2015), the country is rarely affected by cyclones. Thus, the chapter sought to ascertain whether climate change experts' knowledge of IKS in countries highly susceptible to extreme weather events differs from those least affected by such weather events.

Against this background, this chapter demystifies climate change experts' perception of IKS in facilitating climate change adaptation. This is necessary since various knowledge systems are needed to fortify the management and conservation of natural resources (e.g. Thakur et al. 2020), which in turn may facilitate the actualisation of both the Sustainable Development Goals (SDGs) and Agenda 2063. Such analysis may shed light on issues of concern regarding climate change experts' perception of IKS, which could hamper their acceptance of such knowledge in climate change policies and governance discourse.

This chapter proceeds in four parts. The first part succinctly unpacks the literature to highlight the role of IKS in scaling up mitigation and adaptation to climate change in Southern Africa. This sets the scene to analyse the extent to which the various climate change policies in the countries mentioned above make room for IKS, the focus of the second part of the chapter. The third section of this chapter sheds light on the methodology used to obtain primary data from climate change experts in the countries under investigation. The fourth section showcases the results and discusses the findings with particular emphasis on ways it could facilitate the actualisation of SDG 13—climate action—in Southern Africa, while the final section draws decisive conclusions. It is hoped that findings in this chapter will stimulate conversations regarding how climate change experts could play an instrumental role in ensuring that IKS feature more prominently in climate change policies and governance discourses.

6.2 The Role of Indigenous Knowledge in Climate Change Adaptation and Mitigation in Southern Africa

Climate change has had a deleterious effect in Southern Africa. "Trend analysis of temperatures across Southern Africa reveals that annual minimum and maximum temperatures have increased at an average rate of 0.057 °C per decade and 0.046 °C per decade, respectively, between 1901 and 2009. Further analysis reveals that the periods of most rapid warming occurred post-1970, a period for which the rate of increase in both average annual minimum and maximum temperatures is statistically significant at the 95 per cent confidence level. After 1976, minimum temperatures began increasing by 0.27 °C per decade and maximum temperatures by 0.25 °C per decade" (Davis and Joubert 2011, pp. 17–18) Future climate projection suggests that extreme weather conditions will increase, with all global circulation models for B1 and A2 scenarios predicting temperature rises between 1 and 3 °C by 2060 (Tadross et al. 2011). This is expected to decrease the agricultural production of the major cereal crops—maize, sorghum and millet—cultivated in the region (Ajani 2013). Maize, sorghum and millet yields are estimated to decrease by 22, 17 and 17% by 2050 (Schlenker and Lobell 2010). Under the current state of affairs, not only will this scenario amplify existing food and nutrition insecurity in the region, but it will firmly situate rural households in the poverty maze from which they are desperately striving to escape.

Despite the extreme weather challenges confronting rural households in Southern Africa, including Zambia, Malawi and Mozambique, those households are not docile victims to the challenges. They are continuously finding ways heavily underpinned by IKS to outwit the weather challenges to produce their food to obtain their livelihoods (Mubanga and Umar 2014; Nkomwa et al. 2014; Matos et al. 2021). Like most indigenous people in SSA, a key strategy in Southern Africa to produce food effectively is the application of indigenous weather forecasts to determine the types and quantity of food to grow in a particular farming season. The indigenous weather forecasts are not limited to lunar and cloud observations and close monitoring of plants and animal behaviours (Nkomwa et al. 2014; Jiri et al. 2015). These indigenous indicators used to generate forecasts have been applauded by local people for their ability to create weather forecasts with fine spatial resolution, better than most scientific forecasts produce due to the dearth of weather stations in proximity to rural communities (Ebhuoma and Leonard 2020).

It is important to highlight that due to the increased occurrences of climate variability in Southern Africa, farmers call for scientific forecasts to help generate more accurate forecasts (Tall et al. 2015). While this call is widely acknowledged in SSA, some argue that farmers in the region will continue to rely heavily upon indigenous forecasts because their cultural identity underpins it as the knowledge was passed on to households by their forefathers (Tume et al. 2019). This has fortified the call for integration of IKS with scientific forecasts to generate richer weather information. For example, Nyadzi (2020) revealed that utilisation of integrated forecasts resulted in better crop output than using only one forecast. Also, the integration of knowledge will help to preserve indigenous knowledge, which is gradually becoming extinct partly because young people abandon agriculture and migrate from rural spaces to urban areas in search of better livelihood opportunities (e.g. Ebhuoma et al. 2021).

In the context of climate change mitigation, local people in Southern Africa have been known to conserve carbon in soils through practices like zero tilling, mulching and other soil management techniques (Ajani 2013). Natural mulches moderate extreme soil temperatures, suppress diseases and harmful pests, and conserve soil moisture (Makate 2019). Before the advent of chemical fertilisers, local farmers largely depended on organic farming, which also reduces GHG emissions. It is widely recognised that forests play an important role in the global carbon cycle by sequestering and storing carbon (Nkomwa et al. 2014). Local farmers in for example Zambia (Tume et al. 2019), Malawi (Kupika et al. 2019) and Mozambique (Ajani 2013) are known to have practised the fallow system of cultivation, which enabled forest development.

Besides the fact that these well-managed forests provide food and timber resources to the community, they also serve as carbon sinks. Agroforestry is another practice that has been very effective in sequestering carbon from the atmosphere. It is a rational land-use planning system that tries to find some balance in the raising of food crops and forests (Belcher et al. 2015). In addition to the fact that agroforestry techniques can be perfected to cope with the new, drier conditions anticipated under a higher population density, they lead to an increase in the amount of organic matter in the soil, thereby improving agricultural productivity and reducing the pressure exerted on forests (Petzold et al. 2020).

6.3 Climate Change Policy Frameworks in Southern Africa

6.3.1 Zambian Climate Policy Framework

Zambia has a National Policy on Climate Change (NPCC) enacted in 2016 (Republic of Zambia 2016). The NPCC objectives include promoting and strengthening the implementation and adaptation and disaster risk reduction measures to reduce vulnerability to climate variability and climate change and to encourage investments in climate-resilient and low-carbon development pathways to generate co-benefits. The policy also provides incentives for addressing climate change more effectively and fostering research and development to improve understanding and decision-making in response to climate change (Republic of Zambia 2016).

A number of internal and external stakeholders have been identified as key to the implementation of the NPCC (Republic of Zambia 2016), including the traditional leaders and local communities. However, there is no clear policy statement pointing specifically to the promotion and research on IKS for climate change mitigation and adaptation in the country (Republic of Zambia 2016). Reasons for the lack of a clear policy on the promotion and research on IKS on climate change mitigation may include shallowness or non-inclusion of the subject in school curriculums and a lack of support from the central government towards awareness of the IKS practices in the country (Lupele 2020). The climate policy of Zambia seems more inclined to the conventional methods of building resilience than to the local IKS (Republic of Zambia 2016). Notwithstanding, the Zambian government must be applauded for promoting community-based risk management activities and communities' ability to develop physical and social infrastructure that are resilient to the adverse effects of climate change (Republic of Zambia 2016), as this is where IKS will prove instrumental in showcasing its expertise regarding adaptation to climate change.

6.3.2 Malawian Climate Policy Framework

Malawi has a National Climate Change Management Policy (NCCMP) enacted in 2016. The policy, together with supporting legislative frameworks such as the Environment Management Act and the Forest Act, supports the mitigation of and adaptation to climate change by Malawi (Government of Malawi 2016). The successful implementation of the NCCMP required the involvement and participation of all stakeholders, including local communities, NGOs, civil society, faith-based organisations, identified vulnerable groups, the private sector and academia; these stakeholders had to cooperate with the country's government (Malawi Environmental Affairs Department 2016). The policy avows the Malawi government's commitment to fully address climate change issues to reduce the vulnerability of its people, ecosystems and socio-economic development through adaptation and mitigation, technology transfer and capacity building (Government of Malawi 2016). Translating this policy into action will build Malawi's resilience to overcome the challenges of climate change and embrace the available opportunities to enable the country to lay a solid foundation for a sustainable and prosperous Malawi (Government of Malawi 2016; Nkomwa et al. 2014). The policy statement, however, failed to mention the application of IKS. It only acknowledges the importance of consulting with local communities and vulnerable groups to ensure that effective climate change management strategies are put in place (Government of Malawi 2016). Notwithstanding, the space created for consultation with local communities may be interpreted as being tantamount to providing indigenous peoples with a viable platform to dictate their climate change adaptation strategies. In an assessment of the role of IKS in enabling local people—overwhelmingly dependent on agriculture for their livelihood—to build their adaptive capacity to climate change in Malawi, Nkomwa et al. (2014) concluded that effective adaptation to climate change could not be achieved without the use of the IKS.

6.3.3 Mozambican Climate Change Policy Framework

Mozambique ranks high on the climate change index based on annual and seasonal temperature and precipitation indicators (Kupika et al. 2019). In light of these scenarios, climate change poses a major challenge to development in Mozambique. This is further evidenced by recent climate-related disasters. Recognising the need to ensure mitigation and adaptation to climate variability and change to reduce people's vulnerability, the Mozambican government carved out its National Climate Change Adaptation and Mitigation Strategy (NCCAMS) in 2012 (Petzold et al. 2020). The NCCAMS identifies critical areas and actions to diminish the severity of the impacts through adaptation measures and the reduction of climate risks, and to create benefits through mitigation and low carbon development opportunities (Makate 2019).

The Mozambiquan climate policy framework also has the National Adaptation Plan of Action to Climate Change (NAPA) which identifies the country's more urgent needs regarding climate change through a participative assessment process (Nkomwa et al. 2014). The overall purpose of this document as part of the climate policy framework for Mozambique is to improve the country's capacity to deal with the effects of climate change. More specifically, the NAPA pursues four objectives: (1) strengthening the warning system for natural disaster prevention; (2) capacity-building and awareness-raising for farmers to deal with the effects of climate change; (3) reducing these effects in coastal areas; and (4) improving the management of water resources (Sietz et al. 2008). The NAPA addresses climate change issues on a country-wide scale but fails to provide a platform to explore the role of IKS in climate change adaptation and mitigation.

It is worth mentioning that Mozambique's Environmental Strategy for Sustainable Development (ESSD) provides an integrated vision for sustainable environmental management in the country and comes with complete guidelines for stakeholders involved in development programmes (Sietz et al. 2008). The ESSD includes the following four elements: (1) capacity-building to set national and local objectives and implement integrated management systems to preserve natural resources and ecosystems; (2) setting up institutions with the technical capacity required for urban planning, infrastructure development, waste management, and water and sanitation services; (3) providing the technical and financial support required to strengthen the country's capacity to measure, curb and evaluate environmental pollution (including any impacts on health and climate change issues); and (4) keeping demographic growth within the limits of socioeconomic and sustainable development objectives (Habibu et al. 2017). The ESSD directly addresses climate change issues, but then mainly from the perspective of atmospheric (greenhouse gas) pollution, specifying a range of steps to be taken to meet the terms of the United Nations Framework Convention on Climate Change.

Despite this wide range of policies, plans and strategies in place, the country has no clear policy or strategy for upholding IKS in the mitigation and adaptation of climate change. In terms of indigenous knowledge, for example, the NCCAMS only highlights the need to promote the use of indigenous energy resources for clean and renewable energy. In its adaptation and climate risk reduction strategies, it emphasises the need to increase the adaptive capacity of vulnerable people but fails to explicitly recognise these cohorts of people as holders of valuable and actionable knowledge that can facilitate building resilience to climate change. Notwithstanding, the NCCMS acknowledged that "it is now urgent to coordinate responses, ensuring that CC planning processes are integrated at all levels—spatially (at local, community, district, province and national levels) and sectoral" (Government of Mozambique 2012, p. 14).

6.4 Methodology

Primary data were obtained from Zambia, Mozambique and Malawi (Fig. 6.1). The study adopted both a quantitative and qualitative research approach as the distributed questionnaires comprised closed and open-ended questions. Purposive sampling was used to select 30 respondents in the study areas. The e-mails eliciting respondents' participation were sent to workers with at least two years' experience in climate change management in their various organisations.

The participants were identified as individuals with a track record of working in climate change in collaboration with the government of the relevant country. The respondents were selected from ministries responsible for climate change management in Malawi, Mozambique and Zambia. Also, some respondents were environmentalist academics and, according to the key informants, have participated in a number of climate change forums from state universities in Malawi, Mozambique and Zambia. Figure 6.2 shows the position of respondents. Data collection was done during the first quarter of 2021. Questionnaires were distributed via e-mail due to travel restrictions during the Covid-19 pandemic. To ensure the anonymity of respondents, e-mails were sent to Heads of Ministries, Heads of Parastatals and Heads of Departments in universities for further distribution to their subordinates whose names were not disclosed. Respondents were assured of confidentiality and anonymity to avoid bias.

It is important to highlight that the open-ended questions aimed to ensure that the experts answered questions comprehensively in order not to restrict their views. Open-ended questions were distributed as an interview guide in preparation for calls conducted with respondents to ensure they had time to reflect on the questions. Video calls were considered at first but they were not as successful due to poor network. The researchers thus resorted to no video calls.

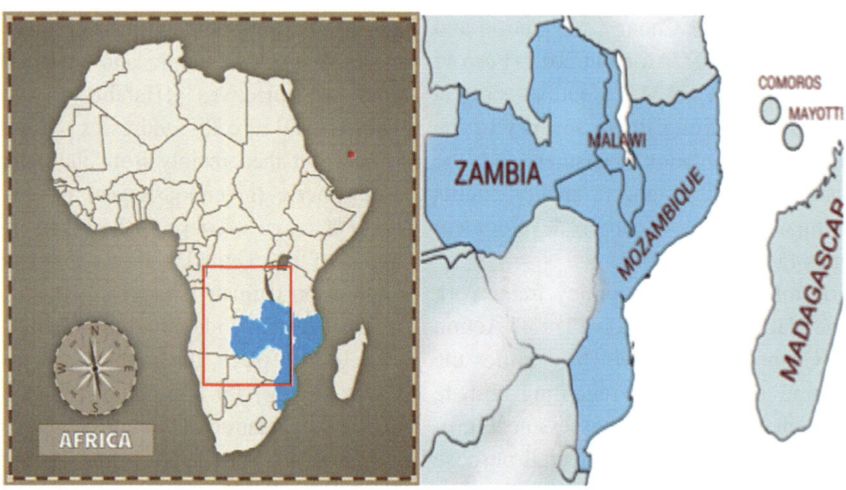

Fig. 6.1 Map of the area of study areas (Zambia, Mozambique and Malawi). *Source* Author (2022)

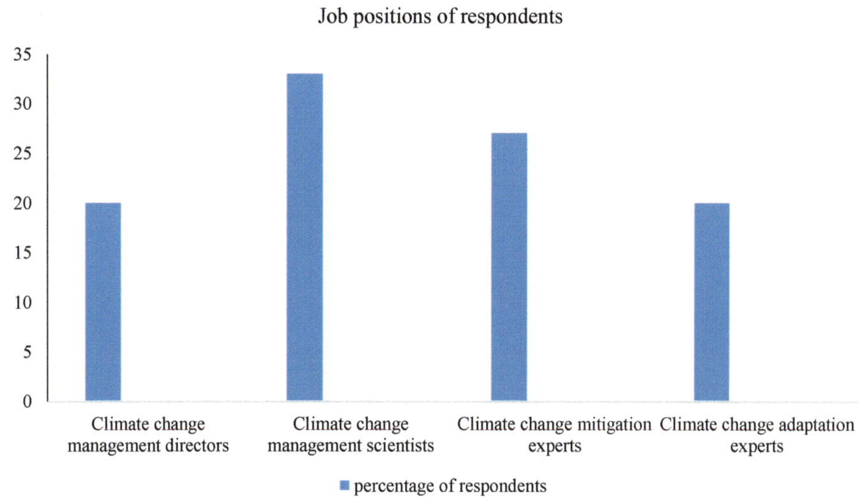

Fig. 6.2 Job positions of respondents. *Source* Survey (2021)

Data obtained from the closed questions were uploaded on the Statistical Package for Social Sciences (SPSS), analysed using the Chi-Square Test of Independence, and presented in the form of tables, graphs and pie charts. Data obtained from the open-ended questions were analysed thematically. Qualitative data have been presented verbatim to avoid misinterpreting the viewpoints of climate experts.

6.5 Results and Discussion

6.5.1 Response Rate

30 climate change experts in Mozambique, Zambia and Malawi were purposively identified and questionnaires were distributed electronically to them due to movement restrictions imposed by the government of the various countries due to the COVID-19 pandemic. However, from the 30 questionnaires distributed, only 15 were completed and returned (Table 6.1).

As shown in Table 6.1, the overall response rate is 50% i.e. 17% from Malawi, 20% from Mozambique and 13% Zambia. Although the response rate is small, this chapter is timely to kick-start the conversation regarding climate change experts' viewpoint on IKS.

6.5.1.1 Level of Education of Respondents

Figure 6.3 indicates the highest level of education of respondents from the three countries. Of the respondents, 20% attained a doctorate, 27% a master's degree, and 53% were undergraduates. Literature indicates that individuals with professional qualifications have a better understanding of the dynamics in areas of expertise (McGrath 2002). This implies that a well-educated human capital workforce is the most valuable component of intellectual capital and enhances organisational performance.

Table 6.1 Response rate

Questionnaires	Malawi	Mozambique	Zambia	Total
Sent	10	10	10	30
Received	5	6	4	15
Response rate (%)	17	20	13	50

Source Survey (2021)

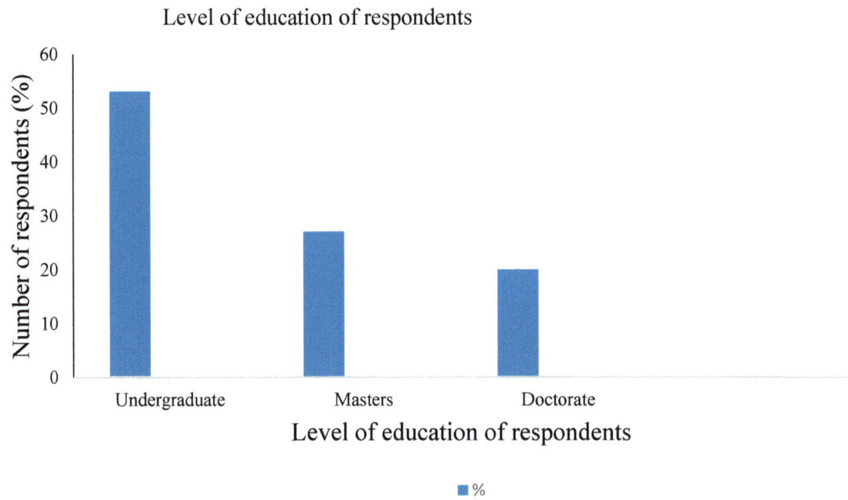

Fig. 6.3 Level of education of respondents. *Source* Survey (2021)

6.5.1.2 Work Experience of Respondents

Table 6.2 shows the work experience of respondents in Mozambique, Malawi and Zambia concerning the role of IKS in climate change mitigation and adaptation. Of the respondents, 27% had 3–7 years working experience, 33% had 7–15 years working experience, and 40% had 15 years and above. The results may imply that skills horned in work experience are resourceful for better understanding of systems and processes of IKS.

6.5.2 Climate Experts' Knowledge on Indigenous Practices Employed in Climate Change Adaptation

IKS has always played a fundamental role in enabling rural households in Zambia, Malawi and Mozambique to obtain their livelihoods in harsh and unprecedented weather changes. As stated by a respondent from Mozambique:

> Mozambique is a country on the shores of the Indian Ocean so it is the recipient several devastating weather phenomenon associated with climate change effects on the ocean fronts. Take for example cyclone Japhet of 2000 and cyclone Idai of 2019. These weather phenomena had great damage on the livelihoods of the survivors, the country's budget was overwhelmed, however, people had to make use of some common indigenous knowledge obtained from their ancestors to adapt and move on with life... Source: Survey (2021).

Results from the closed questions in the questionnaires show the following IKS adaptation practices: traditional farming techniques, livelihood diversification, weather and climate forecasting systems, climate-resilient buildings, dissemination and communication of climate information, knowledge-sharing networks, flood

Table 6.2 Work experience of respondents

Years of experience	Frequency	Percent
3–7 years	4	27
7–15 years	5	33
15 years and above	6	40
Total	15	100

Source Survey (2021)

protection plants, and influencing weather through will and belief in supernatural intervention.

As shown in Table 6.3, according to climate change experts, indigenous people's coping method of choice for climate change was the diversification of livelihood. According to Ogalleh et al. (2012), livelihood diversification is an important climate change coping strategy for local farmers in Southern Africa to avert their inability to obtain a livelihood in the event of a bad harvest.

The second most acknowledged adaptive strategies of indigenous peoples were both weather and climate forecasting systems and constructing climate-resilient buildings. Climate experts' opinion on the importance of weather and climate forecasting in Zambia, Mozambique and Malawi could be attributed to the fact that most (70%) of the rural households in their respective countries rely on food production for their livelihood. This means they rely on indigenous ways of predicting the weather and climate to determine when to plant how much of what type of food in any particular growing season (Lucio 1999; Kalanda-Joshua et al. 2011; Nkomwa et al. 2014; Mbewe et al. 2019). This finding will please indigenous scholars calling for indigenous weather forecasts to be integrated with scientific forecasts to co-produce an integrated forecast, harmonising the best of both worlds. However, due to the relatively small number of respondents, it remains unclear how receptive climate experts in the relevant countries will be to the integration of indigenous and scientific weather forecasts. This can primarily be ascribed to the fact that calls made for the integration of indigenous and scientific weather forecasts have been mainly echoed by social scientists and human geographers (Tume et al. 2019). These scientists seem more receptive to diverse forms of knowledge than physical geographers and scientists whose research philosophy is linked to the positivist approach. The positivist approach entails that forecasts can be verified scientifically, which is not the case with indigenous forecasts. About 50% of the respondents were physical scientists. Because we are in what Sarrasanti et al. (2020) termed the decade of action (2020–2030) to achieve the SDGs, and because the number of respondents in this chapter is so small, further investigation is desperately needed. Such research may unearth entry points to inject measures to ensure wider acceptability of indigenous ways of predicting the weather among physical geographers and scientists.

The second least acknowledged technique in the questionnaires was the dissemination of climate information. Studies show that indigenous peoples communicate both scientific (Paul and Routray 2010) and indigenous (Habibu et al. 2017) forecasts among themselves mainly through word of mouth. Climate scientists, on the contrary, prefer using information and communication technologies (ICTs) like mobile phones, television and radios to communicate weather and climate information (McNaught et al. 2014; Mudombi and Godwell 2014). The scientists' preference for ICTs is linked to widespread reach within the shortest timeframe and cost-effectiveness (Harvatt et al. 2011). However,

Table 6.3 Climate change experts' knowledge of climate change adaptation practices based on indigenous knowledge in Mozambique, Malawi and Zambia

Factor	Frequency	Percentage
Livelihood diversification	4	27
Weather and climate forecasting systems	3	20
Climate resilient buildings	3	20
Dissemination of climate information	2	13
Indigenous plants for flood protection	2	13
Influencing weather through will and faith	1	7
Total	15	100

Source Survey (2021)

indigenous people's preference for word of mouth is due to the fact that many are not educated and may find it difficult to make sense of scientific forecasts which often contains technical jargon that impedes understanding (Hansen et al. 2011). Some also lack ICT devices and a constant power supply (Ebhuoma and Leonard 2020).

Climate experts' finding on the building of climate-resilient buildings corroborates the suggestion in the literature that indigenous peoples are not docile victims of climate change. They consistently find creative ways to fortify their houses to minimise the adverse effects of climate change. Indigenous farmers understand that the destruction of their houses will have a ripple effect on their ability to engage effectively in food production as their resources will be channelled towards rebuilding in the aftermath of a disaster (Simatele and Simatele 2015; Ebhuoma and Simatele 2017). Deliberate efforts by indigenous peoples in Mozambique, Malawi and Zambia to fortify their houses against storms, floods or high temperatures are not limited to raising houses above flood levels and using straw roofs to moderate rising temperatures. As shown in Table 6.3, the respondents underlined the cultivation of indigenous plants to act as a buffer against floods—a form of ecosystem-based adaptation—and indigenous peoples resorting to their will and faith in God as strategies to adapt to climate change; this corroborates findings in Mexico (e.g. Eakin 1999) and Ghana (e.g. Naab et al. 2019).

6.5.2.1 Early Warning Signs Indicating Drought in Malawi, Mozambique and Zambia

Climate change experts from Malawi, Mozambique and Zambia identified some early warning signs of the IKS indicating drought. These indicators include the following: specific birds not making noise at dawn (13%); shrubs/plants not flowering (20%); a long winter season (34%); an abundance of wild fruits (13%); specific birds consistently making a particular noise at dawn (7%); and a lot of fruit from *Pericopsis angolensis* trees (13%) (Table 6.4).

Anecdotal evidence suggests that indigenous forecasts seem abstract to climate policy developers. However, this finding indicates the contrary. In this regard, a respondent from Zambia stated:

> There is need of policies from government that allow experts to take their knowledge on climate change adaptation to the people in rural areas and should be ready to learn from the rural people in return on how they have been and are planning to survive the effects of climate change as was learnt from local traditions… It is known that locals always have their established ways of adapting but there is lack of coherence with policies in the country. Source: Survey (2021).

This finding is crucial in the light of indigenous farmers' acknowledgement that indigenous indicators are not as reliable as in previous decades, partly due to the unprecedented rate of climate change (Habibu et al. 2017; Nkomwa et al. 2014). The growing unreliability of indigenous indicators has led to numerous calls for scientific forecasts to compensate for the limitations of indigenous forecasts (Tume et al. 2019). The shortage of weather stations in proximity to rural communities, due to a lack of resources, poses a challenge to accuracy in forecasting future weather patterns. Therefore, indigenous farmers continue to rely heavily upon indigenous indicators (Habibu et al. 2017). Some scholars have argued that scientific forecasts should start by compensating for the limitations of IKS (Ebhuoma 2020). This will ensure that the cultural identity of indigenous peoples—the weather forecasting skills acquired from their forefathers—are protected, and it will build their trust. Also, it would facilitate a paradigm shift among development practitioners who view IKS as archaic (Nkomwa et al. 2014) as it permits equal contribution from both indigenous and western epistemologies to produce forecasts. As a respondent from Malawi noted:

> It is well known that people in the rural areas have since time immemorial been practising some adaptation methods that are still applicable in this era of climate change. However, little effort has been put by policy makers towards putting these practices on spotlight to bring them on the same importance level as the proven scientific methods… Source: Survey (2021).

Table 6.4 Early warning signs indicating drought in Mozambique, Malawi and Zambia

Indicator	Frequency	Percentage
Specific birds not making noise at dawn	2	13
Shrubs/plants not flowering	3	20
A long winter season	5	34
An abundance of wild fruits	2	13
Specific birds consistently making a particular noise at dawn	1	7
A lot of fruit from a *Pericopsis angolensis* tree	2	13
Total	15	100

Source Survey (2021)

The recognition of indigenous climate change experts' potential to predict droughts in the countries mentioned above by utilising indigenous knowledge may also be linked to the various national climate change policy documents. These documents underline the need to engage this vital knowledge base and promote the participation of local communities and traditional institutions in innovations to tackle climate change (Makate 2019). Thus, it is highly likely that the need to incorporate IKS to ensure it is rigorously engaged with in the pursuance of SDG 13 in Zambia, Mozambique and Malawi may be widely accepted by climate experts.

It is noteworthy to highlight that the chapter's limitation is the shortage of respondents from the three countries who are climate change experts. This is mainly due to COVID-19 that hampered face-to-face meetings. The reliance on e-mails and few completed questionnaires arguably hampered the data collection process severely. It also hindered the authors' attempt to conduct a rigorous comparison of the knowledge on IKS of climate change experts from the different countries. Nonetheless, the findings of this chapter offer valuable evidence regarding climate experts' knowledge of IKS. This evidence may be pivotal to facilitating the integration of IKS with knowledge underpinned by western science to catalyse the actualisation of SDG 13.

6.6 Conclusion and Recommendations

IKS has played a fundamental role in enabling rural households in Zambia, Malawi and Mozambique to build their resilience and adaptive capacity to climate variability and change (Lucio 1999; Kalanda-Joshua et al. 2011; Mbewe et al. 2019). This has resulted in numerous calls for IKS to be integrated with scientific forecasts to produce an integrated forecast, which might facilitate the actualisation of SDG 13. Yet, studies have failed to unpack climate change experts' knowledge regarding the importance of IKS in facilitating climate change adaptation among rural households. Failure to demystify climate change experts' understanding of IKS could have huge ramifications for its integration into climate change policies. The study obtained primary data through the use of questionnaires that comprised of both closed and open-ended questions from 30 purposively selected climate change experts in Zambia, Malawi and Mozambique.

According to climate change experts, adaptation strategies employed by indigenous peoples include livelihood diversification, weather and climate forecasting, constructing climate resilient buildings, dissemination of weather and climate information, cultivation of indigenous plants to act as a buffer for flood protection, and influencing weather through their faith in God. Also, in terms of drought forecasting, climate change experts underlined, among others, specific birds not making noise at dawn and plants not flowering as key indicators used by indigenous peoples to forecast droughts.

Building on the knowledge of climate change experts, it is recommended that a forum for interaction between the scientists and keepers of indigenous knowledge, especially at the grassroots level, be created in Zambia, Malawi and Mozambique. This forum can be helpful to allow IKS keepers to contribute to climate and weather forecasting processes and other climate change adaptation interventions at the local level. It is also recommended that policies should be implemented that recognise the rights and status of indigenous peoples as keepers of knowledge that is both valuable and actionable. Doing this will make indigenous knowledge keepers legitimate joint decision makers in local, national and regional level adaptive and mitigation initiatives, which will likely contribute towards achieving both SDG 13 and Agenda 2063.

References

Ajani E (2013) Use of indigenous knowledge as a strategy for climate change adaptation among farmers in sub-Saharan Africa: implications for policy. Asian J Agric Ext Econ Sociol 2(1):23–40. https://doi.org/10.9734/ajaees/2013/1856

Belcher B, Achdiawan R, Dewi S (2015) Forest-based livelihoods strategies conditioned by market remoteness and forest proximity in Jharkhand, India. World Dev 66:269–279. https://doi.org/10.1016/j.worlddev.2014.08.023

Chanza N (2014) Indigenous knowledge and climate change: insights from Muzarabani, Zimbabwe

Conceição Junior P da, United Nations Human Settlements Programme (2010) Climate change assessment for Maputo, Mozambique: a summary. United Nations Human Settlements Programme (UN-HABITAT)

Davis C, Joubert A (2011) Southern Africa's climate: current state and recent historical changes. In: Davis C (ed) Climate risk and vulnerability: a handbook for southern Africa. https://researchspace.csir.co.za/dspace/bitstream/handle/10204/5737/Davis_2011.pdf?sequence=1&isAllowed=y. Accessed 14 November 2021

Dube K, Chapungu L, Fitchett JM (2021) Meteorological and climatic aspects of cyclone Idai and Kenneth. In: Nhamo G, Dube K (eds) Cyclones in southern Africa. Sustainable development goals series. Springer, Cham. https://doi.org/10.1007/978-3-030-74262-1_2

Eakin H (1999) Seasonal climate forecasting and the relevance of local knowledge. Phys Geogr 20:447–460

Ebhuoma E, Simatele D (2017) Defying the odds: climate variability, asset adaptation and food security nexus in the delta state of Nigeria. Int J Disaster Risk Reduct 21:231–242

Ebhuoma EE (2020) A framework for integrating scientific forecasts with indigenous systems of weather forecasting in southern Nigeria. Dev Pract 30(4):1–13

Ebhuoma E, Leonard L (2020) An operational framework for communicating flood warnings to indigenous farmers in southern Nigeria: a systems thinking analysis. GeoJournal 1–18. https://doi.org/10.1007/s10708-020-10221-4(0123456789(),-volV)(01234567

Ebhuoma OO, Gebreslasie M, Ebhuoma EE, Leonard L (2021) 'The future looks empty': embodied experiences of distress triggered by environmental and climatic changes in rural KwaZulu-Natal, South Africa. GeoJournal 1–17. https://doi.org/10.1007/s10708-021-10426-1

Government of Mozambique (2012) National climate change adaptation and mitigation strategy. https://www.ctc-n.org/sites/www.ctc-n.org/files/resources/mozambique_national_climate_change_strategy.pdf. Accessed 15 November 2021

Government of Malawi (2016) National climate change policy. https://www.reliefweb.int/sites/reliefweb.int/files/resources/NCCM-Policy-Final-16-11-2016.pdf

Habibu T, Chisanga K, Bosco Mvula A (2017) The role of indigenous knowledge in climate adaptation: experiences with farmer perceptions from climate change project in Sedumbwe agricultural camp of Southern Zambia. Int J Sci Res Publ 7(9). www.ijsrp.org

Hamel K, Tong B, Hofer M (2019) Poverty in Africa is now falling-but not fast enough. https://www.brookings.edu/blog/future-development/2019/03/28/poverty-in-africa-is-now-falling-but-not-fast-enough/. Accessed 21 October 2021

Hansen JW, Mason SJ, Sun L, Tall A (2011) Review of seasonal climate forecasting for agriculture in sub-Saharan Africa. Exp Agric 47(2):205–240

Harvatt J, Petts J, Chilvers J (2011) Understanding householder responses to natural hazards: flooding and sea-level rise comparisons. J Risk Res 14(1):63–83

ILO (2019) Indigenous peoples and climate change emerging research on traditional knowledge and livelihoods

Jiri O, Mafongoya P, Chivenge P (2015) Indigenous knowledge systems, seasonal 'quality' and CC adaptation in Zimbabwe. Clim Res 66(2):103–111. https://doi.org/10.3354/cr01334

Kalanda-Joshua M, Ngongondo C, Chipeta L, Mpembeka F (2011) Integrating indigenous knowledge with conventional science: enhancing localised climate and weather forecasts in Nessa, Mulanje, Malawi. Phys Chem Earth Parts a/b/c 36:996–1003

Kupika OL, Gandiwa E, Nhamo G, Kativu S (2019) Local ecological knowledge on climate change and ecosystem-based adaptation strategies promote resilience in the Middle Zambezi Biosphere Reserve, Zimbabwe. Scientifica 2019. https://doi.org/10.1155/2019/3069254

Lucio FDF (1999) Use of contemporary and indigenous forecast information for farm level decisión making in Mozambique. Consultancy report. UNDP/UNSO, p 72

Lupele J (2020) Ministry of lands and national resources national climate change learning strategy: background report

Makate C (2019) Local institutions and indigenous knowledge in adoption and scaling of climate-smart agricultural innovations among sub-Saharan smallholder farmers. Int J Clim Change Strateg Manag (Emerald Group Publishing Ltd.) 12(2):270–287. https://doi.org/10.1108/IJCCSM-07-2018-0055

Matos A, Barraza L, Ruiz-Mallén I (2021) Linking conservation, community knowledge, and adaptation to extreme climatic events: a case study in Gorongosa National Park, Mozambique. Sustainability 13:6478. https://doi.org/10.3390/su13116478

Mavhura E, Manyena SB, Collins AE, Manatsa D (2013) Indigenous knowledge, coping strategies and resilience to floods in Muzarabani, Zimbabwe. Int J Disaster Risk Reduct 5:38–48

Mavhura E (2020) Learning from the tropical cyclones that ravaged Zimbabwe: policy implications for effective disaster preparedness. Nat Hazards 104:2261–2275

Mbewe MA, Phiri A, Siyambango N (2019) Indigenous knowledge systems for local weather predictions: a case of Mukonchi Chiefdom in Zambia. Environ Natl Resour Res 9(1):16–26

McGrath I (2002) Materials evaluation and design for language teaching. Edinburgh University Press, Edinburgh

McNaught R, Warrick O, Cooper A (2014) Communicating climate change for adaptation in rural communities: a pacific study. Reg Environ Change 14 (4):1491–1503

Mudombi S, Godwell N (2014) Access to weather forecasting and early warning information by communal farmers in Seke and Murewa districts, Zimbabwe. J Human Ecol 48(3):357–366

Mubanga KH, Umar BB (2014) Smallholder farmers' responses to rainfall variability and soil fertility problems by the use of indigenous knowledge in Chipepo, Southern Zambia. J Agric Sci 6(6):75–85

Naab FZ, Abubakari Z, Ahmed A (2019) The role of climate services in agricultural productivity in Ghana: the perspectives of farmers and institutions. Clim Serv. https://doi.org/10.1016/j.cliser.2019.01.007

Nkomwa EC, Joshua MK, Ngongondo C, Monjerezi M, Chipungu F (2014) Assessing indigenous knowledge systems and climate change adaptation strategies in agriculture: a case study of Chagaka Village, Chikhwawa, Southern Malawi. Phys Chem Earth 67–69:164–172. https://doi.org/10.1016/j.pce.2013.10.002

Nyadzi E (2020) Best of both worlds: co-producing climate services that integrate scientific and indigenous weather and seasonal climate forecast for water management and food production in Ghana. Ph.D. Thesis, Wageningen University and Research, Wageningen, The Netherlands. https://library.wur.nl/WebQuery/wurpubs/fulltext/515697. Accessed 27 October 2021

Ogalleh SA, Christian RV, Josef E, Michael H (2012) Local perceptions and responses to climate change and variability: the case of Laikipia District, Kenya Sustainability 4(12):3302–3325. https://doi.org/10.3390/su4123302

Paul SK, Routray JK (2010) Flood proneness and coping strategies: the experiences of two villages in Bangladesh. Disasters 34(2):489–508

Petzold J, Andrews N, Ford JD, Hedemann C, Postigo JC (2020) Indigenous knowledge on climate change adaptation: a global evidence map of academic literature. Environ Res Lett (IOP Publishing Ltd.) 15 (11). https://doi.org/10.1088/1748-9326/abb330

Republic of Zambia (2016) Ministry of national development planning republic of Zambia national policy on climate change Ministry of lands, Natural Resources and Environmental Protection

Sarrasanti N, Donkor FK, Santos C, Tsagkari M, Wannous C (2020) It's about time we care about an equitable world: women's unpaid care work and COVID-19. IEEE Eng Manag Rev. https://doi.org/10.1109/EMR.2020.3031313

Schlenker W, Lobell DB (2010) Robust negative impacts of climate change on African agriculture. Environ Res Lett 5(1). https://doi.org/10.1088/1748-9326/5/1/014010

Sietz D, Richard MB, Klein JT, Lotsch A (2008) Mainstreaming Climate adaptation into development assistance in Mozambique: institutional barriers and opportunities. http://econ.worldbank.org

Simatele D, Simatele M (2015) Climate variability and urban food security in sub-Saharan Africa: lessons from Zambia using an asset-based adaptation framework. S Afr Geogr J 97:243–263

Stuart C (2017) Energy and climate change adaptation in developing countries. https://sun-connect-news.org/fileadmin/DATEIEN/Dateien/New/euei_pdf_2017_energy_and_climate_change_adaptation_in_developing_countries_1_.pdf. Accessed 15 November 2021

Tadross M, Davis C, Engelbrecht F, Joubert A, van Garderen EA (2011) Regional scenarios of future climate change over southern Africa. In: Davis CL (eds) Climate risk and vulnerability: a handbook for Southern Africa. Council for Scientific and Industrial Research, Pretoria, South Africa, pp 92. http://researchspace.csir.co.za/dspace/bitstream/handle/10204/5800/Archer2_2011.pdf?sequence=1&isAllowed=y. Accessed 23 October 2021

Tall A, Kaur H, Hansen J, Halperin M (2015) Tanzania summary of baseline studies: country report for the GFCS adaptation program in Africa. In: CCAFS Working Paper no. 124. Wageningen, Netherlands: CGIAR research program on climate change, agriculture and food security (CCAFS). https://ccafs.cgiar.org/publications/tanzania-summary-baseline-studiescountry-report-gfcs-adaptation-program-africa#.XR265eszaUk. Accessed 3 July 2019

Thakur R, Rane AV, Harris G, Thakur S (2020) Future prospective and possible management of water resources in respect to indigenous technical knowledge in South Africa. In: Singh P, Milshina Y, Tian K, Gusain D, Bassin JP (eds) Water conservation and wastewater treatment in BRICS nations, 2020, pp 353–363. https://doi.org/10.1016/B978-0-12-818339-7.00018-7

Tume SJP, Kimengsi JN, Fogwe ZN (2019) Indigenous knowledge and farmer perceptions of climate and ecological changes in the bamenda highlands of Cameroon: insights from the Bui Plateau. Climate 7 (12). https://doi.org/10.3390/cli7120138

World Bank (2021) Global program for safer schools: Malawi. https://gpss.worldbank.org/en/country/malawi. Accessed 21 October 2021

Shorai Kavu holds a B.Tech Honours in Production Engineering and an M.Sc. in Renewable Engineering. Currently, she is a doctor of philosophy student in the Centre for Development Studies in the School of Life Long Learning and Development Studies at Chinhoyi University of Technology, Zimbabwe. Her research interests are climate change mitigation and adaptation, renewable energy for energy access, mitigation and adaptation to climate change, Renewable Energy and Climate Change Policy, local and indigenous knowledge systems, and community livelihoods. I am a member of the United Nations Framework Convention on Climate Change (UNFCCC) Register of Experts–Reviewer of Biennial Reports and National Communications for Developed Countries.

Collen Kajongwe (Ph.D.) is a Senior Lecturer in the Department of Human Resources under the School of Business Management in Chinhoyi University of Technology, Zimbabwe. He holds a Ph.D. in Human Resources, M.Sc. In Peace Leadership and Conflict Resolution and B.Sc. In Human Resources Management. He has published extensively in in a number of Journals and Conferences. His research interests include labour laws, human capital management, management of SMEs, Production and Operations Management and indigenous knowledge systems.

Eromose E. Ebhuoma (Ph.D.) is a postdoctoral research fellow at the Department of Environmental Sciences, College of Agriculture and Environmental Sciences, University of South Africa (UNISA). His research expertise spans climate change vulnerability and adaptation, climate services, climate change policy and governance, local and indigenous knowledge systems, political ecology, rural livelihoods and environmental sustainability in sub-Saharan Africa. He is a steering committee member of the South African Adaptation Network.

Approaching Positionality in Research on Indigenous Knowledge Systems

7

Sinoxolo Magaya and Jennifer M. Fitchett

7.1 Introduction

Indigenous knowledge systems (IKS) have only recently gained traction in geographical research, particularly research on climate change adaptation (Chambers et al. 2019; Elmqvist et al. 2003). The studies on IKS highlight that traditional monitoring methods for tracking environmental change are perceived to have more guaranteed efficacy because they have been used by rural populations to maintain successful food production for many centuries (Mazzocchi 2006; Orlove et al. 2010; Radeny et al. 2019; Rukema and Umubyeyi 2019). Indigenous knowledge (IK) is a tacit and situated form of knowledge reflecting in-depth experiences that are unlikely to have come from formalised structures because they are an inherent part of livelihoods and tradition (Ebhuoma 2017; Fernández Llamazares Onrubia 2015; Kijazi et al. 2013). Therefore, the positionality of both the researcher and the researched is significant for data collection and analysis in any research it relies on (Reid and Sieber 2020). Following the emergence of an initial wave of papers documenting IKS (Chand et al. 2014; Gegeo 1998; Yirga 2010), there is a growing body of literature that engages critically with issues of power, ownership of knowledge and ethical considerations relating to work with indigenous communities (Agrawal 2002; Akena 2012; Chambers et al. 2017; Kwame 2017). What has not yet been documented sufficiently is the importance of considering documentation of and critical reflection on positionality, particularly when working in the context of pressing contemporary issues such as climate change.

Positionality involves people's perceptions of the world from various embodied locations (Martin 2017). In research, it refers to the views and the position a researcher adopts in relation to the study topic (Manohar et al. 2017). Researchers' positionality can inhibit or promote different aspects of the research process (Qin 2016). In IKS research, recognising a researcher's positionality is considered a significant step towards decolonising methodologies (Khupe and Keane 2017). Researchers' reflexivity and understanding of their positionality is argued to promote a comprehensive representation of the voices of the researched (Fisher 2015; Khupe and Keane 2017; Milner 2007). This paper explores the role of researcher positionality in IKS research in the context of climate change adaptation through the experiences of an indigenous insider researcher (first author) from two rural areas in the Eastern Cape and KwaZulu-Natal provinces, South Africa, and a researcher representing the role of the outsider. Both Umzimkulu (KwaZulu-Natal) and Lusikisiki (Eastern Cape)

S. Magaya · J. M. Fitchett (✉)
School of Geography, Archaeology and Environmental Studies, University of the Witwatersrand, Johannesburg, South Africa
e-mail: Jennifer.Fitchett@wits.ac.za

E. E. Ebhuoma and L. Leonard (eds.), *Indigenous Knowledge and Climate Governance*, Sustainable Development Goals Series, https://doi.org/10.1007/978-3-030-99411-2_7

are in the former homeland regions of the provinces and have the highest population densities (Graw et al. 2017). The homelands or Bantustan areas were created during the apartheid era to enforce residential segregation and marginalise the black population from major political and socioeconomic activities (Kwenda et al. 2020; Todes and Turok 2018). Such former homelands are still mostly poverty-stricken and struggle with service delivery (Kwenda et al. 2020). Farms in the eastern part of the Eastern Cape province are mostly owned by households with the lowest income that practice farming mainly to ensure food availability (Graw et al. 2017; Molekwa 2013). Subsistence farmers in rural sub-Saharan Africa often lack adequate financial resources to ensure optimum productivity and to limit excessive damage or adapt to the changes caused by climate stressors (Ebhuoma 2017; Shisanya and Mafongoya 2016). In these rural areas, IK is argued to be one of the most important social assets that has been effective in promoting food production (Mafongoya et al. 2016). Zuma-Netshiukhwi et al. (2013) and Rukema and Umubyeyi (2019) also found that IK for subsistence agricultural management is widely used in rural South Africa to both track seasonality and develop adaptive strategies. This suggests that farming activities and agricultural success in the two poor rural areas under study can, to some extent, be attributed to IKS strategies.

7.2 Researcher's Positionality

The position a researcher adopts is influenced by factors such as age; sexuality; ethnicity; social class; linguistics; geographical and intellectual location; culture; and race (Moser 2008; Qin 2016). These largely represent the background experiences that form a researcher's perceptions and predominantly capture intersectionality of social identities for most researchers (Adu-Ampong and Adams 2020; Bourke 2014; Carstensen-Egwuom 2014). The social and cultural beliefs borne from experiences and intersecting identities have a significant influence on the accumulation and interpretation of information by researchers and ultimately affect knowledge production (Manohar et al. 2017). These attributes influence a researcher's relationality and ability to interpret and assign meaning to actions, language and terminology within a socio-cultural context (Holmes 2020).

The spatial and cultural specificity of IK information cannot be over emphasised (Orlove et al. 2010). IKS are based on knowledge and the know-how accumulated over generations is applied successfully in guiding human societies in their interaction with and observation of their environments (Mbewe et al. 2019; Orlove et al. 2010). As a result, researcher/participant subjectivities are likely to affect IKS research data. Furthermore, in most qualitative research methodologies, knowledge production is based on the meaning participants in a study assign to certain events and the ways in which the researcher interprets the experiences of the respondents (Muhammad et al. 2015). These related concepts are both framed within varying social and cultural contexts and constitute positionality (Khupe and Keane 2017; Qin 2016).

Bourke (2014) noted that presenting more objective results and analysis is the main component when contributing to knowledge. Chavez (2008) argued that neither an insider nor an outsider researcher can achieve pure objectivity because every researcher has an identity and characteristics that inform their perceptions. Therefore, scholarly literature researchers must be aware of their subjectivities to understand the beliefs that influence their observations and data interpretation, particularly in qualitative studies (Bourke 2014; Reid and Sieber 2020). This is important because in most qualitative research methodologies the researcher is the data collection instrument (Xu and Storr 2012). Researchers are not divorced from their beliefs and experiences during the research process. As such, there is always a degree of subjectivity involved in collecting and analysing qualitative data (Holmes 2020; Hoogendoorn and Visser 2012). This applies even more to IKS research because IK is inherently developed through experiences and based on the perceptions and beliefs of communities (Chambers et al. 2017; Reid and Sieber 2020). Therefore, for researchers to contribute to credible and valid knowledge production, they must acknowledge their identities as individuals and their positions within groups that shaped or continue to inform their subjective views (Muhammad et al. 2015; Reid and Sieber 2020). Thus, positionality represents a juncture of objectivity and subjectivity (Qin 2016) and reflection on their positions in relation to a study and its participants would enable researchers to be more cognisant of their biases (Milner 2007).

7.3 Exploring Positionality During the Research Process

This work stems from research towards a master's degree by the first author, with the second author acting as supervisor. From the beginning of the process, the two authors engaged regularly and critically on the issue of positionality. It was presumed that the familiarity of the first author with the two study sites (Fig. 7.1) because she had grown up in the rural communities of Umzimkulu and Lusikisiki would facilitate greater access than would be afforded to many others. At the same time, however, it could influence the data collection and analysis process through both her own preconceptions and the perceptions of the community (Reid 2020). The first author represented the insider researcher, while the second author, who had not previously visited either of the two communities, represented the outsider researcher.

To remain aware of positionality before commencing with fieldwork, we began our research by capturing the first author's life story, her relationship with the community and their beliefs, and her understanding of her positionality (Sultana 2007). That allowed us to reflect on predetermined perceptions relevant to both the research process and the IK of the region when conducting and analysing the results of the interviews (Akena 2012). The following record resulted:

> I was born and raised in Clydesdale location in Umzimkulu. I have a relationship with both study areas, Lusikisiki, my home from my father's side and Umzimkulu, my home from my mother's side. Therefore, my own IK has been constructed by, and is similar to, the communities that I will engage with in the study. However, my IK has been informed mostly by the culture and history of Umzimkulu because I spent more time there, but I identify as an indigenous community member of both areas and I am perceived in that manner by the community members in these places.
>
> At home, we were always involved in different forms of subsistence farming to supplement food sources. Food was always more available during summer and the beginning of autumn (December–March) because my grandmother sowed her garden in September and most of the crops matured from the end of November to the beginning of December. In February to March we would start eating 'autumn foods'. Therefore, my initial perception of the world was largely formed through the lens of poverty, rurality and family unity.
>
> There were no distinguished gender roles between me and my brothers in our household, which was viewed as unconventional by some community members but was never an issue in my household. One of the most important values instilled in me by my family is the importance of education as a tool against poverty, hunger and for the betterment of myself. Although my family has always advocated for education, the contextual inefficiency and cultural detachment of western knowledge have also been highlighted constantly. Therefore, there has always been a clear demarcation between the two knowledge systems and an indication to always maintain these boundaries unless the information unequivocally parallels or compliments each other. However, although the relationship between these epistemologies was not formally established, I sometimes used IK exposure to understand and memorise information from school. For example, I was introduced to the use of kraal manure as a fertiliser at home without extensive explanation on the science behind its practicality. This experience facilitated my understanding of the concept when it was introduced to me at school.
>
> I have always been exposed to IK strategies for agricultural management, daily activities and risk reduction. However, these observations and activities were not explained to me; I copied and deduced from my elders and sometimes I followed instructions to do so. In my household and in accordance with the culture of our community in general, challenging elders and probing for more information than they felt I needed to know were discouraged and a sign of disrespect. These cultural limitations prevented rigorous conversations,

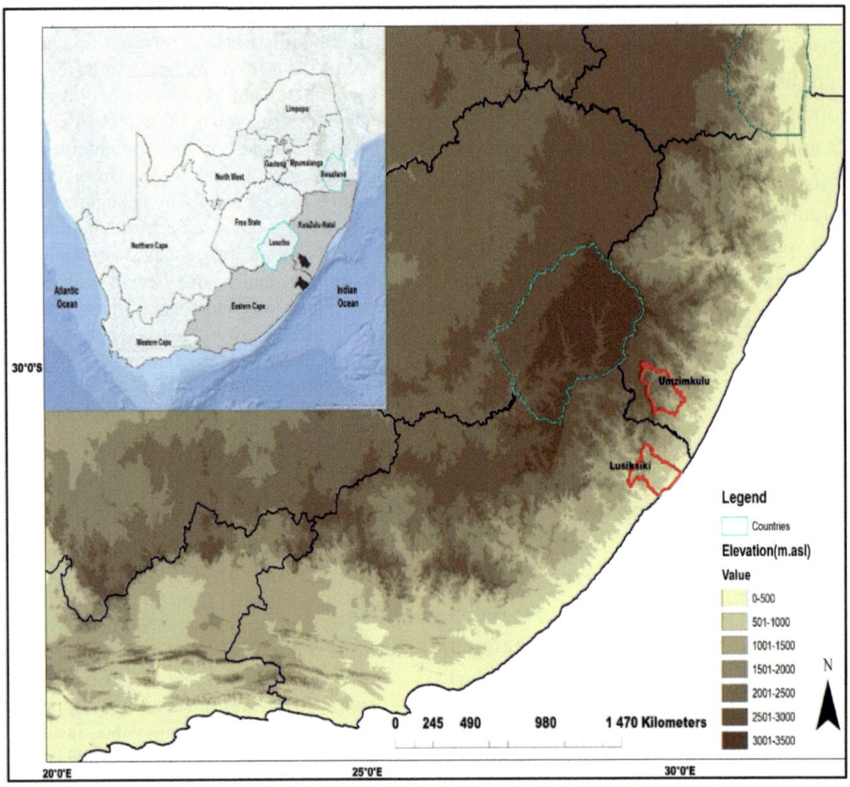

Fig. 7.1 Location of Umzimkhulu and Lusikiski within KwaZulu-Natal and Eastern Cape provinces, South Africa

added to my skepticism of IK and facilitated a futile comparison between IK and western science on my part. I felt that western science offered more background and underlying causes, and thus more certainty. To me, that emphasised the importance of empirical evidence for knowledge.

I was introduced to western science through the natural science subject in grade 4 and an expanded form including landscapes and climatology in geography in grade 10. I was intrigued by natural science in grade 5 when the concept of pollination was introduced, because from that moment the observation of bees and butterflies moving from one plant to the next, something I had seen for most of my life, suddenly acquired meaning. That was when my love and passion for the natural environment developed and it became meaningful to me. From that moment, I always tried to find scientific explanations for my observations, particularly related to plants and animals. Before the introduction of a compartmentalised view of nature, my perception of nature was informed by everyday interactions that varied depending on seasons and the varying activities governed by seasonal weather.

Moffat (2016) emphasises that there is more than one way of knowing and applying knowledge. Being born and raised in a rural community and having engaged in subsistence farming, I have always been aware of this. As such, my ontology has been influenced significantly by both IK and western science information, but this is the first time I merge these systems from a balanced perspective that seeks to draw informed conclusions and contribute to broader knowledge. These two parts of my identity influence my perceptions on IK, agricultural management, climate change and climate change adaptation. I believe that poverty in my childhood added to my recently changed belief of associating IKS with lack of better resources rather than an extensive knowledge resource that sustains cultural identity and offers a significant tool to preserve community values and contribute to environmental management. However, regardless of my earlier perception of IK, I have always been curious to understand the processes behind used indicators and the keen skills displayed by my elders in weather forecasting and risk reduction. IKS have a significant influence on my world views, but I constantly tried to situate the significant part of my roots and ontology within scientific discourse. However, my master's research was borne from my desire to understand the relationship between ringing metal and abating a hailstorm from a physical science and/or climatological perspective. While this research does not address this question, it demonstrates the influence of IK on my constant curiosity to situate the significant part of my roots and ontology within scientific discourse.

The second stage of reflection on positionality took place once the fieldwork had been conducted. We considered the role that positionality had played in access to the community and respondents, in the power dynamics and in the respondents' perceptions (Krenceyova 2014). Reflexivity as an approach that "suggests that, rather than trying to eliminate their effect, researchers should acknowledge and disclose their selves in their work, aiming to understand their influence on and in the research process" (Holmes 2020, p. 3), before and during the research process allowed the insider researcher to identify the aspects that made her similar to and different from the respondents (Adu-Ampong and Adams 2020). An understanding of these was necessary both before and during data collection and throughout the data analysis process as it facilitated the presentation of results representing the views of the participants without such views being distorted by western-influenced beliefs about the natural environment, climate change and IKS (Kwame 2017). It furthermore forced the first author to broaden the interview questions to obtain information considered as 'general' information by the communities. That was done to ensure that valuable IK was not discounted because of it being regarded as not 'real' knowledge, and to avoid the assumption that all community members perceived things the way the researcher and her family did. It was crucial that the researchers' pre-conceived perceptions and understandings were noted to ensure that their biases did not steer data analysis but that the first author's insider identity was used to inform a deeper understanding of information and to facilitate data collection and analysis (Erlingsson and Brysiewicz 2017). In this regard our understanding of the role of positionality was guided by a combination of participant observations, the statements of the respondents and the literature. Of note is that, although we acknowledge biases, we aimed to ensure accurate representation of the participants' views by listening to the audio recordings of interviews repeatedly to facilitate better understanding through the identification of the tone of the respondents' voices. That also enabled recollection of the events and the setting of interviews, thus facilitating contextualisation of the information. We, nonetheless, do not declare complete objectivity.

7.4 Access to Researched Communities

The background and ethnicity of the first author facilitated both physical and intellectual access to the study areas. This is consistent with the argument of Hoogendoorn and Visser (2012), who noted that race, gender and socio-cultural connections dictate fieldworkers' access to data in most qualitative research in South Africa. This underscores some of the challenges faced in accessing research information even beyond the scope of IKS research (Fisher 2015). The insider researcher did not require extended time to establish a relationship or trust to communicate with the participants. That might not have been the case if she were an indigenous outsider; that is white, rich or a poor non-South African black. However, as an indigenous community member, the researcher was also aware that western systems can exploit rural poor communities (Krenceyova 2014; Reid 2020). Therefore, bearing in mind her power as a researcher, she was extremely aware of her outsider position as postgraduate student and the assumed perceptions of participants towards her because of that. Furthermore, the she was mindful of cultural norms and respectful language use around elders, which could limit her exploration of the subject topic. She resolved to remain within the ethical boundaries of both the community and the university and to refrain from manipulative behaviour towards community members and their IK. Culturally, she identifies with community members and understands IK in the study areas. However, her qualifications, age (24 years) and the level of her involvement in farming activities limited her value for the research topic. Most respondents viewed her as one of the children in the community and the following statement showed that their perceptions and comments were based on her age rather than on her level of education or her status as a researcher: "You children, you are not interested in farming though. Young boys do not even know how to milk a cow. Can you even hold a garden hoe and plough? I doubt it." (NMZ 1.3, 1.4).

Nonetheless, all the community members who were approached participated in the interviews and offered information based on their experiences, as such contributing to the dataset. Adu-Ampong and Adams (2020) stated that the significance of an affiliation with a prestigious institution or an established supervisor on a subject varies depending on the setting of an interview and the intended key informants. In this study, the researcher's position based on age was more important than her education level, gender and affiliation with a large, prestigious university. In commenting on her education, participants tried to understand the content of and the reason for her choosing the specific degree. Some also enquired about the cost of education. That was immediately followed by encouragement to take care of her parents whom they assumed were paying for the education. However, overall the insider researcher's familiarity with the communities made interaction with the participants easier, while their views on and enthusiasm for education reflected community support for education in the area.

7.5 Power

Relevant to culture, race, language and ethnicity the first author represented an insider in the communities. However, although community members did not acknowledge her education levels, she understood that her position as a university student and as a researcher who was collecting data, together with the age gap between her and most of the participants, resulted in some community members regarding her as an outsider (Milligan 2016). As a member of the communities, she was aware of the value community members in the two study areas attached to education. Umzimkulu and Lusikisiki communities appreciate and encourage education and the researcher knew that before embarking on the data collection process. Therefore, her position as a researcher made it possible for her to approach and engage elders in a way they would generally have been be skeptical of had she been regarded as merely a child from the community. Furthermore, the absolute regard for her intellect as a postgraduate student was notable among most participants when the study was explained. The importance of the participants' contribution to knowledge creation and research was pointed out to regulate power dynamics and that served to emphasise co-construction of knowledge between the researchers and the participants. The insider researcher consequently had to exhibit only an average understanding of the IK strategies in the areas so as not to overshadow the views of the participants, and the value of the latter's knowledge for the research had to be emphasised constantly. It was important for the researcher to display indigenous insider knowledge and to affirm the concept of co-production of knowledge while also demonstrating genuine interest and ensure trust in IKS. The following is an example of one of her insider conversations in Lusikisiki:

Interviewee: Do you also sow in Umzimkulu?
Researcher: Yes, but there are not as many people who work their gardens anymore.
I: They sow *emasimini* (the big farms)?
R: No. They do use their gardens; people have big gardens that side. Very few people still use *amasimi*.
I: I see. I also don't use *amasimi* anymore. It is mostly because we are old now; so we don't have the energy to work *amasimi*. Then how do you know when it is about to rain that side?
R: Aaahh, the usual, you know. We hear southern ground hornbills and sometimes we see ants.
I: Oh, I see. I do not particularly believe in the ants. The southern ground hornbills are also not a reliable anymore, have you noticed…?

Several scholars have emphasised that insider/outsider positionality is malleable to varying research topics and study areas (Chavez 2008; Fisher 2015; Holmes 2020). Kwame (2017) argued that insider/outsider status should not be regarded as a methodology flaw in research but rather be seen as an opportunity to gain different insights for the co-production of knowledge. In this study, the research process was influenced by the perspectives of both an insider and an outsider on IKS in Umzimkulu and Lusikisiki. As a white woman and an associate professor with no socio-cultural attachment to the study areas, the second author and supervisor of the study represented an indigenous outsider. This suggests that she represented etic perspectives from a non-cultural-specific context (Holmes 2020). Indigenous outsider views are predominantly rooted in realist perspectives that describe differences between cultures from an ontological position that assumes a pre-defined reality (Holmes 2020; Kwame 2017). This perspective facilitated a realist consideration of the collected data and awareness of knowledge that, from an insider's perspective, might have been overlooked on the premise that it is 'just how things are done'. Furthermore, her appraisal of and outsider views on IKS in climate-change research provided an opportunity for the development of the study.

7.6 Participants' Perception of the Insider Researcher

While research is reported through the lens of the researcher, it represents an environment shared by both the researcher and the researched (Bourke 2014). Therefore, the social and cultural identities of the participants are also significant pertaining to results. In this study, the identity and positionality of the first author as an indigenous insider with outsider influence did not exempt her from the scrutiny of some participants regarding her motives for conducting the study in the selected areas. Statements like the following reflected that: "Are you one of these political parties that are always coming here and promising us things?" (EC 2.3); "Who are you doing this research for? It is not like we will see or use this information." (EC 1.4).

Yet other participants challenged her interest in the study and her positionality through statements like the following:

> So, you want information about our farming. I hear you, and I will give it to you, but I doubt you can even carry a garden hoe and plough, or pluck weed. You are here with just your pen and books. What are you going to do with this information then? (EC 2.1).

For the insider researcher as just a young woman, rigorous engagement with elders would generally be limited, particularly after they questioned her sincerity and interest in the topic under study. However, her status as a student/researcher empowered her to probe and seek understanding of the community members' perceptions regardless of their doubts. The participants' perception that the researcher had no real understanding of IKS because she was just a young member of the community facilitated collection of rich data. Although the youth's disinterest in IKS was regarded negatively, the eagerness of the community to ensure that a student succeeded at school preceded any beliefs and perceptions they might have had in that regard. The experience of the insider in the study of Adu-Ampong and Adams (2020) was, however, different. In that study, participants used the insider position of the researchers to withhold information on the premise that, as they were also Ghanaian, they should have had an understanding of some of the issues they were asking about.

7.7 Discussion

Studies often emphasise that IK is rooted in cultural cohesion, that it is place-based and that it can usually be associated with long-settled communities who have strong ties with the natural environment (Chambers et al. 2019; Chand et al. 2014; Gadzirayi et al. 2006). Therefore, varying interpretations of the biophysical environment between and within communities are an inherent attribute of IKS, particularly IK that is important for perceptions of and adaptations to climate change in different areas (Goduka 2012). Community members from Umzimkulu and Lusikisiki demonstrated respect for cultural differences relevant to IKS by highlighting the variations that may exist between areas that are spatially close. For example, in both Umzimkulu and Lusikisiki participants who had acquired their current homes through marriage highlighted differences between IKS in their birth areas and marriage areas, which were mostly less than 10 kms apart. Furthermore, participants often emphasised that their knowledge was not exhaustive of IKS in those areas and was subject to change according to arising needs because of climate change or availability of new resources and knowledge. That suggested that community members in the areas acknowledged the possible limitations of IKS. These farmers could therefore be more receptive to the integration of knowledge systems to ensure food production and maintain livelihoods, unlike the finding of a study in Nigeria where community members believed that God is the ultimate power and that their influence on the natural environment was insignificant (Ebhuoma 2017).

Indigenous rural realities contribute to the creation, implementation and evolution of IK. IK furthermore manifests agency and resilience in response to exploitation, natural stressors and climate variability and change (Valdivia et al. 2010). The recognition of IKS as a viable tool for climate-change adaptation and other environmental conservation efforts increases research interest in this epistemology and the locations where it is produced and applied (Magni 2017). This presents challenges related to exploitation of the knowledge, which is often connected to a community's cultural identity and values (Chambers et al. 2017; Janke 2018). Chambers et al. (2017) and Tong (2017) argued that several challenges relevant to knowledge owned by communities would have to be addressed for proper dissemination of IK ensuring property rights that would protect data from appropriation and commercialization. However, the obscure intellectual property rights for community-owned knowledge make this process challenging and render IKS vulnerable to exploitation, particularly by individuals who have no direct connection with the study areas and who may be seeking the knowledge merely for purposes of commercialisation (Chambers et al. 2017; Masango 2010; Watts 2013). Many indigenous societies consider IK sacred and as a part of their identities. It should therefore not be sold but shared only with individuals who could be trusted to maintain its integrity throughout every stage of data processing and reporting (Janke 2018; Watts 2013). Watts (2013) argued that the disengaged and commercial nature of scientific research dissemination perpetuates the risk of loss of the significance of IK and the connection that communities have with the natural environment, which is the basis for continued development and use of IKS (Janke 2018). It would furthermore lead to loss of significant cultural values and thus weaken this social asset for climate change adaptation in rural areas (Birkmann and Teichman 2010; Parsons et al. 2016).

The perception of climate change and climate change adaptation is influenced by traditional, cultural and religious beliefs and values in Umzimkulu and Lusikisiki. While most Christian community members attributed the prolonged dry spells and extreme heat to punishment from God for their wrongdoings, a traditional healer attributed it to the neglect of ancestors in favour of western development. Nonetheless, all participants associated these climatic consequences with their ill-treatment of the natural environment and stated that if they could use the natural resources in their areas with care and respect, the gods would reward them with more agricultural success and good weather. Although this displays a limited understanding of the underlying scientific complexities related to climate change, it presents a starting point for disseminating more anthropogenic climate-change-related information in these communities. Community members in Umzimkulu and Lusikisiki acknowledged one of the essential components for understanding anthropogenic climate change, namely the influence of human activities on the natural environment (Parry et al. 2007). Acknowledging the participants' positionality and the belief systems within which climate change is viewed could, therefore, facilitate production of knowledge that honours and is aligned with community beliefs and values. That could, in turn, promote adoption of more pragmatic adaptive strategies instead of western-science-based strategies that lack context. It is imperative that IK research on climate change represents the social, cultural and, to some extent, economic condition within which the knowledge is developed, used and transformed (Kassam et al. 2017).

Recognising positionality in IK research compels researchers to acknowledge their biases and to regulate the power dynamics created by their roles as reporters of culturally rooted knowledge (Gao and Mataira 2019). This suggests that researchers have the power to shape the truth and perceptions about the indigenous communities in which they conduct research (Reid and Sieber 2020). The perceptions and beliefs of researchers affect the way they ask questions, understand information and interpret responses (Jacobson and Mustafa 2019). Addressing the connotations of poverty and rurality that the primary researcher associated with IKS, and acknowledging certain biases within themselves in relation to the subject topic and the researched area or people, the researchers of this study gained insight into effective ways of approaching the research process (Holmes 2020). That facilitated access to information and ensured that the researchers presented results that represented the views of the participants accurately. The imposition of incorrect assumptions, distortion and decontextualisation of IKS and related strategies is one of the pervasive challenges in the presentation of traditional methods and worldviews (Briggs 2005). The process of reflexivity ensures that researchers avoid engaging in IKS research without interrogating their positionality, which may lead to a distortion of IK in pursuit of neutrality (Milner 2007). Critically reflecting on their experiences and political, cultural and social identities reduces the influence of implicit biases and facilitates a contextualised understanding of information (Chavez 2008). Moreover, declaring positionality demonstrates to the readers the ontologies within which the research data were analysed (Adu-Ampong and Adams 2020).

The evolving alliance between IK and western science for environmental conservation and climate change adaptation supports an openness for the sharing of knowledge and the exploration of new and more effective measures with significant input from communities (Makondo and Thomas 2018). Collaboration between these knowledge systems could promote effective climate change adaptation and promote taking advantage of already existing strategies at all levels, from household measures to national policies, particularly for the most vulnerable communities (Chambers et al. 2019). In many sub-Saharan African countries, including Nigeria (Ebhuoma 2017), Tanzania (Radeny et al. 2019), Zimbabwe (Mafongoya and Ajayi 2017) and Malawi (Nkomwa et al. 2014), research has demonstrated that IKS provide a reasonable foundation to strengthen adaptive capacity, particularly for small-scale agriculture. Therefore, IKS are considered to be a significant attribute of the bottom-up approach for climate change adaptation in this region (Mapfumo et al. 2016). In Umzimkulu and Lusikisiki, participants demonstrated that IK represents creative planning and action based on observations within the continuously changing natural environment. Therefore it is contextual within an area and sensitive to specific environmental challenges (Magni 2017).

In this study the collection of rich information on IKS for climate change adaptation was eased by the custodians' appraisal of education and the primary researcher's relationship with the communities in the study areas. Furthermore, due to community members' assumption that because of her age the primary researcher was not well versed in IK that is significant for agricultural management, participants were more willing to provide in-depth descriptions of the indicators they used, the various adaptive strategies they had developed and the changes that they had observed, both concerning the climate and the efficiency of IKS strategies. That provided a synopsis of qualitative changes relative to what is considered normal in the natural environment and climate of the area, particularly rainfall characteristics and seasonality (Savo et al. 2016). The descriptions also provided a comprehensive timeframe for the development of adaptation strategies and displayed adaptability and a self-organising process of knowledge systems within the communities (Fernández Llamazares Onrubia 2015). Reflexivity is a very important component in making available credible climate change adaptation information that honours the socio-cultural context within which relevant knowledge has been developed (Bourke 2014; Latai-Niusulu et al. 2020). In line with the custodians' perspectives, the primary researcher's awareness of her own limitations regarding the subject under study, her fluency in the local dialects and her understanding of local social norms and cultural nuances aided the acquisition of rich data and assisted in affirming the validity of the results (Latai-Niusulu et al. 2020). Constant reflexivity on the part of the researchers furthermore influenced the continuous regulation of the already existing power dynamic between the two knowledge systems throughout the data analysis process (Gao and Mataira 2019). Such earnest engagement with reflexivity is essential for the development of contextually relevant climate change adaptation information that, through integration of knowledge systems, can have the quantitative attributes offered by western science (Makondo and Thomas 2018).

7.8 Conclusion

IKS are important for climate change adaptation in many rural areas, particularly for small-scale farmers in sub-Saharan Africa. Custodians of IK are the gatekeepers of the knowledge and decide to share this knowledge based on their interpretation of a researcher's position within their society. This could lead to failure to procure research data if community members deem a researcher or the institution they represent untrustworthy. If the custodians grant access to their information, it is a researcher's responsibility to frame the results in a way that honours the socio-cultural context of the custodians honestly and respectfully. This is particularly important in climate change research, because the ways in which societies experience environmental change are tied to the economic, cultural and overall social conditions of communities. Therefore, IKS research dynamism for climate change mitigation and adaptation is dependent on the relationship between researchers and participants, which is mainly governed by positionalities. This chapter highlighted that a researcher's insider positionality and acknowledging intersecting identities could contribute effectively to knowledge production. Understanding the influence of researcher–participant positionality in research data collection and analysis would promote production of valuable knowledge and strategies that are significant for the integration of knowledge systems aimed at climate change adaptation. This is particularly important in sub-Saharan Africa regions where community members are exposed to contextually irrelevant quantitative information, which is informed by western science and offered by government agencies, that is less effective in ensuring successful food production. Integration of the two knowledge systems would expose communities to more information, which could strengthen their asset portfolios through the conservation of natural assets and an advanced mobilisation of social assets.

References

Adu-Ampong EA, Adams EA (2020) "But you are also Ghanaian, you should know": negotiating the insider–outsider research positionality in the fieldwork encounter. Qual Inq 26(6):583–592. https://doi.org/10.1177/1077800419846532

Agrawal A (2002) Indigenous knowledge and the politics of classification. Int Soc Sci J 54:287–297. https://doi.org/10.1111/1468-2451.00382

Akena FA (2012) Critical analysis of the production of Western knowledge and its implications for Indigenous knowledge and decolonization. J Black Stud 43 (6):599–619. https://doi.org/10.1177/0021934712440448

Birkmann J, Von Teichman K (2010) Integrating disaster risk reduction and climate change adaptation: key challenges—scales, knowledge, and norms. Sustain

Sci 5:171–184. https://doi.org/10.1007/s11625-010-0108-y

Bourke B (2014) Positionality: reflecting on the research process. Qual Rep 19(33):1–9. https://doi.org/10.46743/2160-3715/2014.1026

Briggs J (2005) The use of indigenous knowledge in development: problems and challenges. Prog Dev Stud 5(2):99–114. https://doi.org/10.1191/1464993405ps105oa

Carstensen-Egwuom I (2014) Connecting intersectionality and reflexivity: methodological approaches to social positionalities. Erdkunde 68(4):265–276. https://doi.org/10.3112/erdkunde.2014.04.03

Chambers LE, Plotz RD, Dossis T, Hiriasia DH, Malsale P, Martin DJ, Mitiepo R, Tahera K, Tofaeono TI (2017) A database for traditional knowledge of weather and climate in the Pacific. Meteorol Appl 24(3):491–502. https://doi.org/10.1002/met.1648

Chambers L, Lui S, Plotz R, Hiriasia D, Malsale P, Pulehetoa-Mitiepo R, Natapei M, Sanau N, Waiwai M, Tahani L, Willy A (2019) Traditional or contemporary weather and climate forecasts: reaching Pacific communities. Reg Environ Change 19:1521–1528. https://doi.org/10.1007/s10113-019-01487-7

Chand SS, Chambers LE, Waiwai M, Malsale P, Thompson E (2014) Indigenous knowledge for environmental prediction in the Pacific Island countries. Weather Clim Soc 6(4):445–450. https://doi.org/10.1175/WCAS-D-13-00053.1

Chavez C (2008) Conceptualizing from the inside: advantages, complications, and demands on insider positionality. Qual Rep 13(3):474–494. https://doi.org/10.46743/2160-3715/2008.1589

Ebhuoma EE (2017) Climate change risk communication and asset adaptation of indigenous farmers in the Delta State of Nigeria. Doctoral thesis, University of the Witwatersrand. WIReDSpace. https://hdl.handle.net/10539/25070

Elmqvist T, Folke C, Nyström M, Peterson G, Bengtsson J, Walker B, Norberg J (2003) Response diversity, ecosystem change, and resilience. Front Ecol Environ 1(9):488–494. https://doi.org/10.2307/3868116

Erlingsson C, Brysiewicz P (2017) A hands-on guide to doing content analysis. Afr J Emerg Med 7(3):93–99. https://doi.org/10.1016/j.afjem.2017.08.001

Fernández Llamazares Onrubia Á (2015) Indigenous knowledge of a changing environment an ethnoecological perspective from Bolivian Amazonia. Doctoral thesis, Universitat Autònoma de Barcelona, Barcelona. DDD de la UAB. https://ddd.uab.cat/pub/tesis/2015/hdl_10803_327020/afllo1de1.pdf

Fisher KT (2015) Positionality, subjectivity, and race in transnational and transcultural geographical research. Gend Place Cult 22(4):456–473. https://doi.org/10.1080/0966369X.2013.879097

Gadzirayi CT, Mutandwa E, Chihiya J, Chikosha M (2006) Indigenous knowledge systems in sustainable utilization of wetlands in communal areas of Zimbabwe: case of Hwedza district. Afr J Agric Res 1(4):131–137. https://academicjournals.org/journal/AJAR/article-full-text-pdf/FACBBF736721

Gao IA, Mataira PJ (2019) Insights into positionality and the voicing of indigeneity in research: towards a critical theory of inbetweenness. Dutkansearvvi Dieđalaš Áigečála 3(2):108–124. http://dutkansearvi.fi/wp-content/uploads/2020/02/Dutkensearvi-Vol3Issue2-Mataira-Silan.pdf

Gegeo DW (1998) Indigenous knowledge and empowerment: rural development examined from within. Contemp Pac 10(2):289–315. http://hdl.handle.net/10125/13230

Goduka N (2012) Re-discovering indigenous knowledge – *ulwazi lwemveli* for strengthening sustainable livelihood opportunities within rural contexts in the Eastern Cape Province. Indilinga Afr J Indig Knowl Syst 11:1–19. https://hdl.handle.net/10520/EJC126068

Graw V, Ghazaryan G, Dall K, Delgado Gómez A, Abdel-Hamid A, Jordaan A, Piroska R, Post J, Szarzynski J, Walz Y, Dubovyk O (2017) Drought dynamics and vegetation productivity in different land management systems of Eastern Cape, South Africa—a remote sensing perspective. Sustainability 9(10), Article 1728. https://doi.org/10.3390/su9101728

Holmes AGD (2020) Researcher positionality—a consideration of its influence and place in qualitative research—a new researcher guide. Shanlax Int J Educ 8(4):1–10. https://doi.org/10.34293/education.v8i4.3232

Hoogendoorn G, Visser G (2012) Stumbling over researcher positionality and political-temporal contingency in South African second-home tourism research. Critical Arts 26(3):254–271. https://doi.org/10.1080/02560046.2012.705456

Jacobson D, Mustafa N (2019) Social identity map: a reflexivity tool for practicing explicit positionality in critical qualitative research. Int J Qual Methods 18:1–12. https://doi.org/10.1177/1609406919870075

Janke T (2018) From smokebush to spinifex: towards recognition of indigenous knowledge in the commercialisation of plants. Int J Rural Law Policy 1:1–27. https://doi.org/10.5130/ijrlp.1.2018.5713

Kassam KAS, Avery LM, Ruelle ML (2017) The cognitive relevance of indigenous and rural: why is it critical to survival? Cult Sci Edu 12:97–118. https://doi.org/10.1007/s11422-016-9745-5

Khupe C, Keane M (2017) Towards an African education research methodology: decolonising new knowledge. Educ Res Soc Change 6:25–37. https://doi.org/10.17159/2221-4070/2017/v6i1a3

Kijazi AL, Chang'a LB, Liwenga ET, Kanemba A, Nindi SJ (2013) The use of indigenous knowledge in weather and climate prediction in Mahenge and Ismani wards, Tanzania. J Geogr Reg Plan 6(7):274–279. https://doi.org/10.5897/JGRP2013.0386

Krenceyova M (2014) Who is allowed to speak about Africa? A reflection on knowledge, positionality, and authority in Africanist scholarship. Afr Insight 44:8–22. https://hdl.handle.net/10520/EJC161967

Kwame A (2017) Reflexivity and the insider/outsider discourse in indigenous research: my personal experiences. AlterNative Int J Indig Peoples 13(4):218–225. https://doi.org/10.1177/1177180117729851

Kwenda P, Benhura M, Mudiriza G (2020) Former homeland areas and unemployment in South Africa: a decomposition approach. Discussion Paper 12941. Institute of Labor Economics (IZA). http://hdl.handle.net/10419/215337

Latai-Niusulu A, Nel E, Binns T (2020) Positionality and protocol in field research: undertaking community-based investigations in Samoa. Asia Pac Viewp 61:71–84. https://doi.org/10.1111/apv.12252

Mafongoya P, Rusinamhodzi L, Siziba S, Thierfelder C, Mvumi BM, Nhau B, Lewis Hove L, Chivenge P (2016) Maize productivity and profitability in conservation agriculture systems across agro-ecological regions in Zimbabwe: a review of knowledge and practice. Agr Ecosyst Environ 220:211–225. https://doi.org/10.1016/j.agee.2016.01.017

Mafongoya PL, Ajayi OC (2017) Indigenous knowledge systems and climate change management in Africa. CTA, Wageningen, The Netherlands. https://hdl.handle.net/10568/91189

Magni G (2017) Indigenous knowledge and implications for the sustainable development agenda. Eur J Educ 52 (4):437–447. https://doi.org/10.1111/ejed.12238

Makondo CC, Thomas DS (2018) Climate change adaptation: linking indigenous knowledge with western science for effective adaptation. Environ Sci Policy 88:83–91. https://doi.org/10.1016/j.envsci.2018.06.014

Manohar N, Liamputtong P, Bhole S, Arora A (2017) Researcher positionality in cross-cultural and sensitive research. In: Liamputtong P (ed) Handbook of research methods in health social sciences. Springer, pp 1–15. https://doi.org/10.1007/978-981-10-2779-6_35-1

Mapfumo P, Mtambanengwe F, Chikowo R (2016) Building on indigenous knowledge to strengthen the capacity of smallholder farming communities to adapt to climate change and variability in southern Africa. Climate Dev 8:72–82. https://doi.org/10.1080/17565529.2014.998604

Martin B (2017) Methodology is content: indigenous approaches to research and knowledge. Educ Philos Theory 49(14):1392–1400. https://doi.org/10.1080/00131857.2017.1298034

Masango CA (2010) Indigenous traditional knowledge protection: prospects in South Africa's intellectual property framework? S Afr J Libr Inf Sci 76:74–80. http://www.rci.uct.ac.za/usr/rcips/resources/masango.pdf

Mazzocchi F (2006) Western science and traditional knowledge: despite their variations, different forms of knowledge can learn from each other. EMBO Rep 7 (5):463–466. https://doi.org/10.1038/sj.embor.7400693

Mbewe M, Phiri A, Siyambango N (2019) Indigenous knowledge systems for local weather predictions: a case of Mukonchi Chiefdom in Zambia. Environ Natl Resour Res 9(2):16–26. https://doi.org/10.5539/enrr.v9n2p16

Milligan L (2016) Insider-outsider-inbetweener? Researcher positioning, participative methods and cross-cultural educational research. Comp J Comp Int Educ 46(2):235–250. https://doi.org/10.1080/03057925.2014.928510

Milner HR (2007) Race, culture, and researcher positionality: working through dangers seen, unseen, and unforeseen. Educ Res 36(7):388–400. https://doi.org/10.3102/0013189X07309471

Moffat M (2016) Exploring positionality in an Aboriginal research paradigm: a unique perspective. Int J Technol Incl Educ 5:763–768. https://doi.org/10.20533/ijtie.2047.0533.2016.0096

Molekwa S (2013) Cut-off lows over South Africa and their contribution to the total rainfall of the Eastern Cape Province. Master's thesis, University of Pretoria. UPSpace. http://hdl.handle.net/2263/36767

Moser S (2008) Personality: a new positionality? Area 40 (3):383–392. http://www.jstor.org/stable/40346141

Muhammad M, Wallerstein N, Sussman AL, Avila M, Belone L, Duran B (2015) Reflections on researcher identity and power: the impact of positionality on community based participatory research (CBPR) processes and outcomes. Crit Sociol 41(7–8):1045–1063. https://doi.org/10.1177/0896920513516025

Nkomwa EC, Joshua MK, Ngongondo C, Monjerezi M, Chipungu F (2014) Assessing indigenous knowledge systems and climate change adaptation strategies in agriculture: a case study of Chagaka Village, Chikhwawa, southern Malawi. Phys Chem Earth Parts A/B/C 67–69:164–172. https://doi.org/10.1016/j.pce.2013.10.002

Orlove B, Roncoli C, Kabugo M, Majugu A (2010) Indigenous climate knowledge in southern Uganda: the multiple components of a dynamic regional system. Clim Change 100:243–265. https://doi.org/10.1007/s10584-009-9586-2

Parry ML, Canziani OF, Palutikof JP, Van der Linden PJ, Hanson CE (2007) Climate change 2007: impacts, adaptation, and vulnerability. Working group II contribution to fourth assessment report, vol 4. Intergovernmental panel on climate change. https://www.ipcc.ch/site/assets/uploads/2018/03/ar4_wg2_full_report.pdf

Parsons M, Fisher K, Nalau J (2016) Alternative approaches to co-design: insights from indigenous/academic research collaborations. Curr Opin Environ Sustain 20:99–105. https://doi.org/10.1016/j.cosust.2016.07.001

Qin D (2016) Positionality. In: Naples N, Hoogland RC, Wickramasinghe M, Wong WC (eds) The Wiley Blackwell encyclopedia of gender and sexuality studies. https://doi.org/10.1002/9781118663219.wbegss619

Radeny M, Desalegn A, Mubiru D, Kyazze F, Mahoo H, Recha J, Kimeli P, Solomon D (2019) Indigenous knowledge for seasonal weather and climate

forecasting across East Africa. Clim Change 156 (4):509–526. https://doi.org/10.1007/s10584-019-02476-9
Reid B (2020) Positionality and research: "Two-eyed seeing" with a rural Ktaqmkuk Mi'kmaw community. Int J Qual Methods 19:1–12. https://doi.org/10.1177/1609406920910841
Reid G, Sieber RE (2020) Unavoidable expertise, 'technocratic positionality', and GIScience: eliciting an indigenous geospatial ontology with the Eastern Cree in Northern Quebec. Gend Place Cult 28(4):541–563. https://doi.org/10.1080/0966369X.2020.1811209
Rukema JR, Umubyeyi BS (2019) Indigenous knowledge approach in maintaining a livelihood in the face of disastrous climate change: case of drought in Msinga villages, KwaZulu-Natal. Jàmbá J Disaster Risk Stud 11:1–10. https://doi.org/10.4102/jamba.v11i1.758
Savo V, Lepofsky D, Benner JP, Kohfeld KE, Bailey J, Lertzman K (2016) Observations of climate change among subsistence-oriented communities around the world. Nat Clim Chang 6(5):462–473. https://doi.org/10.1038/NCLIMATE2958
Shisanya S, Mafongoya P (2016) Adaptation to climate change and the impacts on household food security among rural farmers in uMzinyathi District of Kwazulu-Natal, South Africa. Food Secur 8(3):597–608. https://doi.org/10.1007/s12571-016-0569-7
Sultana F (2007) Reflexivity, positionality and participatory ethics: negotiating fieldwork dilemmas in international research. ACME Int J Crit Geogr 6(3):374–385. https://acme-journal.org/index.php/acme/article/view/786
Todes A, Turok I (2018) Spatial inequalities and policies in South Africa: place-based or people-centred? Prog Plan 123:1–31. http://hdl.handle.net/20.500.11910/10967
Tong LA (2017) Aligning the South African intellectual property system with traditional knowledge protection. J Intellect Prop Law Pract 12(3):179–190. https://doi.org/10.1093/jiplp/jpw207
Valdivia C, Seth A, Gilles JL, García M, Jiménez E, Cusicanqui J, Navia F, Yucra E (2010) Adapting to climate change in Andean ecosystems: landscapes, capitals, and perceptions shaping rural livelihood strategies and linking knowledge systems. Ann Assoc Am Geogr 100(4):818–834. http://www.jstor.org/stable/40863604
Watts V (2013) Indigenous place-thought and agency amongst humans and non humans (First woman and sky woman go on a European world tour!). Decolon Indig Educ Soc 2(1):20–34. https://jps.library.utoronto.ca/index.php/des/article/download/19145/16234/
Xu MA, Storr GB (2012) Learning the concept of researcher as instrument in qualitative research. Qual Rep 17(21):1–18. https://doi.org/10.46743/2160-3715/2012.1768
Yirga G (2010) Assessment of indigenous knowledge of medicinal plants in Central Zone of Tigray, northern Ethiopia. Afr J Plant Sci 4:6–11
Zuma-Netshiukhwi G, Stigter K, Walker S (2013) Use of traditional weather/climate knowledge by farmers in the south-western Free State of South Africa: agrometeorological learning by scientists. Atmosphere 4 (4):383–410. https://doi.org/10.3390/atmos4040383

Sinoxolo Magaya (M.Sc., B.Sc. Hons, B.Sc.) recently completed her Master of Science in Geography at the University of the Witwatersrand, with distinction. Her research explored the use of phenological, meteorological and astronomical indicators in forecasting weather and planning seasonal subsistence agricultural practices in two rural communities in South Africa.

Jennifer M. Fitchett (Ph.D., M.Sc., B.Sc. Hons, B.Sc.) is an Associate Professor of Physical Geography in the School of Geography, Archaeology and Environmental Studies at the University of the Witwatersrand. Her research is situated within the discipline of Biometeorology, exploring the impacts of climate change on plants, animals and human communities. Jennifer currently serves as the President-Elect of the Society of South African Geographers, and the African Councilor of the International Society of Biometeorology.

Part II

Integration of Indigenous and Scientific Knowledge Systems to Scale Up Resilience to Climate Change

Making Climate Services Actionable for Farmers in Ghana: The Value of Co-Production and Knowledge Integration

8

Emmanuel Nyadzi, Andy B. Nyamekye, and Fulco Ludwig

8.1 Introduction

The interaction of multiple biophysical, political, and socioeconomic stresses reduces the adaptive capacity of sub-Saharan Africa, making the region extremely vulnerable to climate variability and change (Kotir 2011; Hendrix and Glaser 2007). Temperature and rainfall variation, as well as extreme weather conditions, are projected to intensify, affecting all sectors, especially agriculture in sub-Saharan countries (Schlenker and Lobell 2010). The expected consequences include disrupted and shortened growing seasons and reductions in suitable agricultural lands and yields (Niang et al. 2014; Müller et al. 2011). Smallholder farmers in particular are disproportionately affected by climate variability and change since the majority depend on rain-fed agriculture (Sarr et al. 2015; Niang et al. 2014; Jalloh et al. 2013). In Ghana, extreme weather conditions have caused crop failures leading to yield reduction and food insecurity (Asare-Nuamah 2021; Müller-Kuckelberg 2012). Rainfall variability is a problem for farmers, particularly those in the Northern region. Irrigated and rainfed farms in the Northern part of the country are impacted by these changes because of the difficulties in predicting weather and seasonal climate conditions with implications for food production (Asante and Amuakwa-Mensah 2015; Kranjac-Berisavljevic' et al. 2003). The unpredictability of weather and seasonal climate influences the precision of seasonal and in-season farm-level decisions (Lawson et al. 2019; Asante and Amuakwa-Mensah 2015). For example, farmers have to re-sow seeds several times due to delays in rains which affects germination, increasing the cost of production and straining their livelihoods (Ndamani and Watanabe 2013). Irrigation managers and farmers rely on river discharge information to decide on the frequency, quantity and method of water distribution to farms. However, the unreliability of scientific forecasts to predict the rains and river discharge ahead of a season puts them in a dilemma (Ndamani and Watanabe 2013).

Growing concerns about the impacts of climate variability and change on agriculture have attracted the attention of the national and international community to strengthen weather and climate information services to manage the risk of climate variability and change (Gumucio et al. 2019; Vaughan and Dessai 2014). However, studies have shown that current weather and

E. Nyadzi (✉) · F. Ludwig
Water Systems and Global Change Group, Wageningen University, P.O. Box 47, 6700 AA Wageningen, The Netherlands
e-mail: emmanuel.nyadzi@wur.nl

F. Ludwig
e-mail: fulco.ludwig@wur.nl

A. B. Nyamekye
Food and Agriculture Organization of the United Nations, Rome, Italy

E. E. Ebhuoma and L. Leonard (eds.), *Indigenous Knowledge and Climate Governance*, Sustainable Development Goals Series, https://doi.org/10.1007/978-3-030-99411-2_8

climate information services for agriculture are facing serious constraints in how they are used (if at all). Mis-match between forecast and needs, forecast inaccuracy, language barriers, use of technical forecast terminologies which are difficult to understand, inconsistency and untimeliness of information provision impede their use (Feleke 2015; Onyango et al. 2014). More importantly, climate information services in many parts of the world, including Ghana, are developed in a one-directional manner; where science produces new knowledge and makes it accessible for end-users with no or limited involvement of end-users. This limits the acceptability and use of forecast information (Nyadzi et al. 2021; Okello et al. 2012). The continued extensive reliance on indigenous forecasts for some local farmers across Africa, points to the limitation of scientific forecasts (Gwenzi et al. 2016; Nyantakyi-Frimpong 2013; Zuma-Netshiukhwi et al. 2013).

Indigenous knowledge (IK) emerged as a result of indigenous people observing changes in their environment and adapting to the changing conditions through experimentation (Kassa and Temesgen 2011). Farmers in Ghana like those from elsewhere in Africa have mastered the art of using indigenous ecological indicators such as the behaviour of animals like insects, birds and mammals, and positions of the sun, cloud and moon, wind speed and direction, and physiological changes of vegetation to generate forecasts for farming activities. They practised this even before modern scientific weather and climate forecast systems were developed (Orlove et al. 2010; Changrsquo et al. 2013). Studies have suggested that IK has the potential to enhance farmers' adaptation to climate variability and change, although it could be insufficient because of projected changes in future climate (Nyadzi et al. 2021; Derbile et al. 2016). Therefore, for farmers to benefit from weather and climate forecast information, indigenous and scientific forecasting need to be integrated (Nyadzi et al. 2020; Ziervogel and Opere 2010).

The discourse on climate information services as a tool to support adaptation has stimulated scholars to study it from different angles: improving the skills of scientific forecast models for agriculture decisions (Esquivel et al. 2018; Hansen 2005), increasing co-production of climate information services (Bremer et al. 2019; Vincent et al. 2018), and exploring the value of integrating indigenous and scientific forecasting knowledge (Ebhuoma and Simatele 2019; Zuma-Netshiukhwi et al. 2013). This chapter, therefore, aims at improving the design and operationalisation of climate information services in Northern Ghana through co-production and integration of scientific and indigenous forecasts to generate actionable knowledge that supports farm decision-making. While the value of co-production and knowledge integration to climate information services has been acknowledged in the academic literature (Nyadzi 2020; Bremer et al. 2019) very few projects have implemented this and thus reflected upon its benefits. In this chapter, the contribution of co-production and knowledge integration in the design, information generation and uptake of climate information services was explored. Implicitly, the findings expand on how actionable knowledge anchored in scientific and indigenous knowledge and made available through climate information services can improve farmer decision-making in managing climate uncertainties and contributing to United Nations Sustainable Development Goal 13 (UN SDG 13).

The next Sect. 8.2 illustrates the concepts adopted. Section 8.3 describes the study area and the various methods employed, followed by Sect. 8.4 which presents results obtained. Section 8.5 discusses results and primarily elaborates on the implications of co-production and knowledge integration in improving climate information services.

8.2 Conceptual Framework

The underlying conceptual framework of this study prioritises the integration of scientific knowledge and IK and data through the interaction between researchers and rice farmers within the context of co-production (Fig. 8.1). The framework intends to push for a new generation of climate information services inspired by co-production and knowledge integration.

Central to the framework is the concept of co-production which involves the engagement of producers and users of knowledge to collaboratively engage each other to address complex societal problems (van Kerkhoff and Lebel 2015). Co-production highlights the need to encourage a power balance between producers and users of knowledge (Vincent et al. 2018), and calls for joint efforts between parties in determining the output of their collaboration (Rice 2002). The logic of co-production may seem simple at first sight but practice shows it is rather challenging especially in responding to climate change problems (Nyadzi 2020). In this climate information services project, co-production is operationalised as the active involvement of farmers and researchers in co-designing and co-creating climate information services to address the complex problem of climate variability and change in a way that recognises and uses different knowledge systems and expertise in forecasting, in other words knowledge integration.

Knowledge integration is defined as a combination of specialised knowledge to transform individual knowledge into collective knowledge for considerable results (Okhuysen and Eisenhardt 2002). The debate about integrating indigenous and scientific knowledge is not new and researchers have always recognised that their work is a conscious and critical revision of IK, often considered to be based on superstitions (Rist et al. 2006). We operationalised the concept of knowledge integration as a collective process of synthesising specialised yet differentiated indigenous and scientific knowledge and data possessed by farmers and researchers into a common knowledge that is efficient, flexible and within scope to improve farm decision-making. In doing so, the limitation of each knowledge system could be addressed to produce forecast

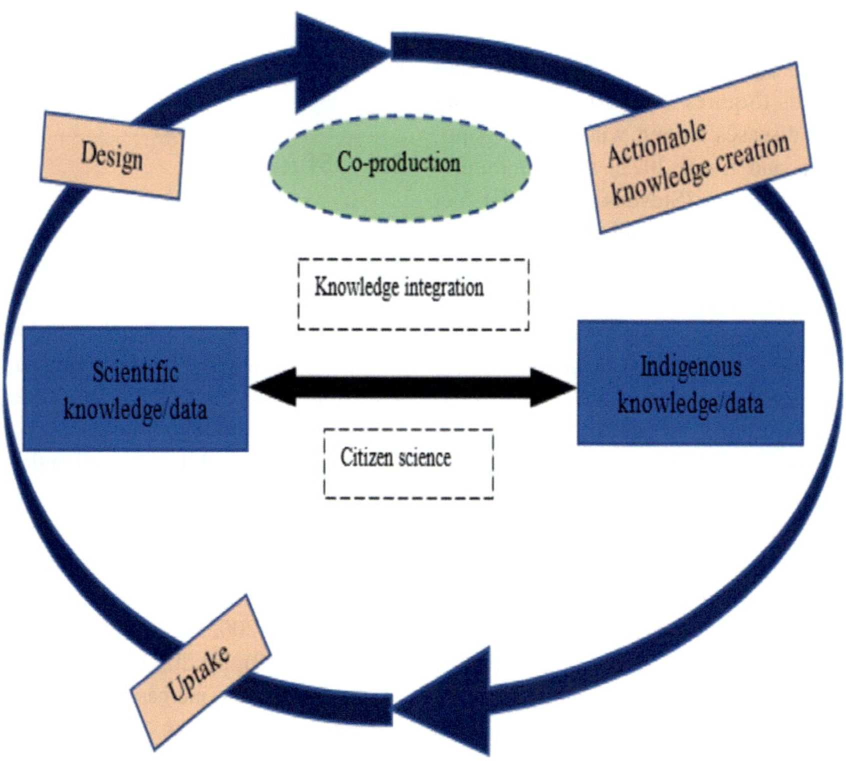

Fig. 8.1 Conceptual framework of the study. *source* Authors' Construct (2021)

information that is reliable and acceptable among farmers. Aside from regular meetings and workshops with farmers, the unique approach adopted in this research is the use of citizen science.

Citizen science is the active involvement of non-scientists of differing degrees in generating new scientific knowledge, from which they also actively stand to benefit either intrinsically (e.g. increased scientific literacy) or extrinsically (e.g. increased social capital) (Buytaert et al. 2014; Bonney et al. 2009). In this chapter, citizen science is defined as the participation and collaboration of farmers in the collection, sharing and interpretation of data (e.g. rain forecast and observational data) generated at a finer resolution which otherwise would be impossible to achieve by scientists only. Generally, citizen science has been successful in advancing scientific knowledge by gathering large amounts of data in different locations for a longer period. Citizen science has been applied in several fields of endeavour, for example, water quality monitoring (Walker et al. 2021), mapping spatially non-continuous permanent rivers (Turner and Richter 2011), examining population distribution and change of birds (Squires et al. 2021), the spread of infectious diseases among wild animal populations (Eisen and Eisen 2021) and community monitoring of poaching (Stevens et al. 2013). Yet application in meteorology and atmospheric science has been relatively limited, for example, precipitation measurement (CoCoRaH 2010). The novelty of applying citizen science in this study is not particularly for advancing theories on citizen science, but to actively involve non-scientists in the process of generating integrated forecasts. Integrating both indigenous and scientific forecasts has the potential of generating actionable knowledge that is useful for farmers (Gagnon and Berteaux 2009).

Actionable knowledge depicts knowledge that specifically supports stakeholder decision-making and consequent actions (Geertsema et al. 2016). It emphasises knowledge being produced from an interactional process and a reference to ultimate use in decision-making (Nana et al. 2013; Meinke et al. 2006; Dewulf et al. 2005). In the domains of information and knowledge research, reference is made not only to scientific knowledge but also to IK systems (Koocheki 2003; Kirchhoff et al. 2013). Drawing from Cash et al. (2003), the study argues that the interactional process of knowledge creation from information involves an interplay of the factors of salience, legitimacy and credibility within rice farming systems. In this study, actionable knowledge is defined as indigenous and scientific knowledge that is locally relevant, trustworthy, and produced in a fair and transparent manner. We define salience as locally relevant, timely, and relatable scientific and indigenous knowledge; credibility as knowledge that is trustworthy and can be based on scientific evidence or trust in the experience (IK) of fellow farmers; and legitimacy as scientific and indigenous knowledge produced in a fair, balanced, and transparent way. How various platforms can contribute to actionable knowledge creation was determined by drawing on how current information systems in rice farming systems enable actionable knowledge creation.

8.3 Study Area

The study this chapter is based on was conducted in the Kumbungu district of the Northern region of Ghana (Fig. 8.2). The Northern region of Ghana, including the Kumbungu district, is a tropical Savanna zone with a single rainfall season (Boogaard et al. 2012; Alhassan et al. 2013). Agricultural production in the area is already negatively affected by the climatic conditions with six to seven months dry season (October/November–March/April) and five months rainy season (April/May to September/October) (Nyadzi 2016; Amikuzuno and Donkoh 2012; Barry et al. 2005). According to Owusu and Waylen (2009), by 2050 the region is projected to experience an increase in rainfall and temperature intensity, an expected reduction in rainfall frequency with consequences on food security in the region (Antwi-Agyei et al. 2012). A large part of the district is rural with about 26% being literate and 74% non-literate (GSS 2014). Considered in this study are the three different groups of farmers in the study area (i.e. irrigated rice production only, rainfed only, and both irrigated and rainfed rice production). The project focused on the rice farming system because of the increasing demand for rice and its susceptibility to climate variability and change and the rising concern among Ghanaian policy makers (Angelucci et al. 2013; Asante and Amuakwa-Mensah 2015; CARD 2010; MOFA 2009). The study was conducted in the North of Ghana because it is ranked as the highest rice producing region (Angelucci et al. 2013) and has the potential to reduce poverty among smallholder farmers (Angelucci et al. 2013; Donkoh et al. 2010), who have limited access to scientific climate information to take adaptive decision-making (Ndamani and Watanabe 2013).

8.4 Methodology: The Climate Information Services Project

8.4.1 Data Collection

This study used multi-data and multi-method approaches to explore the possibility of making weather and seasonal climate forecast information useful for decision-making in rice farming systems. Data was collected through primary and secondary sources. For the primary data collection, a purposive sampling approach was adopted to select participating farmers. The selection was based on farmers experience in using indigenous forecast, farming system practised, forecasting skills and willingness to partake in the study. Primary data was collected through interviews, focus group discussions, workshops and the use of mobile phones. These methods were used to solicit information to contextualise the problem, collect farmers information needs and regularly engage farmers for their feedback (see Nyadzi

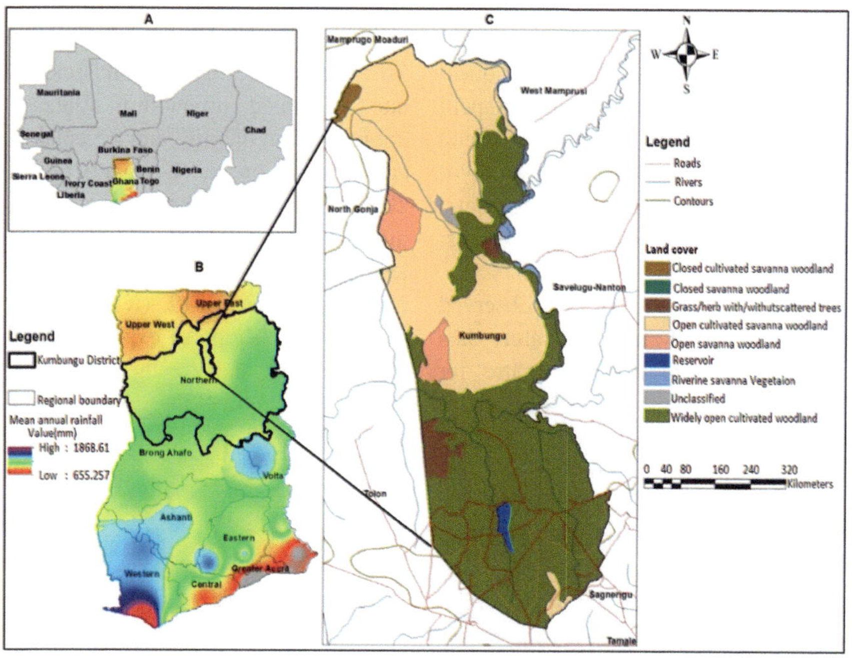

Fig. 8.2 Map of study area: panel "A" shows the location of Ghana in West Africa. "B" shows the map of Ghana (with annual average rainfall from 1981 to 2010) indicating the location of the Northern region and Kumbungu district. "C" shows the detailed map of Kumbungu district. *source* Nyadzi (2020)

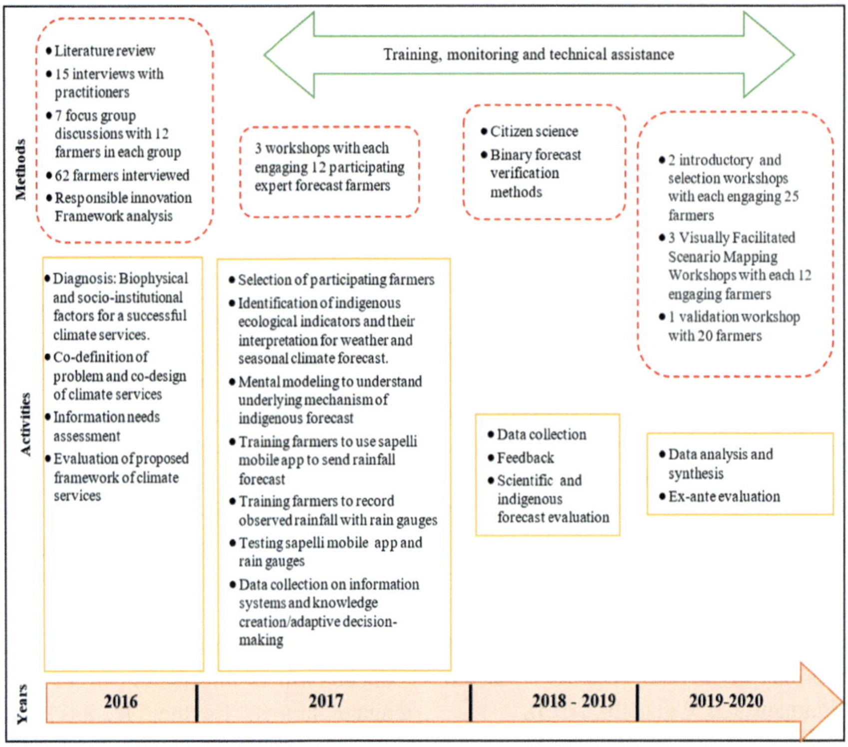

Fig. 8.3 Flow of project; timeline, activities and methods. *Source* Authors self Construct (2021)

2020 and Nyamekye 2020 for details on questions and methods). Mobile phones with Sapelli mobile application were used to record and send forecasts and observed rainfall for a period of time (daily forecast for seven months in 2017 and seasonal forecast for 2017 and 2018) (see Nyadzi et al. 2020).

Secondary data such as forecast and observed daily and seasonal rainfall and temperature data was collected from the European Centre for Medium-Range Weather Forecasts (ECMWF) System 4 and the Ghana meteorological agency (Gmet). This data was used to conduct skills or accuracy tests of the scientific forecast (Nyadzi et al. 2019). Figure 8.3 shows the project set-up and activities.

Details on the methodology for (i) determining the indigenous forecast accuracy using binary or dichotomous forecast verification measures (yes/no occurrence of an event) and the underlying mechanisms behind farmers' forecasting techniques can be found in Nyadzi et al. 2021. (ii) integrating indigenous and scientific forecast into integrated probability forecast (IPF) to improve the reliability and acceptability of forecast information can be found in Nyadzi et al. 2020. (iii) determining how actionability of information services was analysed based on salience, credibility or legitimacy can be found in Nyamekye et al. 2019. (iv) Visually facilitated scenario workshops to explore how uncertainty and seasonal and weather lead times inform farmer decision-making can be found in Nyamekye et al. 2021).

8.5 Results

8.5.1 The Design

8.5.1.1 Essential Features

The study resulted in the development of new generation climate information services largely influenced by co-production and knowledge integration (Fig. 8.4). The architecture of the new generation climate service is built upon two main knowledge and data sources; indigenous and scientific. While the IK system is based upon farmers' expertise in using indigenous ecological indicators for forecasting, the scientific system is primarily based upon the meteorological knowledge of researchers. The information exchange element within the design offers an additional feature that distinguishes it from current climate information services. Farmers can be actively engaged in the co-production process where they can share their forecast information and receive tangible information and advice for adaptive farm decision-making. The ICT part of the design allows easy and automated data handling activities such as indigenous and scientific data collection, processes, analysis, and visualisation based on a predetermined algorithm. The design introduced citizen science as a means of data gathering and information exchange using Sapelli mobile applications to enable co-production. Using the responsible innovation framework results of the analysis show that the proposed climate information services design is anticipatory, inclusive, reflexive and responsive (Nyadzi et al. 2018).

8.5.1.2 Validation of Knowledge Systems: Skills of Indigenous and Scientific Forecast

The study (based on workshops, interviews and FGDs results) revealed that observational changes in IEIs in addition to historical rainfall patterns serve as the fundamental template that allows farmers to form expectations for the coming season. Farmers have an established mental model of how IEIs influence the prediction of different weather and seasonal climate events. These IEIs include behaviours of ducks, birds, ants, butterflies, mosquitoes, dogs and cows and reptiles (snakes); the appearance of clouds, stars, lightning, moon, flowers on baobab trees; and direction of wind and degree of hotness of the weather. The observable changes in IEIs have their generally held interpretations depending on which event is to be predicted and whether for a short or longer time scale (see Nyadzi et al. 2020 for the interpretations of these IEIs). On average, both farmers and GMet can accurately forecast one out of every three daily rainfall events. At the seasonal scale, one out of every three farmers was able to accurately make onset predictions while two out of every five farmers

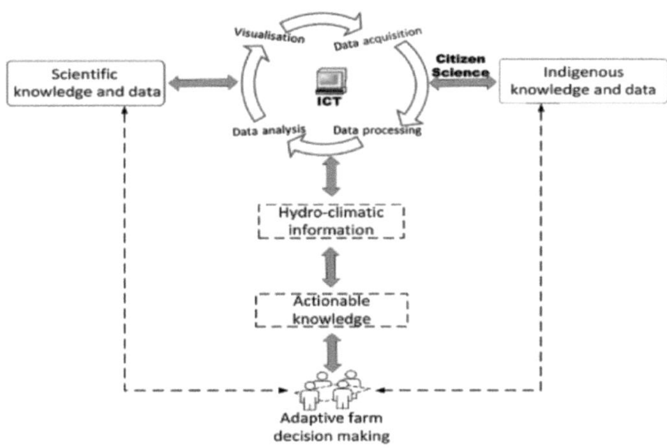

Fig. 8.4 Fundamental architecture of hydro-climatic information services as developed by the study. *Source* Nyadzi et al. (2018)

can get rainfall amount and cessation right. Similarly, GMet was able to predict rainfall amounts accurately in one out of every three communities and one out of every four communities for onset but was unable to accurately predict cessation for the communities. The study also showed that indigenous forecast is not intuitive but a skill rationally developed given age and experience. This result, therefore, informs the next step which aims at finding a quantitative method to integrate farmers' indigenous forecasts and scientific forecasts (from GMet).

8.5.1.3 Knowledge Integration: Combining Indigenous and Scientific Forecast

Results of the forecast verification show that the Integrated Probability Forecast (IPF) method combines the strengths of indigenous forecast (IF) and scientific forecast (SF) and subsequently improves their reliability. IPF performed generally better than any of the individual forecasts. Specifically, IPF showed improved reliability at both daily and seasonal timescales although IF performed better at seasonal timescales. Furthermore, results of the interviews show that IPF method had far greater acceptability potential among farmers (93% of farmers accept) because it combines IF and SF into a single forecast, resolves the issues of contradicting forecast information, requires less meeting time and improves forecast reliability (See Nyadzi et al. 2022). Finally, as a proof of concept, the study demonstrated that it is possible to combine IF and SF into a single objective forecast despite the potentially significant differences between them. The study also showed that using the IPF method to integrate IF and SF improves the reliability and acceptability of the resultant forecast information among farmers.

8.5.2 Actionable Knowledge Creation

Based on the results of the interviews and FGDs the study found four types of information systems: community radio, commercial radio, farmer-to-farmer, and mobile-based-only platforms. Although these systems had similarities, differences exist in their set-up, particularly how they engage and interact as well as provide information towards actionable knowledge creation. The farmer-to-farmer system is better at enabling the creation of actionable knowledge compared to other systems because of its ability to diffuse information extensively and create salient knowledge that focuses on local contexts and aligns information and makes it usable. Farmer-to-farmer systems also enable the legitimacy of knowledge through open participation and sharing through open space informal meetings. Also, farmer-to-farmer systems advanced the credibility of outcomes of deliberations by drawing on both indigenous and scientific information. Community radio systems strongly support cross-fertilisation of knowledge by engaging experienced farmers in knowledge brokerage where both scientific and indigenous knowledge are considered during studio discussions. Mobile-based-only platforms have also gained traction in enabling actionable knowledge creation in food systems. However, the study revealed that existing mobile-based platforms are limited, in that they do not recognise the significant impact IK has on the credibility of knowledge produced. Despite the success in breaking physical barriers to communication by using mobile technology, the focus on national and regional levels and only scientific forecasts limits the salience of emerging knowledge with an emphasis on credibility rather than legitimacy and salience.

Results of the visually facilitated scenario workshops show the level of risks that farmers are willing to take based on the information they receive. Forecasts indicating a higher probability of rainfall occurring are not always favourable for farmers, with the possibility of farmers changing dates for carrying out intended activities such as land preparation, weed control and fertiliser application because of their sensitivity to water conditions. It was evident that a 1-month lead time informs farmers' decision-making better than a 2- or 3-month lead time. Regarding weather forecasts, a 1-week lead time is preferred over a 3-day or a 1-day lead time for in-season decision-making.

8.5.3 Uptake

An initial diagnosis shows the disconnect between climate information service providers and farmers, resulting in low uptake of forecast information. The study(based on the validation workshop) found that a two-directional model of climate information services where farmers and researchers are active in all stages of the process has the potential to increase uptake. The one-way approach (where scientists produce and end-users use), which is still present in existing information systems is limited in uptake, as the production of knowledge has little or no involvement of end-users. In many cases, local people are not involved in data gathering and interpretation. Participation is often limited to a couple of workshops where the aim of the information system is communicated, giving researchers the choice to determine the problem, gather and interpret the scientific data and plan the approach of information delivery with little consideration for farmers own IK and specific information need (see Nyadzi et al. 2018; Ouedraogo et al. 2018).

Results of the validation workshop demonstrated that with proper training and through citizen science, farmers are capable of recording local rainfall with tailor-made rain-gauges and in providing community level forecasts using IK. Further, they also provided some perspective into the interpretation of local data which might be overlooked by researchers. Analysis shows that indigenous forecasting is finer in resolution and more valuable at the community level than GMet forecasting which is coarse and issued at a regional level. Therefore, rather than trying to improve on inherently uncertain scientific forecasts, local knowledge on forecasting can be harnessed and incorporated to improve forecast information (see also Kniveton et al. 2015). When end-users are involved through co-production, ownership of uncertainty is extended to the wider community of users and uptake is improved.

8.6 Discussion and Conclusion

The case study has been used to demonstrate that co-production and knowledge integration significantly improve the design, actionable knowledge creation and uptake of climate information services. A key reflective outcome of the study is that end-users' needs cannot be completely understood through producers' one-size-fits-all lens. Rather, context must be analysed paying attention to the preferences of end-users and how they interact with climate information for decision-making. Moreover, the complex nature of local climate change problems requires multiple perspectives and entry points, thus the need for incorporating the knowledge and experiences of all actors, especially those often marginalised. The study revealed that co-production ensures the development of climate information services that is anticipatory, inclusive, reflexive and responsive, thus generating actionable knowledge (see also Nyadzi et al. 2018). Co-production also results in high level capabilities that are crucial to the success of climate information services, influencing the development of solution-oriented policies for climate action in sub-Saharan Africa. Firstly, they use not only the explicit knowledge but also the tacit one, which resides with end-users who are often vulnerable. Bringing together different expertise and knowledge from researchers and farmers in co-defining and framing the problem, co-designing and co-producing the solutions is an effective approach to creating solutions to help farmers survive the impact of climate change. Secondly, the process enhances the effectiveness and usefulness of research as it aligns its outcome to community choices and needs (Galuszka 2018). Thirdly, in policy circles, actors at all levels must be involved in policy design with a leaning towards a bottom-up approach to policy-making. Moreover, including farmers in the design and delivery of climate information services offers the opportunity to address the relevant gaps and increase impact, particularly policy reforms and actions necessary to address common problems at all levels of society (see also Durose et al. 2012). Nevertheless, the above reasons do not in any way seek to underrate the difficulties with meaningful co-production and knowledge integration processes.

Co-production in climate information services delivery should include relevant institutions (see also Galuszka 2018; Collodi et al. 2017; Moser 2016). While farmers remained central in the co-production process, other important institutions and stakeholders such as extension services and Ghana meteorological agency were seldomly contacted and involved in the climate information services development. Not fully and constantly engaging these institutions has the potential to affect the sustainability of a project especially when mainstreaming it into policy-making and governance. This is especially relevant in Ghana and most parts of sub-Saharan Africa where institutional bureaucracies and regulatory norms could thwart the development of climate information services.

One key feature of a successful co-production lies in the flexibility of knowledge integration. This involves constant reconfiguration and modification of existing knowledge based on experiences of successes, failures and lessons learnt to form new knowledge. This is achievable through a collaborative learning process with emphasis on methods that centre on participants. Engaging farmers throughout the study addresses the challenges of credibility, legitimacy, scale, cognition, familiar institutional practices and complex decision-making currently affecting the uptake of climate information services that focus on scientific information only as detailed by Patt and Gwata (2002) and Nyamekye et al. (2019). Moreover, involving local farmers could address the limitation that comes with scientific downscaling methods in translating forecast information from coarse to finer resolutions due to model inadequacies and quality data available in local communities (Caffrey and Farmer 2014).

Knowledge production underpins the decision-making process of farmers and the outcomes of a farming decision depend on the actionability of the knowledge. According to Antonacopoulou (2009), researchers need to continuously reflect on the actionability of knowledge that is co-produced, by examining the processes of knowledge integration itself. Operationalising actionable knowledge in the context of farm decision-making and as knowledge that goes beyond a unitary perspective to embrace consultation and deliberation by different actors was very beneficial for the climate information services project. Knowledge integration has the potential of creating actionable knowledge that supports transformative adaptive decisions among farmers. Knowledge for developing climate information services should not be based on researchers expertise alone but also include farmers and other actors. Perry and Atherton (2017) argues that the production of actionable knowledge is in itself a social activity in which multiple actors, whether researchers or not, hold relevant knowledge that is useful in characterising and addressing complex challenges. Therefore, knowledge integration and co-production are the surest pathways to creating a co-learning opportunity for developing sustainable climate policy instruments towards achieving SDG 13. These findings contribute to climate action as captured by United Nations' Sustainable Development Goal (UN SDG)13, allowing indigenous communities in sub-Saharan Africa including Ghana to contribute meaningfully to developing early warning systems to anticipate and combat climate change impacts.

Co-production and knowledge integration processes have the inherent potential for capacity building and empowerment. Farmers and researchers learnt from the unique features of each knowledge system. For instance, farmers learn how to systematically measure rainfall using rain gauges while researchers learnt about indigenous ecological indicators and their interpretation for forecasting rainfall. However, we observed that when care is not taken and proper protocol not followed, there is the tendency for a one-directional learning process where researchers will discredit IK and push scientific knowledge onto farmers. Thus, good practice of co-production and knowledge integration could result in an equitable, horizontal sharing of skills and expertise that intrinsically informs each side to re-question and re-define and shape their capacities (see also Collodi et al. 2017). Therefore capacity building remains vital in the successful delivery of climate information services. Mistra Urban Futures (2016) recognises the potential value of capacity building in sustainability research and advocates its incorporation into the co-production and knowledge integration process. This chapter, therefore, calls for a learning framework to distinguish whether these benefits are directly due to co-production and knowledge integration efforts or due to a general trend within a deliberative society. Doing this will shape our understanding of framing the process for co-production and knowledge integration for effective and efficient climate services delivery.

References

Alhassan M, Loomis J, Frasier M, Davies S, Andales A (2013) Estimating farmers' willingness to pay for improved irrigation: an economic study of the bontanga irrigation scheme in Northern Ghana. J Agric Sci 5(4). https://doi.org/10.5539/jas.v5n4p31

Amikuzuno J, Donkoh SA (2012) Climate variability and yields of major staple food crops in Northern Ghana. Afr Crop Sci 20(2):349–360

Angelucci F, Asante-Poku A, Anaadumba P (2013) Analysis of incentives and disincentives for rice in Ghana: technical notes series. MAFAP, FAO, Rome, Italy

Antonacopoulou EP (2009) Impact and scholarship: unlearning and practising to co-create actionable knowledge. Manag Learn 40(4):421–430

Antwi-Agyei P, Fraser EDG, Dougill AJ, Stringer LC, Simelton E (2012) Mapping the vulnerability of crop production to drought in Ghana using rainfall, yield and socioeconomic data. Appl Geogr 32(2):324–334. https://doi.org/10.1016/j.apgeog.2011.06.010

Asante FA, Amuakwa-Mensah F (2015) Climate change and variability in Ghana: stocktaking. Climate 3 (1):78–99. MDPI AG. https://doi.org/10.3390/cli3010078

Asare-Nuamah P (2021) Climate variability, subsistence agriculture and household food security in rural Ghana. Heliyon 7(4):e06928

Barry B, Obuobie E, Andreini M, Andah W, Pluquet M (2005) Comprehensive IWMI., Assessment of Water Management in Agriculture (Comparative Study of River Basin Development and Management. International Water management institute

Bremer S, Wardekker A, Dessai S, Sobolowski S, Slaattelid R, van der Sluijs J (2019) Toward a multifaceted conception of co-production of climate services. Climate Services 13:42–50

Bonney R, Cooper CB, Dickinson J, Kelling S, Phillips T, Rosenberg KV, Shirk J (2009) Citizen science: a developing tool for expanding science knowledge and scientific literacy

Boogaard H, Hoek S, Jacobs C, van't Klooster K, Supit I, van der Wijngaart R, Winsemius H (2012) Weather risk mapping for Northern Ghana. Wageningen University and Research, Alterra

Buytaert W, Zulkafli Z, Grainger S, Acosta L, Alemie TC, Bastiaensen J, De Bièvre B, Bhusal J, Clark J, Dewulf A, Foggin M Zhumanova M (2014) Citizen science in hydrology and water resources: opportunities for knowledge generation, ecosystem service management, and sustainable development. Front Earth Sci 2:26.https://doi.org/10.3389/feart.2014.00026

Caffrey P, Farmer A (2014) A review of Downscaling methods for climate change projections. Tetra Tech ARD. http://www.ciesin.org/documents/Downscaling_CLEARED_000.pdf

CARD (Coalition for African Rice Development) (2010). Mapping of Poverty Reduction Papers (PRSPs), Sector Strategies and Policies related to rice development in Ghana. http://www.fao.org/3/a-at546e.pdf

Cash DW, Clark WC, Alcock F, Dickson NM, Eckley N, Guston DH, Jäger J, M. R. (2003) Knowledge systems for sustainable development. Proc Natl Acad Sci 100 (14):8086–8091

Changrsquo LB, Ngana J (2013) Indigenous knowledge in seasonal rainfall prediction in Tanzania: A case of the South-western Highland of Tanzania. J Geogr Reg Plan 3(4):66–72

CoCoRaH S (2010) Community collaborative rain, hail & snow network. Colorado State University. https://www.cocorahs.org/. Accessed 9 Aug 2019

Collodi J, Di Vicenz S, Murphy R, Visman E (2017) Linking preparedness, response and resilience (LPRR). Disasters and emergencies preparedness programme. Kings College and Christian AID, UK

Derbile EK, Jarawura FX, Dombo MY (2016) Climate change, local knowledge and climate change adaptation in Ghana. In: Adaptation to climate change and variability in rural West Africa. Springer, Cham, pp 83–102

Dewulf A, Craps M, Bouwen R, Abril F, Zhingri M (2005) How indigenous farmers and university engineers create actionable knowledge for sustainable irrigation. Action Res 3(2):175–192

Donkoh S, Awuni J, Namara R (2010). Improving the efficiency of inland valley rice production in Northern Ghana. J Ghana Sci Assoc 12(2). https://doi.org/10.4314/jgsa.v12i2.62818

Durose C, Beebeejaun Y, Rees J, Richardson J, Richardson L (2012) Towards co-production in research with communities. Connected Communities, pp 1–15

Ebhuoma EE, Simatele DM (2019) 'We know our Terrain': indigenous knowledge preferred to scientific systems of weather forecasting in the Delta State of Nigeria. Climate Dev 11(2):112–123

Eisen L, Eisen RJ (2021) Benefits and drawbacks of citizen science to complement traditional data gathering approaches for medically important hard ticks (Acari: Ixodidae) in the United States. J Med Entomol 58(1):1–9

Esquivel A, Llanos-Herrera L, Agudelo D, Prager SD, Fernandes K, Rojas A, Valencia JJ, Ramirez-Villegas J (2018) Predictability of seasonal precipitation across major crop growing areas in Colombia. Climate Services 12(March):36–47. https://doi.org/10.1016/j.cliser.2018.09.001

Feleke HG (2015) Assessing weather forecasting needs of smallholder farmers for climate change adaptation in the Central Rift Valley of Ethiopia. J Earth Sci Clim Change 6(10):1–8

Gagnon CA, Berteaux D (2009) Integrating traditional ecological knowledge and ecological science: a question of scale. Ecol Soc 14(2). https://doi.org/10.5751/ES-02923-140219

Galuszka J (2018) What makes urban governance co-productive? Contradictions in the current debate on co-production. Plan Theory 1–18

Geertsema W, Rossing WA, Landis DA, Bianchi FJ, Van Rijn PC, Schaminée JH, Tscharntke T, Van Der Werf W (2016) Actionable knowledge for ecological intensification of agriculture. Front Ecol Environ 14 (4):209–216

Ghana Statistical Service (GSS) (2014) 2010 Population and Housing Census: District Analytical Report. Kumbungu district. Ghana Statistical Service. Accra. http://www2.statsghana.gov.gh/docfiles/2010_District_Report/Northern/Kumbungu.pdf. Accessed 21 Feb 2021

Gumucio T, Hansen J, Huyer S, van Huysen T (2019) Gender-responsive rural climate services: a review of the literature. Taylor and Francis Ltd., Climate and Development. https://doi.org/10.1080/17565529.2019.1613216

Gwenzi J, Mashonjowa E, Mafongoya PL, Rwasoka DT, Stigter K (2016) The use of indigenous knowledge systems for short and long range rainfall prediction and farmers' perceptions of science-based seasonal forecasts in Zimbabwe. Int J Clim Change Strategies Manage 8(3):440–462. https://doi.org/10.1108/ijccsm-03-2015-0032

Hansen JW (2005) Integrating seasonal climate prediction and agricultural models for insights into agricultural practice. Philos Trans Roy Soc b: Biol Sci 360 (1463):2037–2047. https://doi.org/10.1098/rstb.2005.1747

Hendrix CS, Glaser SM (2007) Trends and triggers: climate, climate change and civil conflict in Sub-Saharan Africa. Polit Geogr 26(6):695–715

Jalloh A, Nelson GC, Thomas TS, Zougmoré R, Roy-Macauley H (eds) (2013) West African agriculture and climate change. https://doi.org/10.2499/9780896292048

Kassa B, Temesgen D (2011) Indigenous knowledge in research and extension. Perspective of agricultural extension. Regional Msc program on: Agricultural information and communication management (Aicm), pp 91–94. https://www.researchgate.net/publication/289326509_PERSPECTIVE_OF_AGRICULTURAL_EXTENSION

Kirchhoff CJ, Carmen Lemos M, Dessai S (2013) Actionable knowledge for environmental decision making: broadening the usability of climate science. Annu Rev Environ Resour 38(1):393–414. https://doi.org/10.1146/annurev-environ-022112-112828

Kniveton D, Visman E, Tall A, Diop M, Ewbank R, Njoroge E, Pearson L (2015a) Dealing with uncertainty: integrating local and scientific knowledge of the climate and weather. Disasters 39(Suppl 1):S35–53. https://doi.org/10.1111/disa.12108

Koocheki AA (2003) Indigenous knowledge in agriculture with particular reference to saffron production in Iran. In: Paper presented at the I international symposium on saffron biology and biotechnology, p 650

Kotir JH (2011) Climate change and variability in Sub-Saharan Africa: a review of current and future trends and impacts on agriculture and food security. Environ Dev Sustain 13(3):587–605

Kranjac-Berisavljevic' G, Blench RM, CR (2003) Rice production and livelihoods in Ghana. Multi-agency partnerships (Maps) for technical change in West African agriculture. https://www.odi.org/sites/odi.org.uk/files/odi-assets/publications-opinionfiles/3990.pdf

Lawson ET, Alare RS, Salifu ARZ, Thompson-Hall M (2019) Dealing with climate change in semi-arid Ghana: understanding intersectional perceptions and adaptation strategies of women farmers. GeoJ. https://doi.org/10.1007/s10708-019-09974-4

Meinke H, Nelson R, Kokic P, Stone R, Selvaraju R, Baethgen W (2006) Actionable climate knowledge: from analysis to synthesis. Climate Res 33(1):101–110

Mistra Urban Futures (2016) Co-production in action: towards realizing just cities, Gothenburg

MOFA (Ministry of Food and Agriculture) (2009) National Rice Development Strategy (NRDS). https://www.jica.go.jp/english/our_work/thematic_issues/agricultural/pdf/ghana_en.pdf

Müller-Kuckelberg K (2012) Climate change and its impact on the livelihood of farmers and agricultural workers in Ghana. Preliminary research results. General agricultural workers' union of GTUC. friedrich ebert stiftung (Fes), pp 1–47

Müller C, Cramer W, Hare WL, Lotze-Campen H (2011) Climate change risks for African agriculture. Proc Natl Acad Sci USA 108(11):4313–4315. https://doi.org/10.1073/pnas.1015078108

Nana YCF, Asuming-Brempong S, Nantui MF (2013) Analysis of cocoa-based agricultural knowledge and information systems in the eastern region of Ghana. Russ J Agric Socio-Econ Sci 14(2)

Ndamani F, Watanabe T (2013) Rainfall variability and crop production in northern Ghana: the case of Lawra district. ResearchGate. https://doi.org/10.13140/2.1.2343.7125

Niang I, Ruppel OC, Abdrabo MA, Essel A, Lennard C, Padgham J, Urquhart P (2014) Africa (Part B: Re). Cambridge University Press, Cambridge, pp 1199–1265

Nyamekye AB (2020) Towards a new generation of climate information systems: information systems and actionable knowledge creation for adaptive decision-making in rice farming systems in Ghana (Doctoral dissertation, Wageningen University)

Nyamekye AB, Dewulf A, Van Slobbe E, Termeer K (2019) Information systems and actionable knowledge creation in rice-farming systems in Northern Ghana. Afr Geogr Rev 1–18. https://doi.org/10.1080/19376812.2019.1659153

Nyamekye AB, Nyadzi E, Dewulf A, Werners S, Van Slobbe E, Biesbroek RG, Ludwig F et al (2021) Forecast probability, lead time and farmer decision-making in rice farming systems in Northern Ghana. Clim Risk Manag 31:100258

Nyantakyi-Frimpong H (2013) Indigenous knowledge and climate adaptation policy in Northern Ghana. https://www.africaportal.org/publications/indigenous-knowledge-and-climate-adaptation-policy-in-northern-ghana/

Nyadzi E (2016) Climate variability since 1970 and farmers' observations in Northern Ghana. Sustain Agric Res 5(526-2016-37880)

Nyadzi E (2020) Best of both worlds: co-producing climate services that integrate scientific and indigenous weather and seasonal climate forecast for water management and food production in Ghana (Doctoral dissertation, PhD Thesis, Wageningen University and Research, Wageningen, The Netherlands)

Nyadzi E, Nyamekye AB, Werners SE, Biesbroek RG, Dewulf A, Van Slobbe E, Long HP, Ludwig F (2018) Diagnosing the potential of hydro-climatic information services to support rice farming in northern Ghana. NJAS-Wageningen J Life Sci 86:51–63

Nyadzi E, Saskia Werners E, Biesbroek R, Long PH, Franssen W, Ludwig F (2019) Verification of seasonal climate forecast toward hydroclimatic information needs of rice farmers in Northern Ghana. Weather, Climate, Soc 11(1). https://doi.org/10.1175/WCAS-D-17-0137.1

Nyadzi E, Werners SE, Biesbroek R, Ludwig F (2020) Techniques and skills of indigenous weather and seasonal climate forecast in Northern Ghana. Climate Dev 1–12

Nyadzi E, Ajayi OC, Ludwig F (2021) Indigenous knowledge and climate change adaptation in Africa: a systematic review. CAB Reviews Perspectives in Agriculture Veterinary Science Nutrition and Natural Resources 2021(16):029. https://www.cabi.org/cabreviews/review/20210181994

Nyadzi E, Werners SE, Biesbroek R, Ludwig F (2022) Towards weather and climate services that integrate indigenous and scientific forecasts to improve forecast reliability and acceptability in Ghana. Environ Dev 100698

Okello JJ, Kirui O, Njiraini GW, Gitonga Z (2012) Drivers of use of information and communication technologies by farm households: the case of smallholder farmers in Kenya. J Agric Sci 4(2). https://doi.org/10.5539/jas.v4n2p111

Okhuysen GA, Eisenhardt KM (2002) Integrating knowledge in groups: how formal interventions enable flexibility. Organ Sci 13(4):370–386+457. https://doi.org/10.1287/orsc.13.4.370.2947

Onyango E, Ochieng S, Awiti A (2014) Weather and climate information needs of small-scale farming and fishing communities in western Kenya for enhanced adaptive potential to climate change. Proc Sustain Res Innov Conf 4:187–193

Orlove B, Roncoli C, Kabugo M, Majugu A (2010) Indigenous climate knowledge in southern Uganda: The multiple components of a dynamic regional System. Clim Change 100(2):243–265. https://doi.org/10.1007/s10584-009-9586-2

Ouedraogo I, Diouf NS, Ouédraogo M, Ndiaye O, Zougmoré RB (2018) Closing the gap between climate information producers and users: assessment of needs and uptake in Senegal. Climate 6(1). https://doi.org/10.3390/cli6010013

Owusu K, Waylen P (2009) Trends in spatio-temporal variability in annual rainfall in Ghana (1951–2000). Weather 64(5):115–120. https://doi.org/10.1002/wea.255

Patt A, Gwata C (2002) Effective seasonal climate forecast applications: examining constraints for subsistence farmers in Zimbabwe. Glob Environ Chang 12(3):185–195

Perry B, Atherton M (2017) Beyond critique: the value of coproduction in realising just cities? Local Environ 22 (1):36–51

Rice MP (2002) Co-production of business assistance in business incubators: an exploratory study. J Bus Venturing 17(2):163–187. https://doi.org/10.1016/S0883-9026(00)00055-0

Rist S, Dahdouh-Guebas F, Dahdouh-Guebas RF, S. A. (2006) Ethnosciences-A step towards the integration of scientific and indigenous forms of knowledge in the management of natural resources for the future. Environ Dev Sustain 8:467–493. https://doi.org/10.1007/s10668-006-9050-7

Sarr B, Atta S, Ly M, Salack S, Ourback T, Subsol S, George DA (2015) Adapting to climate variability and change in smallholder farming communities: a case study from Burkina Faso, Chad and Niger. J Agric

Extension Rural Dev 7(1):418–429. https://doi.org/10.5897/JAERD14.0595

Schlenker W, Lobell DB (2010) Robust negative impacts of climate change on African agriculture. Environ Res Lett 5(1). https://doi.org/10.1088/1748-9326/5/1/014010

Squires TM, Yuda P, Akbar PG, Collar NJ, Devenish C, Taufiqurrahman I, Wibowo WK, Winarni NL, Yanuar A, Marsden SJ (2021) Citizen science rapidly delivers extensive distribution data for birds in a key tropical biodiversity area. Global EcolConserv e01680

Stevens M, Vitos M, Lewis J, Haklay M (2013) Participatory monitoring of poaching in the Congo basin. http://tes-ne.com

Turner DS, Richter HE (2011) Wet/dry mapping: using citizen scientists to monitor the extent of perennial surface flow in dryland regions. Environ Manag 47 (3):497–505. https://doi.org/10.1007/s00267-010-9607-y

van Kerkhoff LE, Lebel L (2015) Coproductive capacities: rethinking science-governance relations in a diverse world. Ecol Soc 20(1). https://doi.org/10.5751/ES-07188-200114

Vaughan C, Dessai S (2014) Climate services for society: origins, institutional arrangements, and design elements for an evaluation framework. Wiley Interdisc Rev: Clim Change 5(5):587–603. https://doi.org/10.1002/wcc.290

Vincent K, Daly M, Scannell C, Leathes B (2018) What can climate services learn from theory and practice of co-production? Climate Services 12:48–58. https://doi.org/10.1016/j.cliser.2018.11.001

Walker DW, Smigaj M, Tani M (2021) The benefits and negative impacts of citizen science applications to water as experienced by participants and communities. Wiley Interdisc Rev: Water 8(1):e1488

Ziervogel G, Opere A (2010) Integrating meteorological and indigenous knowledge-based seasonal climate forecasts in the agricultural sector. https://idl-bnc-idrc.dspacedirect.org/bitstream/handle/10625/46185/132676.pdf?sequence=1&isAllowed=y

Zuma-Netshiukhwi G, Stigter K, Walker S (2013) Use of traditional weather/climate knowledge by farmers in the South-western Free State of South Africa: agrometeorological learning by scientists. Atmosphere 4 (4):383–410

Emmanuel Nyadzi (Ph.D., M.Tech, B.Sc.) is a postdoctoral researcher in Water Systems and Global Change group at Wageningen University, Netherlands. He is an applied climatologist with a strong interest in interdisciplinary research, using multi-approach and multi-data research frameworks to answer complex climate change questions. His recent work focused on developing water and climate information services (predictability, impact, risk, and vulnerability) and Climate change adaptation for and with local or indigenous people.

Andy B. Nyamekye (Ph.D., M.Sc., B.Sc.) is a Digital Capacity Specialist at the United Nations Food and Agriculture Organizations, Rome, Italy. His research borders on technology, innovation and sustainability governance. He is interested in investigating how developing countries can leverage technology and innovation in solving societal problems. As a multi-disciplinary researcher, his recent works employ multiple lenses such as responsible innovation, adaptive decision making, adaptive governance, and informational governance.

Fulco Ludwig (Ph.D., M.Sc., B.Sc.) is a full professor and a multidisciplinary researcher at the Water Systems and Global Change group at Wageningen University, Netherlands. He has a strong background in climate impact modelling and data analyses and has extensive field experience in Africa, Australia, Asia and USA. He focuses on studying global change impacts and adaptation in relation to water, food and energy resources. He is also interested in combining global and local information systems to improve climate services.

Highlighting Avenues to Reconcile Indigenous and Scientific Knowledge to Address Climate Change and Land Degradation in Lesotho: A Case Study of the Maseru District

9

Mulalo Rabumbulu and Itumenleng Masithela

9.1 Introduction

Many indigenous and marginalised communities around the world have been residing and continue to reside in environments that are highly vulnerable to climate change. Small islands and arid, arctic and high-altitude regions are examples of such environments (Abate and Kronk 2013; Obi et al. 2021; Rabumbulu and Badenhorst 2017). The Kingdom of Lesotho is situated at a high altitude. It is the only country in the world that lies entirely at an elevation of 1 000 m, with its height above sea level ranging from 1 300 m to 3 482 m. Because of its high altitude, Lesotho's citizens are more vulnerable to climate change (United Nations Lesotho 2019). An adverse consequence of climate change in Lesotho is increased land degradation, an issue accelerated by population growth (Maro 2011).

M. Rabumbulu (✉)
Department of Geography, Environmental Management and Energy Studies, University of Johannesburg, Kingsway Campus, Auckland Park, 2006, PO Box 524, Johannesburg, Republic of South Africa
e-mail: mrabumbulu@uj.ac.za

I. Masithela
Occupational Safety Department, University of Johannesburg, Kingsway Campus, Auckland Park, 2006, PO Box 524, Johannesburg, Republic of South Africa
e-mail: imasithela@uj.ac.za

Desertification is a characteristic feature of land degradation (Rabumbulu and Badenhorst 2017). It can be classified "as soil erosion caused by a severe depletion of vegetative matter, bush encroachment, shrub invasion of grasslands and a general decrease in rangeland productivity" (Rabumbulu and Badenhorst 2017, p. 219). Degradation of this kind, as is the case in Lesotho, is a major issue because it reduces the biological and economic productivity of any land. Most of Lesotho's land is already degraded. However, such degradation is highly prevalent on especially agricultural land in the country. Thus, although increasing land degradation has become a global environmental issue, it is particularly problematic in Lesotho because most of the population relies on subsistence farming for survival (Maro 2011). If not attended to immediately, this degradation will exacerbate the country's already severe poverty levels due to decreased crop productivity (Maro 2011). The growing population combined with the need for intensified agricultural production therefore reinforces the need to implement effective coping and adaptation strategies to deal with climate change and land degradation.

The United Nations (UN) Convention to Combat Desertification, which focuses specifically on African countries experiencing land degradation, has identified poor agricultural practices as the primary contributor to land degradation (specifically in the form of soil erosion) in Lesotho. The removal of nutrient-rich

E. E. Ebhuoma and L. Leonard (eds.), *Indigenous Knowledge and Climate Governance*, Sustainable Development Goals Series, https://doi.org/10.1007/978-3-030-99411-2_9

topsoil reduces soil depth and water-holding capacity, making plant growth difficult. However, to attribute Lesotho's land degradation to mainly poor agricultural practices may be misleading. Researchers such as Hongslo et al. (2009) and Rabumbulu and Badenhorst (2017) have identified many other complex and interrelated causes of land degradation, including socio-economic, biophysical, climatic and land-use activities. Thus, although the jury is still out on the extent to which unsustainable human activities such as poor agricultural practices exacerbated by biophysical factors and climate variability contribute to land degradation, it could be argued that land degradation is caused by both human and natural forces.

Farmers, as the primary users of agricultural land, can be viewed as key players in land management in Lesotho. Basdew et al. (2017) and Nezomba et al. (2017) have shown that indigenous knowledge (IK) can assist scientists in developing novel ways to cope with and adapt to climate change. Thus, the primary goal of this chapter is to analyse the indigenous indicators used by Lesotho farmers in the Maseru district in the eastern part of the country to adapt to climate change and land degradation. Also, the chapter seeks to understand farmers' perspectives on the integration of scientific and IK strategies in land management against degradation. Since farmers are the primary managers of land who can, potentially, both cause and mitigate land degradation, it would be beneficial to concentrate on their perspectives. Furthermore, because land degradation and climate change are closely intertwined, assessments of Lesotho's interventions could be instrumental in advancing the discourse and the ventures needed to achieve sustainable development goal (SDG) 13 of the UN. SDG 13 focuses on taking immediate action to combat climate change and its consequences (Romero-Daza et al. 2009). Addressing climate change is critical in Lesotho because climate change and environmental degradation could pose a significant threat to Lesotho's ability to actualise other SDGs. These include protecting the environment and promoting sustainable development (SDGs 11, 12 and 15), addressing poverty and hunger (SDGs 1 and 2) and stimulating economic growth (SDG 8).

9.2 Climate Change, Land Degradation and SDG 13 in Lesotho

Climate change refers to changes over decades in rainfall patterns and sea levels and an increase in the frequency and severity of extreme weather events such as storm surges, floods, droughts and heat waves (Galloway McLean 2010).Increased temperatures, rainfall variability and unpredictable and extreme weather events are also among the projected climate changes for Lesotho (United Nations Lesotho 2019). Over the past 10 years, Lesotho has been severely affected by heavy rains, hailstorms, heavy snow and flash floods. According to the UN resident representative in Lesotho, most parts of the country experienced a lower than average crop harvest in the 10 years preceding 2018 compared with previous times due to delayed rains and unexpected snow, frost, extreme temperatures (high and low) and dry spells. This issue is not peculiar to Lesotho as extreme climate events, especially in the last decade, have put global food production systems in jeopardy (Ahmed 2020).

Changes in climatic conditions cause soil erosion and fertility depletion that exacerbate land degradation (Rabumbulu and Badenhorst 2017). Overgrazing is another factor contributing to land degradation in Africa. This occurs because most African farmers typically have huge livestock herds competing for scarce pastures (Ebhuoma et al. 2021). Overgrazing also contributes to soil erosion and soil salinity (Galloway McLean 2010), making the land unsuitable for future agricultural production.

When land is degraded, soil carbon and nitrous oxide are released into the atmosphere, making land degradation a significant contributor to climate change (International Union for Conservation of Nature 2015). Because of land degradation, most rangelands furthermore lose ecosystem services and functions that results in altered nutrient, global carbon and hydrological cycles (Eldridge and Delgado-Baquerizo 2017). In both low- and high-income countries, land degradation affects all terrestrial biomes and agro-ecologies (Nkonya et al. 2016).

Land is one of the few productive assets owned by the rural poor in Lesotho. Hence, most of them work in agriculture (Barbier and Hochard 2018; Donkor et al. 2020; Maro 2011). However, the country's rapidly rising population that increases demand and pressure on land is exposing its citizens ever more to degradation of their land. Arable land is mostly endangered due to poor land management in the quest to grow more food to meet the high demand (Abdi et al. 2013). When subjected to such intense pressure, land in semi-arid and dry, sub-humid areas can take on a variety of forms. In Lesotho, some of the most visible signs of land degradation are the loss of biological productivity, deteriorating rangelands, drying wetlands and poor crop and animal production (Maro 2011). Land degradation furthermore impacts on human development and health. Rabumbulu and Badenhorst (2017) found that controlling land degradation can help to address developmental constraints such as food insecurity and poverty effectively. As such, land degradation has become an environmental, economic and social problem that necessitates ongoing monitoring to ensure a sustainable environment and socio-economic well-being.

The SDGs of the UN should serve as a model for all its members in the fight to protect the environment and ensure sustainability. The Kingdom of Lesotho, led by the King and Prime Minister, has long been committed to implementing both climate change and development agendas geared towards ensuring sustainability. This commitment dates back to the UN Conference on Environment and Development that took place in Rio de Janeiro, Brazil, from 3 to 14 June 1992. Lesotho signed the UN Framework Convention on Climate Change, which was seen as a blueprint for international environmental action (Nkonya et al. 2016), in 1992 and has since made significant progress in taking action on climate change and sustainable development with the development in 1996 of a national climate change programme. Lesotho has also ratified both the Kyoto Protocol and the Paris Agreement in 2000 and 2017, respectively. As such, the country has incorporated the SDG agenda and other international development commitments in its national development plan towards achieving the UN 2030 Agenda for Sustainable Development. Table 9.1 reflects the environmental and climate-related policies, strategies, programmes and regulations that have been developed to ensure that the Kingdom of Lesotho is on course to attain sustainable development.

Because climate change poses a significant threat to indigenous peoples and communities (Donkor et al. 2019), most rural farmers rely on IK, which is central to their cultural resilience, to devise adaptive responses to such change as it occurs (Fereja 2016). Consequently, there is an urgent need for a policy-led integrated approach that would improve farmers' knowledge and skills while avoiding the one-sided scientific approach commonly used to address climate change (Phong et al. 2017). The following section considers how existing IK in Lesotho could be combined with scientific knowledge to address the issue of land degradation and climate change.

9.3 Indigenous and Scientific Knowledge

IK refers to cumulative knowledge, beliefs and a way of doing things that are passed down from one generation to another through proverbs; riddles; folktales; songs; legends and myths; culture; religion; stories; informal education; communication; tradition; and apprenticeship (Abate and Kronk 2013; Obi et al. 2021; Torrents-Ticó et al. 2020). It is frequently passed down among indigenous peoples of a specific geographic area, and the knowledge is quite diverse, covering topics such as agriculture; medicine; security; botany; zoology; food technology; art and craft skills; linguistics; education; natural resource management; hazard management, and so on. It is used to guide human societies in their numerous interactions with their surroundings (Berkes 2012; Nakashima et al.

Table 9.1 Summary of international and national agreements, policies, strategies and programmes developed to address climate change, as well as Lesotho's implementation of the SDG 13 agenda

International agreement and national policy and regulation	Ratification/start date
United Nation convention for climate change	1992
National environment policy	1998
Kyoto protocol	2000
Agriculture sector strategy	2003
Lesotho food security policy	2005
National forest policy	2008
Transport policy	2006
National action plan for food security	2007–2017
Water and sanitation policy	2007
National disaster risk reduction policy	2007
Environment act	2008
National conservation agriculture strategy framework	2012–2017
National health policy	2012
Mountain integrated conversation agriculture project (MICA)	2012–2014
National range resources management policy	2013
Draft renewable energy policy	2013
Long term water and sanitation strategy	2014
Environmental education strategy towards 2014	2014
National range action plan	2015
National strategic resilience framework	2015–2025
Energy policy	2015–2025
Paris agreement	2017
National climate change policy	2017–2027
Soil and water conservation policy	2021

Source Authors' review of Lesotho policy documents (2021)

2018). Other terms with similar meanings that are frequently used interchangeably with the term IK include *indigenous technical knowledge*, *rural knowledge*, *ethnoscience*, *traditional knowledge*, *traditional ecological knowledge*, *local knowledge*, *farmers' knowledge*, *folk knowledge* and *indigenous science* (Obi et al. 2021).

On the other hand, science is defined as a systematic organised inquiry into understanding the natural world and its phenomena (Bwambale et al. 2020). Science is about gaining or imparting a more in-depth and often useful understanding of the world. Scientific knowledge is generally defined by the method through which it is accumulated. In general, knowledge is considered scientific if it has been accumulated through systematic study and is organised according to general principles.

In an attempt to distinguish between the two knowledge systems, some people make the mistake of describing IK as old fashioned, backwards, static or unchanging. In reality, the two knowledge systems are working toward the same goal: gaining a better understanding of how the world works. Even though their approaches appear to differ, both systems have structured ways of acquiring new knowledge. Scientific knowledge is renewed when technology improves, while IK is renewed through learning-by-doing, experimenting and knowledge building (Berkes 2012). Put simply, IK holders have always been able to adapt and modify their actions in response to environmental change. Similar to how scientists use their understanding of scientific principles and natural laws, indigenous communities' adaptive capacity is based on both their in-depth knowledge of and their ongoing interaction with their land. Their excellent understanding of and interaction with their immediate environments enable them to read critical signs from the environment and detect unusual changes (Berkes 2012).

Even though the two knowledge systems serve the same purpose, scientific knowledge has always received more attention, especially in the twenty-first century (Hermann et al. 2020), which could be described as a technologically focused period. The role of science in reducing land degradation is not disputed. However, the argument here is that platforms that permit complementary or integrated usage of both science and IK should be provided. It is important to note that the scientific approach has been the subject of debates due to the narrow approach that scientists have taken. The limitation stems from the fact that science does not provide solutions to *all* the problems that farmers face (Abate and Kronk 2013) and a recent study (Basdew et al. 2017) found that the use of local knowledge is critical, particularly in the context of farming, conservation and natural resource management.

Because using a single knowledge system has failed to manage local issues (Nkuba et al. 2020; Torrents-Ticó et al. 2020), the integration of indigenous and scientific knowledge systems has received burgeoning attention in the last decade (Ebhuoma 2020; Nkuba et al. 2020). Although the integration of the two knowledge systems is expected to produce better results in terms of solving local issues, it is important to note that merging them would come with its own set of unique challenges. For example, research (Nakashima et al. 2018; Nkuba et al. 2020; Obi et al. 2021) has shown that in basing climate change decision making and action on the best available knowledge, both scientific knowledge and IK introduce new challenges. Scientific observations and understandings may sometimes agree with those of indigenous peoples; at times they may differ but be complementary; and at yet other times they may be completely contradictory. As such, it could be challenging to make a joint decision when merging the two knowledge systems.

While literature in the fields of agronomy, soil science and natural sciences has largely shaped the scientific debate on soil fertility depletion and management in Africa (Nezomba et al. 2017; Payton et al. 2003), there are a few examples where the integration of the two knowledge systems has produced positive results. For example, the Vam Ray coast in Vietnam was eroding severely and the initial plan of constructing sea dykes and long fences of eucalyptus poles to protect Avicennia marina seedlings failed. To avoid repeating the same errors, development planners combined scientific knowledge of erosion control with IK of coastal changes, sedimentation and tidal regimes. The main strategy was to transform traditional Melaleuca fences into sea-mud-accumulation fences. The fences were upgraded into silt-trap fences (single-line Melaleuca silt-trap fences—three types—and double-line Melaleuca silt-trap fences—four types), which were constructed and tested for their effectiveness under different circumstances (Phong et al. 2017). The strategies developed as a result of the integration of the two knowledge systems included the construction of a nursery, which is a place or facility used for growing seedlings for a period before moving them to be planted permanently. That allowed for a large supply of seedlings and the creation of a fence that could trap the silt, resulting in the accumulation of mud that protected the seedlings and slowed erosion rates (Phong et al. 2017).

A further example is a study conducted in Tanzania and Uganda during which the knowledge of soils and land resources of both farmers and scientists was considered as important for promoting sustainable land management (Payton et al. 2003). Scientists used geographic information system software to enter and overlap soil characteristics from both local and scientific soil knowledge. Maps of the area drawn by the farmers were also remapped into photographs. The maps that combined indigenous and scientific knowledge were both effective and beneficial in the attempts to control and prevent land degradation (Payton et al. 2003). Another excellent example is the development of neem biopesticides in Togo and Niger that resulted from the integration of both knowledge systems. For centuries, local people used the neem tree to store grains and provide pest protection. The neem seed kernel was extracted and used to create a pesticide that contains about 1% of the bioactive found in this plant, enhancing protection for up to 120 species (Özger et al. 2013). Studies by Özger et al. (2013), Payton et al. (2003) and Verma et al. (2016) furthermore demonstrated unequivocally that incorporating local knowledge into strategies aimed at improving agriculture and natural resource management yielded effective results.

9.4 Material and Methods

9.4.1 Regional Setting

Lesotho is a landlocked country surrounded by South Africa and comprises highlands, plateaus, hills and mountains. The Kingdom is divided into 10 districts and four ecological zones, namely lowlands, foothills, highlands and the Senqu River valley zone (Letsie 2005). The lowlands zone (1 400–1 800 m above sea level) is the most densely populated and farmed region. It covers approximately 12% of the country's area and provides most of the country's fertile and arable soils. The foothills or intermediate zone at the base of the Drakensberg and Maloti Mountain ranges is 1 800–2 100 m above sea level, has fertile soil and covers about 15% of the land area. However, slope angles pose a threat to sustainable crop farming. The highland zone is at the highest altitude in the country (2 100–3 482 m above sea level). This is the dominant land space, accounting for nearly 61% of the country's total land area. The Senqu River valley zone (1 200–1 400 m above sea level) accounts for about 4% of Lesotho's land area. The Maseru district, which was the focus area of this study, lies to the east (Fig. 9.1). The study included three villages in the district that represent three of the country's four major ecological zones. The livelihoods of the Basotho (people living in Lesotho), like those of most Africans, are primarily dependent on natural resources. The land uses in Lesotho are the rangelands for livestock grazing, arable land for crop production, forested land (indigenous and afforested) and settlements (Maro 2011). The land-use patterns are mostly communal in the rangelands and semi-private on cultivated lands.

Lesotho's soils are sedimentary and basaltic in origin, formed primarily by underlying sandstone and basalt parent materials in the lowlands and mountains, respectively. The country's geological formations are part of the Karoo Supergroup's Beaufort and Stormberg series (Fig. 9.2). Mineral soils cover a large part of the country, with the exception of areas above 3 000 m in elevation that have organic soils. Mineral soils include sandy, loamy and clayey soils (Schmitz and Rooyani 1987).

9.4.2 Data Collection and Analysis

The main data collection techniques used in the study were (i) interviews, (ii) observations and (iii) documents. The respondents in this study were chosen using a non-probabilistic purposive sampling method, also known as judgmental sampling. Purposive sampling is the deliberate selection of respondents based on their ability to elucidate a specific theme, concept or phenomenon (Creswell 2007). Participants from the three villages in the Maseru district that represent three of the four major ecological zones were chosen on the basis of their farming experience and knowledge. The three villages are Morija (lowlands zone), Nazareth (foothills zone) and Semonkong (highlands zone). In total, 30 farmers were interviewed (Table 9.2). Government officials were also interviewed to gather information about government plans and interventions. Thematic content analysis was used to analyse data from the semi-structured interviews.

Fig. 9.1 Ecological zones of Lesotho. *Source* Molahlehi et al. (2013)

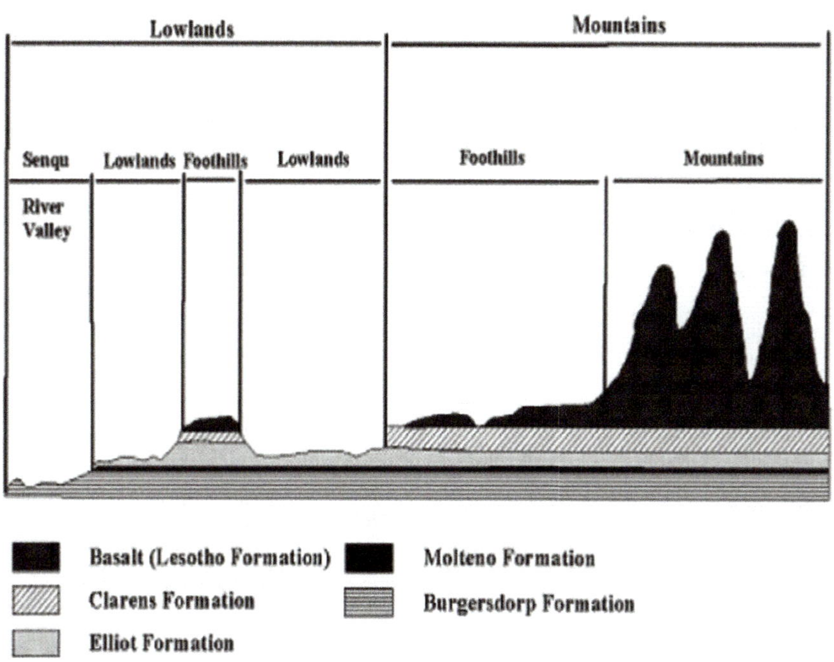

Fig. 9.2 Geological formations of Lesotho. *Source* Majara (2005)

Table 9.2 Demographic data of participants

Ecological zone	Village	Gender	Number of participants
Lowlands	Morija	Female	5
		Male	5
Foothills	Nazareth	Female	6
		Male	4
Highlands	Semonkong	Female	3
		Male	7

Source Fieldwork (2020)

9.5 Results and Discussion

This section is divided into three parts. The first discusses the reconciliation of indigenous and scientific knowledge on land degradation, while the second discusses climate change and weather. The final part covers Basotho land management and coping strategies.

Table 9.3 gives a summary of indigenous indicators of land degradation, climate change and land management strategies used by Maseru farmers. As indicated in the introduction, bio-physical factors also play a major role when it comes to land degradation. One of these factors is soil properties. Some soils are simply more vulnerable to degradation than others due to differences in structure, texture and mineralogical composition. *Lehlabathe* (sandy soil) is highly erodible, whereas *letsopa* (clay soil) is less prone to erosion.

9.5.1 Indigenous Indicators of Land Degradation in Lesotho and Their Reconciliation with Science

Ethnopedology is the study of local knowledge (about the land) and is based on the theory that people who are familiar with soils have a better understanding of the indicators of degraded and

Table 9.3 Summary of indigenous weather forecasts, land degradation and land management strategies used by Maseru farmers

Indigenous indicators	Interpretation
Indigenous indicators of weather and land degradation	
Crescent moon facing down	Basotho sees the crescent moon as a bowl/basin and believes that when the bowl (moon) is facing down, God will pour down rain
Crescent moon facing up	The bowl facing up is thought to be collecting and containing the rain, indicating that it will not rain anytime soon
Thick dark clouds that build up from north	This suggests that it will rain heavily
Thin clouds together with wind	This indicates that a cold day is on the way
Many red ants in straight line (in fields)	This is interpreted as a forewarning of impending heavy rain that will last for several days. The ants are migrating to find shelter away from the rainy area
Mahaqasi/mankabelane(birds that come in numbers)	The birds are said to be dancing and celebrating the arrival of rain
Donkey braying at night	This indicates that it will be cold the next morning

(continued)

Table 9.3 (continued)

Indigenous indicators	Interpretation
Changes in soil colour	Farmers regard *Selokoe*(dark brownish soil) as rich and beneficial to crops. If the soil becomes lighter, they interpret it as a sign of decreased fertility
Indigenous practices to combat land degradation and adapt to climate change	
Digging of gutters around the field	Farmers frequently dig small trenches around their farm to collect excess water and channel it down the slope. This aids in reducing the impact of flooding and erosion on farms
Burning of fields	To reduce pest invasion, weeds that attract pests are burned. The ash residue is then used to enrich soils with potassium and calcium, thereby benefiting crops
Lesokoana(relay race to attract rain during dry season/drought)	The race is run by young girls from one village to the next, and as they sprint and sweat, it is believed that they are attracting rain
Use drought tolerant and heat resistance seeds/crops	This is yet another disaster preparedness measure implemented by Basotho farmers in response to changing climatic conditions
Hail	Basotho believe that their collective behaviour can influence hail. They do not bury their dead until after 2:00 pm and they do not hang their laundry until noon. They believe that if they do not follow these rules, it will bring hail. To stop the hail, the community hires a witch doctor (*Sangoma*) to perform specific rituals
Mantloane(keyhole garden)	This is a gardening technique used to adapt to climate change. It aids in water retention by allowing for easy absorption and accumulation of water. The garden in Lesotho is typically walled with local stone or brick, which retains daytime heat and alleviates low night-time temperatures. The elevated structure also facilitates access for the chronically ill or elderly. The basic garden is 1 m high and 2 m in diameter, but sizes can vary. Growers have 'arms-length access' across the garden with a keyhole-like 'cut out' or walkway design. 'Grey' water from household washing is added through a central composting basket, watering the garden and infusing the soil with nutrients on a continuous basis

Source Fieldwork (2020)

healthy soils (Nezomba et al. 2017). Local knowledge is beneficial because it enables the community or farmers to monitor soil degradation. Another advantage of IK, particularly in poor African countries in the global south such as Lesotho, is that it is less expensive and has local relevance (Basdew et al. 2017; Nezomba et al. 2017). Farmer knowledge of soils and their management is critical in developing more sustainable agricultural systems. The majority of farmers in Lesotho, like most African countries, lack such formal soil science training. However, despite the majority of farmers in this study lacking formal education, they typically had a good understanding of their soils and of the crops that are better suited to specific locations and types of soil (Apanovich and Lenssen 2018; Payton et al. 2003).

Farmers were interviewed and asked how they classified soils locally and what indicators they used in Maseru (local knowledge) to determine land degradation. Farmers distinguished soils by naming them after observed and experienced unique properties and they classified soils based on texture (e.g., sandy or clay) and colour (e.g., *selokoe*—dark soil). They used the classification to determine the soils' fertility and erodibility. In other words, farmers used changes in soil texture and colour to detect degradation. This is similar to previous research (Manyevere et al. 2020; Rabumbulu and Badenhorst 2017) on IK in soil and land degradation classification. However, the indigenous classification system has the limitation of being purely qualitative, relying on the physical characteristics of soils that can be observed with the naked eye, while quantitative measurements are actually required to obtain more accurate results. Integrated indigenous and scientific knowledge would therefore improve farmers' ability to recognise the constraints of each soil unit, which could then also be used as a guide for precision farming. That would, furthermore, assist farmers in detecting other changes (degradation) that cannot be detected by solely examining texture or color.

It is also worth noting that the farmers mentioned other types of degradation, such as increased fluvial erosion and changes in vegetation cover. Although they noticed these changes, it became clear during discussions that they did not necessarily recognise them as signs of land degradation but rather as indicators of changes in climate or weather patterns. For example, a female farmer in Nazareth said:

> Lesotho is turning into a desert; there is no rain; it last rained in March or April; it did not really rain, so having six months with no rain, crop farming cannot happen.

In view of the above, the incorporation of science into existing knowledge could serve as a tool for farmers to see the connections between the various factors that contribute to land degradation (Basdew et al. 2017; Manyevere et al. 2020).

9.5.2 Indigenous Indicators of Climate Change and Weather in Lesotho

The Basotho have many methods to forecast weather (precipitation and temperature) and they use animal behaviour, moon orientation and differences in cloud coverage as indicators in this regard (see Table 9.3). Donkeys braying at night, swarms of larks and swallows, an increase in the number of red ants in a straight line in the fields and the call of the hammerhead bird (*Mamusi-unoke*) are all examples of indigenous indicators. Donkey braying at night is thought to be a forewarning of impending cold weather and other animal behaviour, such as the presence of flocks of birds, is thought to predict the arrival of rain. Several other studies (Abate and Kronk 2013; Al Mamun and Al Pavel 2014; Bacud 2018; Cameron 2019; Obi et al. 2021) also revealed that indigenous communities use animal behaviour to predict natural events such as rainfall, drought, tsunamis, volcanism and earthquakes. Some coastal communities in Nigeria, for example, use the migration patterns of certain birds, particularly swallow-tailed kites, to forecast floods, as noted by Obi et al. (2021, p. 5): "The disappearance of kites in the sky indicates that the flood is approaching, whereas their appearance shows that the flood is retreating." As such, coastal communities regard the migratory behaviour of birds as a sign that flooding may be about to end. Furthermore, observations that birds are beginning to perch only on tall trees is an indication of flooding. The findings from the current study and those of the studies mentioned demonstrate that indigenous indicators are effective in predicting short-term weather and seasonality. Unfortunately, such indigenous indicators do not forecast long-term weather events. Predicting long-term weather events would enable farmers to prepare better for forthcoming farming seasons, since some of them already find the unpredictable shifts in seasons very frustrating. A female farmer in Morija said the following:

> In the past, you would find that in December there are crops that are already growing, but due to the present conditions, in December we are only starting to plough.

Thus, there is a need to develop methods for predicting long-term climate change, and the integration of scientific knowledge with IK could assist in bridging the gap. Integrated knowledge systems could also help to improve the effectiveness or accuracy of some of the indigenous indicators (Table 9.3) currently in use. Also, Table 9.3 shows that Basotho farmers are superstitious about certain natural events such as drought and hail storms and that they apply superstitious beliefs when dealing with such events. As such, they believe that traditional games and ceremonies can be used to summon rain, end droughts and prevent hailstorms. Farmers should therefore be exposed to awareness campaigns and receive ongoing training and education in the causes of climate change so that they could respond appropriately to changing environmental conditions.

9.5.3 IK in Land Management and Adaptation Strategies to Deal with Land Degradation and Climate Change in Lesotho

As can be expected, the Basotho people have developed a number of coping strategies to assist them in dealing with changing environmental (climate change and climate variability) conditions and land degradation. This study found that all farmers in the Maseru district practised crop rotation and rotational grazing. Crop rotation is the practice of growing different crops on the same land in a systematic and recurring sequence (Rabumbulu and Badenhorst 2017). Farmers stated that planting the same crop on a regular basis causes salt to build up in the soil. A male farmer in Nazareth said:

> In order to succeed in farming, you should not plough (sic) the same crop all the time. You must rotate crops with years. If this year I plough (sic) maize, the following year it can be beans, for example.

He went on to say that some seeds secrete "elements" that cause salt accumulation. By changing the crops, farmers could alternate between crops that excrete salts and those that absorb salts, ensuring that there were no excess salts in the soil. Crop rotation could furthermore support pest control while increasing nutrient and water efficiency (Özger et al. 2013).

To increase the productivity of their land, all the farmers said that they preferred to use manure (livestock dung), while some farmers in the foothills also used ashes. All farmers agreed that working manure into their fields provided the best relief strategy for nutrient depletion. In this regard, a participant from Semonkong village said:

> We used to use both fertiliser and manure in the past. The extension workers advise us to use manure because it can hold the soil and it has weight. With compact soil and increased soil weight, erosion gets reduced as it is not easy for wind, rain or animals to carry away the soil. Fertilisers have chemicals that can also affect the weather and the soil.

Unlike farmers in the highlands who used only manure, farmers in the lowlands and foothills said they were frequently forced to mix manure and fertilizer to boost production due to a lack of manure for their large fields. These areas also had fewer livestock than the highlands, making it difficult to collect enough manure for the fields. While both fertilisers and manure could assist in increasing productivity, studies like that of Lal and Stewart (2010) agreed with farmers that manure is more environmentally friendly and helps to conserve soil.

The rise in the frequency and severity of droughts is one of the most serious challenges that Lesotho farmers face because of climate change. According to the farmers, they are aware of heat and drought-resistant seeds or crops, that they can use. Unfortunately, most farmers were hesitant to use these because they looked unfamiliar. While a number of farmers already made use of some scientific advances like genetically modified seeds and greenhouses, the issue was their negative view of such developments. It could thus be argued that a closer look at how the two knowledge strands should work together rather than simply replacing one method with the other could help to change the minds of the farmers. That could assist in achieving SDG 13 because it would result in reduced land degradation, which would reduce the amount of carbon dioxide released and, as a result, global warming would be reduced.

Farmers in the highlands, where livestock rearing is prevalent, said rotational grazing helped to maintain pastures. One farmer emphasised the importance of rotating their grazing to prevent grass depletion. However, one farmer pointed out that even though they rotated grazing, the presence of too many animals and too few pastures defeated the purpose, with overgrazing ultimately becoming unavoidable. Overgrazing has been linked to decreased rain infiltration, accelerated run-off and soil erosion (Abbasi et al. 2016; Ebhuoma et al. 2021). The high grazing intensity is also reducing grass diversity in the highlands.

To reduce the amount of fertile soil loss in Lesotho's mountainous terrain, including loss from fluvial erosion, farmers resorted to creating furrows to channel floods away from the fields. Terraces were also used to facilitate ploughing on steep slopes (Fig. 9.3). While the practice of using fire to control pests and prevent overgrazing on farms is beneficial to deal with pests, with the ash residue enriching the soil with potassium and calcium that benefit crops, it also contributes to excessive soil erosion (see Fig. 9.3).

Keyhole gardening is another innovative strategy implemented by the farmers to prevent soil loss and conserve water. Such gardens are normally enclosed by a stone wall and consist of a raised, circular bed of soil, about two metres in diameter and one metre in height (Romero-Daza et al. 2009). Metal containers, aloe and animal bones are some of the items in the bottom layer of the enclosure that comprises numerous layers. Soil, ash and manure are among the elements making up the other layers. The seeds are planted in a final mixture of soil and manure (see Fig. 9.4). Such keyhole gardens encourage water penetration and upward movement. Since these systems are currently being built on a very small scale, scientific experience in the form of modern, inexpensive engineering techniques could perhaps be applied to try to develop them on a larger scale. Keyhole gardens are one of the government-supported initiatives for dealing with dry spells. Our participants also indicated that the Lesotho government was promoting pro-poor production technologies, and indigenous methods such as keyhole gardens are typically pro-poor.

Fig. 9.3 Terrace farming and burning of fields in foothills and highlands. *Source* Author (2020)

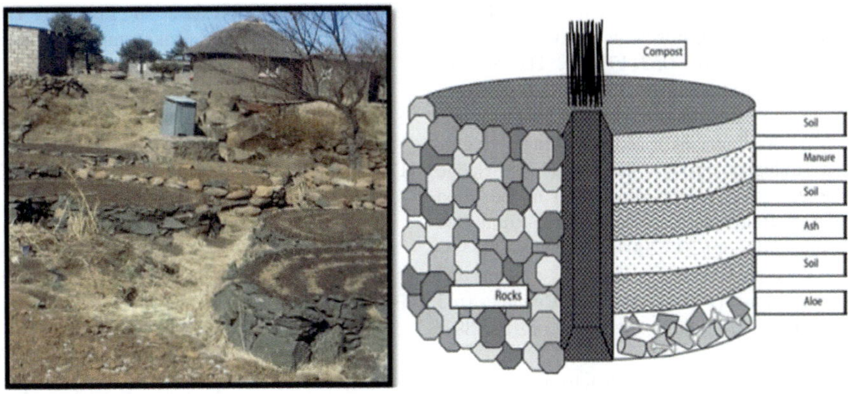

Fig. 9.4 Keyhole garden (Mantloane) that is useful to retain water longer (Left). Sketch showing the structure of a keyhole garden (Right)

9.6 Conclusion

Basotho farmers forecast weather by observing animal behaviour, moon orientation and differences in cloud coverage. However, based on the responses of the farmers interviewed, it appears that they have no way of forecasting long-term climate change. While they are aware of changing environmental conditions and the unfortunate issue of land degradation and have developed various coping and adaptation strategies, their efforts to mitigate the effects of climate change appear futile. This chapter suggests avenues through which IK and scientific knowledge could be combined to produce more locally relevant climate change adaptation solutions in Africa.

Integration of the two knowledge systems provides significant hope for improvement. Although there can be no doubt that land degradation and climate change are intertwined, interventions to address climate variability are challenging since it is almost impossible to change the extent and occurrence of natural phenomena. Hence, climate change mitigation requires intervention on a global scale. Lesotho has incorporated climate change measures in its national policies, strategies and plans to strengthen the country's resilience and adaptive capacity to climate-related hazards and natural disasters. It has put in place policy and institutional frameworks to address and manage climate change and it is clear that Lesotho is committed to accelerating the implementation of SDG 13 and other SDGs. To fully achieve SDG 13, Lesotho should focus on the implementation and evaluation of the policies established. Integrating indigenous and scientific knowledge could also assist in building community resilience by developing effective, low-cost adaptation measures.

Acknowledgements We would like to express our gratitude to the following groups of people: 1) the farmers interviewed for access to their farms, their time and their participation, 2) Lesotho government officials for sharing their information, 3) officials from the Lesotho Ministry of Agriculture and Food Security for arranging meetings with agricultural extension workers through resource centers in the three villages; and last but not least, the editors and two anonymous reviewers for comments that improved the original paper considerably.

References

Abate RS, Kronk EA (2013) Climate change and indigenous peoples: the search for legal remedies. Edward Elgar Publishing

Abbasi T, Abbasi T, Abbasi SA (2016) Reducing the global environmental impact of livestock production: the minilivestock option. J Clean Prod 112:1754–1766. https://doi.org/10.1016/j.jclepro.2015.02.094

Abdi OA, Glover EK, Luukkanen O (2013) Causes and impacts of land degradation and desertification: case study of the Sudan. Int J Agric Forest 3(2):40–51. http://article.sapub.org/10.5923.j.ijaf.20130302.03.html

Ahmed M (2020) Introduction to modern climate change. Andrew E. Dessler: Cambridge University Press, 2011, 252 pp, ISBN-10: 0521173159. The Science of the Total Environment 734:139397. https://www.ncbi.nlm.nih.gov/pmc/articles/PMC7227569/#

Al Mamun MA, Al Pavel MA (2014) Climate change adaptation strategies through indigenous knowledge system: aspect on agro-crop production in the flood prone areas of Bangladesh. Asian J Agric Rural Dev 4:42–58. https://doi.org/10.22004/ag.econ.198381

Apanovich N, Lenssen AW (2018) Cropping systems and soil quality and fertility in south-central Uganda. Afr J Agric Res 13(15):792–802. https://doi.org/10.5897/AJAR2018.13056

Bacud ST (2018) Integration of indigenous and scientific knowledge in disaster risk reduction: resilience building of a marginalized sampaguita growing community in the Philippines. Procedia Eng 212:511–518. https://doi.org/10.1016/j.proeng.2018.01.066

Barbier EB, Hochard JP (2018) Land Degradation and Poverty. Nature Sustain 1(11):623–631. https://doi.org/10.1038/s41893-018-0155-4

Basdew M, Jiri O, Mafongoya PL (2017) Integration of indigenous and scientific knowledge in climate adaptation in KwaZulu-Natal, South Africa. Change Adapt Socio-Ecol Syst 3:56–67. https://doi.org/10.1515/cass-2017-0006

Berkes F (2012) Sacred ecology, 3rd edn. Routledge

Bwambale B, Nyeko M, Muhumuza M, Kervyn M (2020) Questioning knowledge foundation: what is the best way to integrate knowledge to achieve substantial disaster risk reduction? Int J Disaster Risk Reduction 51:1–12. https://doi.org/10.1016/j.ijdrr.2020.101850

Cameron A (2019) Indigenous knowledge and science: the case of *heteropyxis natalensis* as a lightning tree. Indilinga–African J Indigenous Knowl Syst 18 (2):129–144. https://www.ajol.info/index.php/indilinga/article/view/195212

Creswell JW (2007) Qualitative inquiry and research design: choosing among five approaches, 2nd edn. Sage

Donkor FK, Howarth C, Ebhuoma E, Daly M, Vaughan C, Pretorius L, Mambo J, MacLeod D, Kythreotis A, Jones L, Grainger S, Golding N, Anderson JA (2019) Climate services for development: the role of early career researchers in advancing the debate. Environ Commun 13(5):561–566. https://doi.org/10.1080/17524032.2019.1596145

Donkor FK, Mearns K, Ojong E, Tantoh H, Ebhuoma E, Hadisu A, Mavuso S, Mbewe P, Mabeza C, Leclerc A (2020) Attitudinal changes towards agriculture through the generational lens and impact on engagement in related activities: case study from a mountainous area. In: Squires V, Gaur M (eds) Food security and land use change under conditions of climatic variability. Springer

Ebhuoma EE (2020) A framework for integrating scientific forecasts with indigenous systems of weather forecasting in southern Nigeria. Dev Pract 30(4):472–484. https://doi.org/10.1080/09614524.2020.1723494

Ebhuoma OO, Gebreslasie M, Ebhuoma E, Leonard L (2021) 'The future looks empty': embodied experiences of distress caused by environmental and climatic changes in rural KwaZulu-Natal, South Africa. GeoJ. https://link.springer.com/article/https://doi.org/10.1007/s10708-021-10426-

Eldridge DJ, Delgado-Baquerizo M (2017) Continental-scale impacts of livestock grazing on ecosystem supporting and regulating services. Land Degrad Dev 28(4):1473–1481. https://doi.org/10.1002/ldr.2668

Fereja GB (2016) The impacts of climate change on livestock production and productivities in developing countries: a review. Int J Res-GRANTHAALAYAH 4 (8):181–187. https://doi.org/10.29121/granthaalayah.v4.i8.2016.2578

Galloway McLean K (2010) Advance guard: climate change impacts, adaptation, mitigation and indigenous peoples—a compendium of case studies. United Nations University. https://i.unu.edu/media/tfm.unu.edu/publication/242/UNU_Advance_Guard_Compendium_2010_final_web.pdf

Hermann S, Dioul AA, Sall I (2020) Beyond bioproductivity: engaging local perspectives in land degradation monitoring and assessment. J Arid Environ 173:1–5. https://doi.org/10.1016/j.jaridenv.2019.104002

Hongslo E, Rhode R, Hoffman T (2009) Landscape change and ecological processes in relation to land-use in Namaqualand, South Africa, 1939 to 2005. South Afr Geogr J 91(2):63–74. http://hdl.handle.net/11427/24849

International Union for Conservation of Nature (2015) Land degradation and climate change. The multiple benefits of sustainable land management in the drylands. ISSUES BRIEF. https://www.iucn.org/resources/issues-briefs/land-degradation-and-climate-change

Lal R, Stewart BA (2010) Soil quality and biofuel production. CRC Press

Letsie M (2005) Utilisation of Maqalika reservoir as a source of potable water for Maseru city in Lesotho [Master's thesis, Durban Institute of Technology]. DUT Open Scholar. http://hdl.handle.net/10321/147

Majara N (2005) Land degradation in Lesotho: a synoptic perspective [Master's dissertation, University of Stellenbosch]. SUNScholar. https://scholar.sun.ac.za/handle/10019.1/2541

Manyevere A, Mnkeni PNS, Laker MC, Muchaonyerwa P (2020) The use of indigenous knowledge systems in soil classification, appraisal and soil productivity rating for maize cropping: a case of the Xhosa speaking people in Raymond Mhlaba municipality, South Africa. Indilinga Afr J Indigenous Knowl Syst 19(1):96–109. https://hdl.handle.net/10520/EJC-1ee400abca

Maro P (2011) Environmental change in Lesotho: an analysis of the causes and consequences of land-use change in the Lowland Region. Springer. https://www.sahistory.org.za/sites/default/files/archive-files3/pendo_maro_environmental_change_in_lesotho_an_abook4you.pdf

Molahlehi L, Steyn JM, Haverkort AJ (2013) Potato crop response to genotype and environment in a subtropical highland agro-ecology. Potato Res 56(3):237–325. https://doi.org/10.1007/s11540-013-9241-1

Nakashima D, Krupnik I, Rubis JT (2018) Indigenous knowledge for climate change assessment and adaptation. Cambridge University Press

Nezomba H, Mtambanengwe F, Tittonell P, Mapfumo P (2017) Practical assessment of soil degradation on smallholder farmers' fields in Zimbabwe: integrating local knowledge and scientific diagnostic indicators. CATENA 156:216–227. https://doi.org/10.1016/j.catena.2017.04.014

Nkonya E, Mirzabaev A, Von Braun J (2016) Economics of land degradation and improvement—a global assessment for sustainable development. Springer. https://link.springer.com/book/https://doi.org/10.1007/978-3-319-19168-3

Nkuba MR, Chanda R, Mmopelwa G, Kato E, Mangheni MN, Lesolle D (2020) Influence of indigenous knowledge and scientific climate forecasts on arable farmers' climate adaptation methods in the Rwenzori region, Western Uganda. Environ Manage 65(4):500–516. https://doi.org/10.1007/s00267-020-01264-x

Obi R, Nwachukwu MU, Okeke DC, Jiburum U (2021) Indigenous flood control and management knowledge and flood disaster risk reduction in Nigeria's coastal communities: an empirical analysis. Int J Disaster Risk Reduction 55https://doi.org/10.1016/j.ijdrr.2021.102079

Özger S, Pohl D, Karaca I (2013) The use of neem extracts as biopesticide. Turk J Biol Control 4(2):165–178. https://www.researchgate.net/publication/264156413

Payton RW, Barr JJF, Martin A, Sillitoe P, Deckers JF, Gowing JW, Hatibu N, Naseem SB, Tenywa M, Zuberi MI (2003) Contrasting approaches to integrating indigenous knowledge about soils and scientific soil survey in East Africa and Bangladesh. Goederma 111(3–4):355–386. https://doi.org/10.1016/S0016-7061(02)00272-0

Phong NT, Luom TT, Parnell KE (2017) Developing a framework for integrating local and scientific knowledge in internationally funded environment management projects: case studies from Kien Giang Province, Vietnam. Local Environ. The Int J Justice Sustain 22 (11):1298–1310. https://doi.org/10.1080/13549839.2017.1342617

Rabumbulu M, Badenhorst M (2017) Land degradation in the west-central Free State: human-induced or climate variability, the perceptions of Abrahamskraal-Boshoff district farmers. S Afr Geogr J 99(3):217–234. https://doi.org/10.1080/03736245.2016.1231623

Romero-Daza N, Himmelgreen DA, Noble CA, Turkon D (2009) Dealing with the global food crisis in local settings: Nonintensive agriculture in Lesotho, Southern Africa. Ann Anthropol Pract 32(1):23–41. https://doi.org/10.1111/J.1556-4797.2009.01026.X

Schmitz G, Rooyani F (1987) Lesotho geology, geomorphology, soils. National University of Lesotho, Roma

Torrents-Ticó M, Fernández-Llamazares AI, Burgas D, Cabeza M (2020) Convergences and divergences between scientific and indigenous and local knowledge contribute to inform carnivore conservation. Ambio 50:990–1002. https://doi.org/10.1007/s13280-020-01443-4

United Nations Lesotho (2019) The Kingdom of Lesotho voluntary national review on the implementation of the agenda 2030 report. https://sdgs.un.org/documents/lesotho-2019-vnr-report-english-24489

Verma P, Vaughan K, Martin K, Pulitano E, Garrett J, Piirto DD (2016) Integrating indigenous knowledge and western science into forestry, natural resources, and environmental programs. J Forest 114(6):648–655. https://doi.org/10.5849/jof.15-090

Mulalo Rabumbulu (MA, BA Honours, BA, ADHE) is a lecturer in the Department of Geography, Environmental Management and Energy Studies in the Faculty of Science at the University of Johannesburg. She is a physical geographer, specialised in geomorphology. Her research interests are quaternary geomorphology, environmental management, land degradation and tourism geography.

Itumenleng Masithela (MA, BA Honours, BA) is a safety practitioner in occupational safety at the University of Johannesburg. She is an environmentalist, specialised in land degradation and climate change. Her research interests are environmental and natural resource management, climate change, sustainability, environmental justice and effects of climate change on the environment.

How Corporate Scientific Expertise and Privatised Seeds Are Destroying Indigenous Knowledge Systems for Food Security and Climate Change Mitigation: The Case of Sub-Saharan Africa

10

Llewellyn Leonard

10.1 Introduction

Climate change presents a forceful challenge to food security and sustainable development especially in Sub-Saharan Africa, which is food insecure (Makate 2020; Mafongoya and Ajayi 2017; Westengen et al. 2019). Climate change is expected to make agricultural development in Sub-Saharan Africa and across the continent more difficult as weather patterns become less favourable and increase the instability of crop yields (Woetzel 2020). In the wake of global climate change and its resultant challenges, it is critical that the knowledge of local and indigenous people, often referred to as indigenous knowledge (IK), become more profoundly recognised as an important source of climate knowledge and adaptation strategies (Westengen et al. 2019; Jiri et al. 2017; Mafongoya and Ajayi 2017; Mekbib et al. 2017). IK is knowledge that is place-based and deep-rooted in local cultures that are mostly associated with long-settled communities that have strong ties to their natural environments (Ponge 2013; Kronik et al. 2010; Orlove et al. 2010), which can regenerate local food systems while increasing socio-environmental sustainability and resilience (Food and Agricultural Organisation 2009). IK is noted to have the potential to transform the technocratic-community engagement front in the current discussions where climate science and policy regimes are increasingly being interrogated for their sustainability. It is useful to highlight adaptation governance, which centres on participation, equality and justice in decision-making about interventions to contain climatic events. It is recognised that if adaptation is to build community resilience and adaptive capacity, adaptation governance informed by IK and the experiences of locals, cannot be ignored (Chanza and de Wit 2016). It ranges from enhancing the understanding of climate impacts at local levels by increasing meaningful community participation in science (Chanza and de Wit 2016). This collective engagement will be important if Sub-Saharan African countries are to realised Sustainable Development Goal 13 to take urgent action to combat climate change and its impacts.

There are many threats to IK to mitigate the risks of climate change on food security. Such threats may include weak institutional capacities and poor governance, limited financial and technical resources necessary for climate change adaptation, and weak policies that favour toxic fertilizers and agrochemicals, which can make communities more vulnerable (Makate 2020).

L. Leonard (✉)
Department of Environmental Sciences, School of Ecological and Human Sustainability, College of Agriculture and Environmental Sciences, University of South Africa (UNISA), Johannesburg 1709, South Africa
e-mail: llewel@unisa.ac.za

E. E. Ebhuoma and L. Leonard (eds.), *Indigenous Knowledge and Climate Governance*, Sustainable Development Goals Series, https://doi.org/10.1007/978-3-030-99411-2_10

Although there has been a fair amount of literature that has attempted to examine the intersection between IK and scientific knowledge domains (see Arsenault et al. 2019; Makondo and Thomas 2018; Ford et al. 2016), there has been the limitedly explored threat in academic debates about the influence of capitalist political economies and associated scientific expertise promoting privatised seeds that have worked to destroy IK, despite such seeds working to encroach food security for indigenous peoples (Bellevue 2017). These unjust practices include re-engineering seeds (namely, making them drought resistant or having early ripening tendencies, thereby homogenising plant species and jeopardising the diversity of crops crucial to indigenous people), and then privatising seeds and controlling how food is produced. For example, according to Serumaga (2014), global agribusiness interests have silently engrained themselves in the information environment over the past decade, which has been rooted in European Enlightenment thinking looking to science for a solution and asserting a process that commands to discuss the matter on its 'scientific' and technical merits only. What is happening is a displacement of small-scale farming in favour of large-scale agriculture; the removal of IK and foods and the scientific modification and reliance on genetically modified 'expertise' (Serumaga 2014), which have implications for climate change mitigation and food security for local communities in Africa. According to Scoones and Thompson (2011), roughly 70–80% of the seeds planted in Africa are from IK and farm-saved seed, with privatised and genetically modified seeds amounting to approximately 38% in Southern Africa (excluding South Africa, which is the only country with a fully developed commercial seed system).

This displacement of IK and foods can be traced back to many decades. The Green Revolution was an endeavour by First World countries to export chemical United States style agriculture to the Third World since the late 1960s. Northern policymakers were concerned that rampant hunger created the basis for First World science to "help" the Third World by giving it the means to produce more food. The industrialisation of agriculture contributed to the replacement of farms with corporations. The same companies (for example, Aventis, Novartis, Syngenta, Monsanto, and DuPont) that endorsed chemical-based agriculture also introduced genetically engineered food and agriculture by converting seeds into product-delivery systems. The crops produced by Monsanto's Roundup-Ready brand seeds, for example, only tolerate the company's Roundup brand herbicide. These multinationals promise to create a world free of pesticides by equipping each crop with its own 'insecticidal genes' (Mittal & Rossert 2001).

In the contemporary Green Revolution narrative, the low adoption rates of privatised seed from multinationals, especially in Sub-Saharan Africa, is a major reason for the low rates of agricultural productivity seen on the continent. In this paradigm, the role of the private sector as a producer and distributor of seed is seen as key (Agricultural Centre for Biodiversity 2015). Additionally, for crop production to adapt to climate change and for food security, it is fundamental that farmers have access to well-adapted seeds. Unfortunately, private stakeholders will pursue the best market opportunities for their most profitable seeds produced, whilst, IK seed varieties may be more suitable and can increase food security and be used for climate change mitigation in Africa (Westengen et al. 2019).

For centuries, African small farmers have produced local seed varieties and managed, selected, improved, reproduced, stored, planted and swapped seeds with other neighbouring farmers in the local vicinity, at no cost, using their specific inter-generational knowledge and skills (Grain 2018; Agricultural Centre for Biodiversity 2015). This IK produces biodiverse, ecologically resilient seeds that can adapt to the changing climate along with many other challenges (for example, poor soil). This food production requires genetically biodiverse seeds that are selected by farmers every season to suit local ecosystems and can adjust through the farmers' active management to pressures such as climate change (Grain 2018). In parts of the world and in

Africa, corporations such as Monsanto, have patented seeds, driving farmers to buy these seeds and leading to a loss of IK and dictating how food is produced (see Bega 2019 as an example). Additionally, the multinationals have introduced pesticides that complement the seeds sold to farmers to increase profits. These pesticides also alter ecosystems that are vital for indigenous knowledge system applications.

By reviewing the recent work on privatised science and seeds in Sub-Saharan Africa, the effects on indigenous knowledge systems and impact on food security are explored. The link between privatised and genetically altered seeds and links with food security and climate change mitigation have not been comprehensively explored in academic debates. There seems to be a disconnect between scientific knowledge and IK when it comes to proposing and addressing issues of food security for climate change mitigation. Questions remain—should IK formations and locally selected seed variants be the only available option for food security for climate change mitigation? Or should a fusion of science with IK be a better option? If the latter is a better option, how should such a partnership be developed? This chapter attempts to address these questions by exploring the case of Sub-Saharan Africa and how this has played out across the landscape by drawing on different cases. Indirectly, this chapter aims to contribute to the United Nations Sustainable Development Goal 13 surrounding climate action so that communities and civil society generally may work to hold governments accountable for injustices related to food insecurity and the loss of IK formations. The chapter is divided into six sections, including this section. The second section briefly describes the foundation by exploring the roots of Sub-Saharan African food injustice and dispossession. The third section will explore science and the privatisation of seeds and implications for IK, food security and climate change mitigation. The fourth section will explore how IK can assist in mitigating the impacts of climate change on food security. The fifth section presents the debate for reflexive scientisation, integrated knowledge systems and against the privatisation of seeds and food. The final section engages in the discussion and conclusion of the chapter.

10.2 The Roots of Sub-Saharan African Food Injustice and Dispossession

Agricultural and social scientists have been aware of the existence of IK since colonial times (Khanyama-Phiri 2017). This was during the first interaction between Europeans and indigenous communities in Southern Africa. Jan Van Riebeeck, the founder and first Dutch commander of the European settlement at the Cape of Good Hope, and official for the Dutch East India Company, saw the Dutch (and Germans) settle in the area in 1652 (Speake 2003). The British gained control of the Cape of Good Hope in 1795, with colonial rule expanding northwards and eastwards. Due to British expansion, there emerged a long conflict between the Dutch and English, which saw the Dutch leave the Cape and settle in the Orange Free State, and Transvaal (Hajdu 2006). According to Brewer (2005), although black South Africans attempted to combat colonisation, the process of colonisation saw Africans lose their lands, crops, animals and their homes. Beinart et al. (1989), also note that during the pre-industrial era, the expansion of agricultural activities was closely dependent on the dispossession of the Khoisan and later Xhosa communities, which also saw the core labour force on most settler farms up to the early nineteenth-century being made up of slaves. This 'primitive accumulation' involved the expropriation of people's means of independence and forcing them into labour. The brutal history of primitive accumulation had significant support from the early masters of classical political economy, who saw the traditional system of self-dependence in the countryside as an impediment to capitalist production, which needed labour (Perelman 2007). The dispossession of African populations from lands for their food production gave rise to, and mirrored, relations of power and privilege.

Europeans were ignorant of the long established and successful ways in which indigenous black Africans ensured their survival by protecting the soils, plants and creatures forming a basic part of the texture and meaning of rural, non-industrial existence (Anderson and Grove 1987). As Adams and McShane (1996) note, humanity lived for thousands of years without depleting its ecosystems. Europeans disrupted this established balance excluding those whose lives were most affected (Anderson and Grove 1987). Anderson and Grove (1987) note, prescriptions for the environment from soil to water conservation have come mainly from Europeans. These 'experts' often lack critical sociological insights into the potential impacts of their programmes on local African populations. In their efforts to mirror Europe in terms of physical structure and social, political and economic organisations, white people became masters of the Southern African environment, dominating the environment with all its resources, possibilities and limitations. This laid the foundations and ideology for the further contemporary environmental and food injustices discussed below, where a prescription for food production has come from Western scientific domains, and in many cases, has led to privatisation of food systems and has not contextualised IK for food production and how it works to mitigate climate change.

10.3 Science and the Privatisation of Seeds: Implications for IK, Food Security and Climate Change Mitigation

The private seed industry has made dramatic gains in Sub-Saharan Africa in recent years (African Centre for Biodiversity 2015). Since 2006, numerous funding partners have become key actors in the agricultural space in Sub-Saharan Africa by establishing the Alliance for a Green Revolution in Africa (AGRA). Some of these funders include the Bill and Melinda Gates Foundation, United States Aid, United Kingdom Aid, United Nations Environment, The Rockefeller Foundation and the African Union, to name a few. Some of the private sector partners include Bayer (previously Monsanto), Syngenta, Microsoft and Nestle, who provide agribusiness linkage for smallholder farmers to access yield-enhancing inputs, post-harvest management, logistics and markets. AGRA has since established 119 African seed companies and is noted to work with Farmers Organisations, and Youth and Women Organisations on shared interests to increase the adoption of models and technologies for increased agricultural production and productivity; to create an enabling policy environment for inclusive agricultural transformation at national and state (districts/county levels); and to increase housed income for smallholder farmers while catalysing inclusive agricultural transformation (namely, seed, fertilizer, markets and agriculture finance) (AGRA 2020).

According to the African Centre for Biodiversity (2015), the expansion of the privatised hybrid maize use is problematic for farmers and IK systems. Privatised seeds need to be repurchased annually and the use of other agrochemicals is generally more costly and gives farmers less flexibility in their farming practices. Input subsidy programmes benefit multi-national seed companies. However, according to Wise (2020), it has been fourteen years since the Bill and Melinda Gates and Rockefeller foundations launched the AGRA to bring Africa its own Green Revolution in agricultural productivity. Armed with high-yield commercial seeds, fertilizers and pesticides, AGRA had the ambition of doubling productivity and incomes by 2020 for 30 million small-scale farming households, while reducing food insecurity by half in 20 countries. However, AGRA has not produced a productivity surge. Rather, many climate-resilient, nutritious crops have been displaced by the expansion of supported crops such as maize. The number of undernourished in AGRA's 13 focus countries, has increased 30% during the organisation's well-funded Green Revolution campaign. However, a 2020 report published by several stakeholders including the Rosa-Luxemburg-Stiftung and the INKOTA-netzwerk entitled "False

Promises: The Alliance for a Green Revolution in Africa", paints a different picture of AGRA and its foot prints in Zambia, Tanzania, Kenya and Mali. The research was also conducted by researchers from Tufts University. Some of the findings include that in countries in which AGRA operates, there has been a 30 percent increase in the number of people suffering from hunger. Maize, heavily promoted by Green Revolution programmes, showed a 29% yield growth, well short of AGRA's goal of 100%. Perhaps, the significant finding was that the commercialisation of seeds caused further erosion of food security and nutrition for poor small-scale food producers, where AGRA incentives for priority crops drove land use towards maize and away from more nutritious and climate-resilient traditional crops like millet and sorghum. Seeds for traditional crops, which were formerly easy and cheap to secure via farmers exchange, have to be paid for by farmers. Unfortunately, the study further found that the programme has resulted in negative environmental impacts, including acidification of soils under monoculture cultivation, with fossil fuel based synthetic fertilizers negatively impacting aspects of climate change mitigation and adaptation. Overall, farmers have lamented about imposing Western technologies that were unsuitable for the continent's soils, farmers, and food systems, and the lack of consultation with farmers about the interventions.

Considering the impact of privatised seeds on African farmers, it is therefore surprising that the African Union (AU) has been one of the funders of AGRA. The AU was formed in 1963, and by July 2011, it had 54 member states to promote the unity and solidarity of African states and to coordinate and intensify their cooperation and efforts to achieve a better life for the peoples of Africa (United States Aid 2017). Additionally, the AU centres on Africans controlling their destiny and recognising equality and justice, including the communal characteristic and drive to embrace Africa's culture and common heritage (AU, n.d.). Social scientists warn that privatised and genetically modified seeds are being pushed by international donors without fully taking into account the agriculture practices of poor farmers and the specific crop characteristics that they most want and need (Paliwal 2019). Under capitalist political economics, there is a tendency for knowledge domains to become a commodity so that innovations and technologies protected under intellectual property rights give those with particular knowledge domination opportunities for profit (Jessop 2007), which suggests moving away from a collective IK domain system for the common good. On the other hand, Nakashima et al. (2000) note that IK, which is transmitted orally, is holistic as knowledge is strongly rooted to the earth and values all knowledge forms and human experiences. This type of IK has been generally marginalised by Western structures and forms of domination (Barnhardt and Kawagley 2005).

10.4 How IK Can Assist in Mitigating the Impacts of Climate Change on Food Security

Understanding that crops are subject to erratic weather events, indigenous communities and farmers have customarily used IK to favour the farming of varied crop diversities and have domesticated, enhanced and preserved many hundreds of crop types over time for food security (Kalanda-Joshua et al. 2011). A major challenge for crop production, especially due to commercialisation, is due to, not only privatised and 'tampered' seeds, but farmers' use of high level chemicals, reduced diversity of the cropping systems and controlling nature to a high degree (Ponge 2013). On the other hand, IK can be used to tackle the emerging problems of soaring food prices and climate change, and tap into vast knowledge of previous variations in climate and for mitigation and adaptation strategies for ensuring food security (Food and Agricultural Organisation 2009). For example, Godfrey et al. (2009) examined the role of indigenous knowledge in enhancing household food security in Uganda by focusing on the Mukungwe sub-county, Masaka district. The authors found that many households mulched their crops using local materials and used locally concocted pesticides to control pests both in the field and in the storehouse. Additionally, some households developed a technique of burying certain foods like fresh cassava tubers in moistened soil to last longer. These practices enhanced household food security and lessened the need to buy food.

With IK, the promotion and use of underutilised crops can also provide adaptation opportunities for increasing productivity for enhanced climate change resilience through crop diversification and genetic variation. Access to a diversity of local crops is necessary for farming communities. Hence, on-farm conservation of natural genetic resources via IK can assist in addressing the problems of climate change. Their adaptability and resilience to stresses have the potential to provide farmers with coping strategies in the face of climate change. The genetic diversity created through such practices has traditionally allowed farmers to cope and adapt to adverse climatic conditions, and this process will continue to serve that function in future (Jiri et al. 2017). For example, in the mountains and foothills of Lesotho, farmers have been experimenting and carrying out farm trials with different crops since land productivity has been low and subject to fluctuations. Trials help farmers to establish which crops can cope best with the shifting sowing season with climate change constraints overcome by using a simple technique based on intercropping and crop diversification, with the added localised application of organic fertilisers such as manure and ashes. Some crops have also been shown to act as natural repellents to certain insects, and the intercropping practice has contributed to pest control. The rotation of crops assists in breaking the life-cycle of insect pests and regular weeding throughout the year has been proven to control pests and diseases (Mekbib et al. 2017).

Farmers grow a wide-range of crops to mitigate climatic and marketing risks (Lynam 2011). Most rural areas in Africa have always experienced climate variability, and farmers have production systems of crops, perennial fruits and trees around the homestead or in the vicinity of rivers and forests, and incorporate a variety of staple crops for managing risks. Farming systems incorporating perennial fruit trees such as mangoes, not only provide options for household food supply but household nutritional diversity as well. These systems are designed to spread risks and vulnerability in the face of unpredictable climatic events. Additionally, in countries such as Botswana and Zambia, IK has been using intra-cropping diversity for varieties of pearl millet and sorghum, which have remained more or less constant throughout the dry periods for the last few decades (Jiri et al. 2017). IK is also used for changing settlement patterns and food and agricultural practices to adapt to, or mitigate, such climate change impacts as droughts, floods, sea level rises, changes in temperature, flora, fauna or water conditions, and outbreaks of pests and diseases (Food and Agricultural Organization 2009). For example, African indigenous small-scale farmers have over centuries of knowledge formations, and have developed local strategies of adaptation to climate change for food security (Ajani et al. 2013). These adaptation strategies include knowledge of behaviours of living entities such as insects, and fauna and flora, which are used as indicators of climate change, and of the appropriate plant varieties and agricultural practices (for example, mixed cropping practices) to cope with climate changes. Crops are harvested at different periods of the year to ensure food security (Agrawal 2003). The introduction of modern scientific systems and knowledge formations for food security in Africa has not stopped most people to still rely on their local IK systems for food security and to adapt to changing climatic conditions (Morrison and Ostry 2008). With indigenous production methods and farming systems, due to their ability to plant more than one crop using locally adapted seeds, farmers who use traditional seeds own the bulk of what they produce and are food sovereign (Deutsche Welle 2018).

10.5 Theoretical Framework: Reflexive Scientisation, Integrated Knowledge Systems and the Privatisation of Seeds and Food Insecurity

Reflexive scientisation can be a way in which Western scientific knowledge systems do not dominate indigenous knowledge formations that hamper food security and climate change adaptation. Reflexive scientisation has been extensively applied to IK. The concept reflexivity concerns self-confrontation and non-knowledge (Beck 2009) and where scientific expertise and dominate institutions are no longer treated as infallible but is both internally and externally scrutinised, what Beck calls "reflexive scientisation". According to Beck (1992), risks that move to the centre stage (namely, in this case, Western private science and seed production working to destroying IK for food security and climate change adaptation), may dissolve trust in science and their dominant institutions. When facing risks, as highlighted in cases across Sub-Saharan Africa, space is opened for people (namely, indigenous farmers, civil society and even African governments) to scrutinise science and dominant institutions and develop counter-expertise (namely, sub-politics). This means that farmers and civil society work to protect IK and need to counter the privatisation of seeds and the resultant food insecurity created. Here, it is relevant to draw on the framework provided by Ottinger and Cohen (2011) concerning how those promoting certain scientific systems can alter their practices in response to the public outcry about the way things are done, with scientific experts responding to the demands of affected users, which makes choices available to them to multiply and broaden the expert's room to manoeuvre. Disruption can be twofold, with transformation occurring because of the expert's deliberative and strong engagement with society or because of dominant institutional practices by outside pressure to incorporate concerns. Kunseler (2016) suggests that manoeuvres can increase the options available to inform scientific advice. Unfortunately, those with scientific expertise (namely, Aventis, Novartis, Syngenta, Monsanto, and DuPont) who are devoted to the redesigning and privatisation of seeds logic can fail to perceive failures at the ground in communities when they promote privatised seeds as the solution to food insecurity and for climate change adaptation. Thus, replacing the modernist logic with a reflexive one is often difficult and viewed as subversive for the knowledge-power nexus implicated in institutionalised forms of scientific advice (Kunseler 2016).

According to Leonard and Lidskog (2020), an interplay exists between reflexive scientisation, risk communication, knowledge integration and trust. The problem of trust can be viewed as fundamental in the undercurrents that occur between private institutions that may thrust their engineered seeds onto farms that have carried out IK techniques to combat climate variation and food insecurity for decades. Dominate institutions and their scientific systems may be viewed as inflexible, with pre-established skill sets (Ottinger and Cohen 2011). Trust is not only a relational process based on experience, but is also based on an incalculable process (for example, a participatory process) that is viewed as credible by all parties and whereby public concerns are genuinely taken into account (Leonard and Lidskog 2020). Horlick-Jones (2004) finds a need for 'alternative expertise' that includes the everyday experience and interest of people. This is important in the context of many Sub-Saharan African countries where local communities have engaged in indigenous farming for centuries and have indigenous farming scientific expertise. This alternative expertise should provide insights into information about the limits of Western scientific knowledge, their communication strategies and opportunities for reassessing their dominant knowledge formations, and also how they interact with society to solve problems (Kunseler 2016; Sankowska and Soderlund 2015; Engdahl and Lidskog 2014).

Sullivan and Lloyd (2006) refer to 'community orientated environmental sciences', which is a community/science interface to increase the effectiveness of interventions, and strives to integrate peer review science with community actors' knowledge systems to educate, mobilise and defend constituencies. Although different stakeholders may emphasis different solutions and stories, leading to frame differences (Kunseler 2016), knowledge integration requires the combination of knowledge. Whilst trust may facilitate the exchange of knowledge, it may not automatically facilitate the integration of knowledge, which requires reflexivity to facilitate knowledge amalgamation (Sankowska and Soderlund 2015). Within the context, conceptions of reflexive scientisation, knowledge integration and trust can be important lessons for outside interventional actors when engaging with local communities and farmers in Sub-Saharan Africa. Unlike the practices of multinationals in Sub-Saharan Africa, who develop and privatise seeds, and who have prescribed these seeds as solutions to food insecurity and climate adaptation, to the exclusion of IK systems, reflexive scientisation can be used to rather support IK systems to better adapt to climate change and improve food security. This will ensure that IK systems are preserved and can be passed on to future generations The approach here is to ensure that Western scientific and corporate interventions in Sub-Saharan Africa work to enable mutual learning and deliberation with communities to support and further enable IK systems and practices.

The perspective on knowledge integration must therefore ensure that knowledge is used to support society and that seeds are not re-engineered and privatised for profits. As Ponge (2013) highlights, focusing on Kenya, it is clear that IK on its own cannot be a solution to crop production shocks. Therefore, it is reasonable to advocate for a 'marriage of convenience' between the two knowledge systems for effective output and eventual sustainability. However, it is important to appreciate that the two are complementary in their strengths and limitations and by working jointly, each may achieve what neither could alone. Therefore, there is a need to value indigenous peoples' knowledge and practices and build on these assets by supporting an association that blends traditional knowledge and practices with modern scientific approaches. For example, according to the ISSD Africa (n.d), agrobiodiversity (also known as Agro-ecology), are efforts that can strengthen national seed sectors to use crop diversity for sustainable agricultural development and resilience. This system offers farmers an awareness of nature and natural processes, which promote the soil-building practices that Green Revolution practices often undermine. Various food crops are grown in the same field. Organic fertilizers, not fossil-fuel-based synthetic fertilizers, are used in the fields. Biological pest control decreases pesticide use. Researchers work with farmers to increase the output of their seeds rather than substituting them with commercial hybrid seeds. The seeds are free to use and grow. Providing farming communities with information and options on seeds and varieties is proven to boost the demand and use of quality seeds, as well as promote crop diversity and support community seed banks. A number of digital tools can be utilised to provide more information in the hands of farmers. Thus, Western science may be used in this way to support the efforts of indigenous farming practices. As Shonhai (2016) also highlights, although IK needs technological support, the control of the seeds must remain with farmers and communities who must also participate in policy formulation if they are to be able to control, share and breed seeds.

10.6 Discussion and Conclusion

Climate change is expected to make agricultural development in Sub-Saharan Africa and in other places more problematic as weather patterns continue to change and increase the uncertainty of crop yields. In the wake of these challenges, IK will remain an important source of climate knowledge to address challenges and apply appropriate local context adaptation strategies. Despite the importance of IK systems for climate mitigation and food security, this chapter suggests that a number of impediments to IK have become apparent in Sub-Saharan Africa as in other places. This is particularly in the last decade, due to Western corporate scientific expertise and the privatisation of seeds working to supersede IK systems. These include power imbalances and difficulties originating from twofold tensions between Western private science and corporations confined within a capitalist political economy to secure profits, and secondly, with indigenous knowledge systems remaining decontextualised. The private seed industry has made dramatic gains in Sub-Saharan Africa in recent years with numerous funding partners becoming key actors in the agricultural space. On the other hand, IK has provided adaptation opportunities for increasing productivity for enhanced climate change resilience through crop diversification and genetic variation. The genetic diversity created through such practices has traditionally allowed farmers to better cope and adapt to adverse climatic conditions. Unfortunately, international corporations market privatised seeds as a better solution to food insecurity and climate change adaptation. Local communities, due to a lack of awareness, may adopt the privatised seed logic over IK; however, they may regret it later on.

The chapter highlighted that unfortunately, there seems to be a lack of adaptation governance to protect IK systems and guard against the commercialisation of farming practices and privatisation of seeds and farming systems. It has been surprising that the AU has been one of the funders of the AGRA, which has worked to remove IK systems and replace it with a commercial and privatised system. This is despite that the AU was formed to promote the unity and solidarity of African states and to coordinate and intensify cooperation and efforts to achieve a better life for African communities. The practices of the AU to support privatisation of seeds and fully support Western scientific knowledge and commercial farming, goes against the AU's principles of Africans controlling their destiny and recognising equality and justice, and preserving culture and common heritage, such as IK systems. The AU has not moved towards an adaptation governance strategy centred on participation, equality and justice in decision-making about interventions to mitigate climatic events. The paper highlighted that if adaptation is to build community resilience and adaptive capacity realised, then adaptation governance informed by IK and experiences of locals cannot be ignored.

Therefore, meaningful community participation in any endeavours must be prioritised if countries are to enable Sustainable Development Goal 13 and take urgent action to combat climate change and its impacts. What this means is that African governments, via the AU, must genuinely incorporate IK into development and the decision-making process so that indigenous knowledge systems are contextualised for adapting to climate change for food security. This will ensure that there is no imposition of Western knowledge systems that are unsuitable for the continent's soils, farmers, and food systems. This will also enable indigenous communities to protect IK systems so that food security and climate mitigation are better operationalised. Any modern scientific interventions (such as AGRA) must work to support IK systems rather than replacing them. Evidence suggests that AGRA has not produced a productivity surge to combat food insecurity as it initially claimed. Rather, many climate-resilient, nutritious crops have been displaced by the expansion in supported crops such as maize. The number of undernourished people in AGRA's 13 focus countries, has increased and proliferated food insecurity, which has been supported and enabled by poor African visionary leadership and a lack of appropriate policy visions to support IK systems. On the other hand, with IK, farmers have been able to use organic seeds to act as insect repellents and utilise wisdom from experience to aid this, and, through intercropping practices, control pests. Thus, it will be up to civil society networks and farming communities to engage in education and awareness about the importance of IK systems and to push governments and the AU for de-commodification and the de-privatisation of nature. To realise the United National Sustainable Development Goals, particularly, Goal 13 on climate action will also require that the Sub-Saharan and African populations hold their governments accountable and are supported via a pan-African network, to protect and preserve IK.

In addition to calls for a move away from privatisation of seeds and the commercialisation of farming systems, the chapter highlighted how corporations have also used science to manipulate seeds as a way of mitigating climate impacts. However, these developments have not produced the desired results. Although the importance of science is acknowledged in being used to improve upon developments in society, reflexive scientisation is proposed as an outlet that is supportive of indigenous knowledge systems that incorporate participatory democracy. Farmers are the primary agents in farming systems and must be treated as equal partners in any attempts to support their farming practices. Reflexive scientisation can ensure that scientific knowledge systems do not dominate indigenous knowledge formations to hamper food security and climate change adaption. Unfortunately, scientific corporate expertise (namely, Aventis, Novartis, Syngenta, Monsanto, and DuPont) that have worked to engineer and privatise seeds, have failed to perceive failures at the local level, which includes the destruction of IK systems and crop diversity to combat climate change and food insecurity. There is a need to replace their modernist logic with a reflexive one that rather uses science to support IK and for improved food security and climate mitigation.

Therefore, engagement in 'alternative expertise' that emphasises IK is important in the context of Sub-Saharan Africa where local communities have engaged in indigenous farming for centuries. This engagement will provide insights into information about the limits of modern scientific expertise and their dominant knowledge formations, and provide appropriate local mitigation measures to enhance food security. Alternative expertise can contribute to 'food democracy', which is defined by Friedrich et al. (2019) as 'the possibility for all social groups to participate in, negotiate and struggle over how societies organise agricultural production, thereby ensuring that food systems fulfil the needs of people and [a] sustain (re)productive nature into the future'. Such an engagement can work to increase trust between local communities with scientists. Thus, reflexive scientisation can be used to support IK systems to better adapt to climate change and for improved food security. An association that blends traditional knowledge and practices with modern scientific approaches must be supported. Finally, if corporations want to assist farmers, then investment will be key. As Woetzel (2020) notes, modernising Africa's agriculture in the face of a changing climate will require significant investment. Investments in irrigation can increase the likelihood that farmers maintain yields even when the weather is unfavourable. Improvements in the selection of local seed production systems would offer farmers different varieties of seed that are appropriate to changing conditions.

References

African Union (n.d.). https://au.int/en/overview. Accessed 22 April 2021

Alliance for a Green Revolution in Africa (2020). https://agra.org/focus-countries/. Accessed 25 March 2021

Anderson D, Grove R (1987) Conservation in Africa. Cambridge University Press, Great Britain

Arsenault R, Bourassa C, Diver S, McGregor D, Witham A (2019) Including indigenous knowledge systems in environmental assessments. Global Environ Pol 19(3):120–132

Barnhardt R, Kawagley AO (2005) Indigenous knowledge systems and Alaska native ways of knowing. Anthropol Educ Quart 36(1):8–23

Beck U (1992) Risk society. SAGE Publications, London, California and New Delhi

Beck U (2009) World risk society. Polity, Cambridge

Bega, S. (2019). Agriculture Minister says no to Monsanto's "drought tolerant" maize seed, *Independent Online*, 4 October.

Beinart W, Delius P, Trapido S (1989) Putting a plough to the ground: accumulation and dispossession in rural South Africa, 1850–1930. Raven Press, Braamfontein

Bellevue C (2017) GMOs, international law and indigenous peoples. Pace Int Law Rev 30(1):1–42

Brewer A (2005) Under Siege: apartheid in South Africa, Codman Academy Research

Report. http://codmanacademy.org. Accessed 23 March 2013

Chanza N, de Wit A (2016) Enhancing climate governance through indigenous knowledge. South African J Sci 112(3/4), Art. #2014–0286, 7. http://dx.doi.org/10.17159/ sajs.2016/20140286

Deutsche Welle (2018) Resistance to genetically modified seeds in Africa, 23 July. https://mg.co.za/article/2018-07-23-resistance-to-genetically-modified-seeds-in-africa/. Accessed 14 Jan 2021

Engdahl E, Lidskog R (2014) Risk, communication and trust. Public Underst Sci 23(6):703–717

Food and Agricultural Organization (2009) FAO and traditional knowledge. http://www.fao.org/3/i0841e/i0841e.pdf. Accessed 25 March 2021

Ford J, Cameron L, Rubis J, Maillet M, Nakashima D, Willox A, Pearce T (2016) Including indigenous knowledge and experience in IPCC assessment reports. Nat Clim Chang 6:349–353

Friedrich B, Hackfort S, Boyer M, Gottschlich D (2019) Conflicts over GMOs and their contribution to food democracy. Polit Gover 4:165–177

Godfrey J, Lugangwa E, Obua J, Kambugu R (2008) Role of indigenous knowledge in enhancing household food security: a case study of Mukungwe, Masaka District, Central Uganda. Sabinet African J. Published Online: 1 June 2008. https://hdl.handle.net/10520/EJC61539.

Hajdu F (2006) Local worlds: rural livelihood strategies in eastern cape, south Africa. Thesis (PhD), Linkoping University, Department of Water and Environmental Studies, Tema Institute

ISSD Africa (n.d) https://issdafrica.org/about/. Accessed 12 April 2021

Jessop B (2007) Knowledge as a fictitious commodity. In Buğra A, Ağartan K (eds) Reading Karl Polanyi for the 21st century. Market Economy as a Political Project. Basingstoke: Palgrave, pp 115–134

Jiri O, Mafongoya R, Musundire R (2017) The use of underutilised crops and animal species in managing climate change risks. In: Mafongoya PL, Ajayi OC (eds) Indigenous knowledge systems and climate change management in Africa. CTA, Wageningen, The Netherlands

Kanyama-Phiri G, Wellard K, Snapp S (2017) Agricultural systems, 2nd edn. Elsevier

Kronik J, Verner D, Mearns R, Norton A (2010) The role of indigenous knowledge in crafting adaptation and mitigation strategies for climate change in Latin America. Social Dimensions of Climate Change, The World Bank, Washington DC, p 145

Kunseler E (2016) Revealing a paradox in scientific advice to governments. Palg Commun 2:16029. https://doi.org/10.1057/palcomms.2016.29

Leonard L, Lidskog R (2020) Examining reflexive scientisation and scientific expertise in South Durban, South Africa. J Risk Res. https://doi.org/10.1080/13669877.2020.1805638

Lynam J (2011) Plant breeding in sub-Saharan Africa in an era of donor dependence. IDS Bull 42(4)

Makate C (2020) Local institutions and indigenous knowledge in adoption and scaling of climate-smart agricultural innovations among sub-Saharan smallholder farmers. Int J Climate Change Strat Manag 12 (2):270–287

Mekbib S, Olaleye A, Johane M, Wondimu T (2017) Indigenous knowledge to address the challenges of climate change: case of Machobane farming system in Lesotho. In: Mafongoya PL, Ajayi OC (eds) Indigenous knowledge systems and climate change management in Africa, CTA, Wageningen, The Netherlands

Mittal A, Rosset P (2001) Genetic engineering and the privatization of seeds, dollar and sense magazine, March/April issue (234). http://www.dollarsandsense.org/archives/2001/0301toc.html. Accessed 10 Jan 2021

Makondo C, Thomas D (2018) Climate change adaptation. Environ Sci Policy 88:83–91

Morrison GR, Ostry A (2008) Developing and utilizing a database for mapping the temporal and spatial variation in the availability of “local foods” in British Columbia. Environ J 36(1):19–31

Nakashima D, Prott L, Bridgewater P (2000) Tapping into the world’s wisdom, UNESCO Sources, 125, July-August, p 12

Orlove B, Roncoli C, Kabugo M, Majugu A (2010) Indigenous climate knowledge in southern Uganda. Clim Change 100:243–265

Ottinger G, Cohen B (2011) Techno science and environmental justice: expert cultures in a grassroots movement. MIT Press, United States

Paliwal A (2019) How a fierce debate over GMO could determine the future of agriculture in Africa, Fast and Company, 8 February

Perelman M (2007) Articulation from feudalism to neoliberalism. J Devel Stud 2 Africanus 37(2):22–38

Ponge A (2013) Integrating indigenous knowledge for food security. https://elibrary.acbfpact.org/acbf/collect/acbf/index/assoc/HASH0125/f03e1475/4cc7afe4/0524.dir/Integrating%20Indigenous%20Knowledge%20for%20Food%20Security.pdf. Accessed 20 April 2021

Sankowska A, Soderlund J (2015) Trust reflexivity and knowledge integration. Human Relat 68(6):973–1000

Shonhai V (2016). Analysing South African indigenous knowledge policy and its alignment to government’s attempts to promote indigenous vegetables, Ph.D. thesis, University of KwaZulu-Natal

Sullivan J, Lloyd R (2006) The forum theatre of Augusto Boal. Local Environ 11(6):627–646

Wise T (2020) Failing Africa's farmers: new report shows Africa's green revolution is "failing on its own terms". Institute for agriculture and trade policy. https://www.iatp.org/blog/202007/failing-africas-farmers-new-report-shows-africas-green-revolution-failing-its-own-terms. Accessed 22 Feb 2021

Woetzel J, Pinner D, Samandari H, Engel H, Krishnan M, McCullough R, Melzer T, Boettiger S (2020) How will African farmers adjust to changing patterns of precipitation? McKinsey and Company, 18 May

Llewellyn Leonard (Ph.D.) is Professor at the Department of Environmental Science, School of Ecological and Human Sustainability, University of South Africa. Before joining academia, he worked for a human rights environmental organisation working to support vulnerable communities exposed to environmental risks to shape policy and has worked to link communities in Africa, Asia, Latin America and Europe to engage in solidarity and networking. Previously he served as Vice Dean: Research at the University of Johannesburg, College of Business and Economics. His research interest include environmental justice, governance, democracy and human rights, civil-society-state-industry relations, urban risks, sustainable development; risk society and political economy/ecology to name a few. He sits on the editorial board of the Southern African Geographic Journal.

Part III

Integration of Indigenous Knowledge Systems in Climate Change Governance and Planning

Prospects for Strengthening Adaptation Governance Through Indigenous Knowledge Systems

11

Nelson Chanza, Walter Musakwa, and Anton de Wit

11.1 Introduction

Climate change remains a vexing governance issue. This is because of the complexity of the climate change phenomenon itself (Deere-Birkbeck 2009; Termeer et al. 2013), the impacts associated with it (Li 2016; IPCC 2018; Ishtiaque 2021) and the rigidity of existing institutions to accommodate the policies and activities required to tackle climatic challenges in an integrated manner (Meadowcroft 2009; Cadman 2013; Huitema et al. 2016). Given the global nature of climatic impacts, involvement of all actors from different sectors and at different levels of governance is critical for effective climate change responses (Ishtiaque 2021). Meadowcroft (2009) reveals that the international and national administrative architecture, which evolved to address other sorts of problems, are now challenged to handle emerging requirements of climate change mitigation and adaptation. There is now growing acknowledgement that the changing climatic conditions require transformation in institutions at multiple scales (Knieling and Leal Filho 2013; Huitema et al. 2016). This envisaged transformation has put the notion of climate governance in the spotlight. Frohlich and Knieling (2013) define climate governance as a broad spectrum of action and coordination around the management of climate change to acceptable societal expectations. Climate change governance requires continuous interrogation as the mitigation and adaptation policy regimes evolve. Many authors agree that the governance required in climate change management should be about disclosures on greenhouse gas (GHG) and collective effort to curtail emissions, while paying attention to the justice about the negative impacts or benefits associated with climatic impacts (Cadman 2013; Frohlich and Knieling 2013; Ross et al. 2013; Gupta 2016; Huitema et al. 2016; Ishtiaque 2021). It is also important to understand how the governance processes can be used to drive sustainable development goal (SDG) 13 that calls for urgent action to tackle climate change. Frohlich and Knieling (2013) point out that the complex interrelationships between stakeholders and societal coordination processes require proper governance. This means that the governance of climate change incorporates a multitude of structural and regulatory forms across a variety of different stakeholders.

N. Chanza (✉) · W. Musakwa
Department of Urban and Regional Planning, University of Johannesburg, Johannesburg 2006, South Africa
e-mail: nchanza@gmail.com

W. Musakwa
e-mail: wmusakwa@uj.ac.za

A. de Wit
Department of Geosciences, Nelson Mandela University, Port Elizabeth 6031, South Africa
e-mail: anton.dewit@mandela.ac.za

E. E. Ebhuoma and L. Leonard (eds.), *Indigenous Knowledge and Climate Governance*, Sustainable Development Goals Series, https://doi.org/10.1007/978-3-030-99411-2_11

Debate on how to translate the concept of climate governance into practice continues to evolve alongside global interest to address climate change impacts. Many scholars argue that the concept is a theoretical fantasy that cannot be put into praxis (Van Vossole 2012; Cadman 2013; Gupta 2016) because of its complexities. In essence, the practical challenges associated with operationalising the concept arise from the multi-level and multi-sectoral nature of climate governance (Jagers and Stripple 2003; Li 2016; Ishtiaque 2021). Van Asselt (2007) and Cadman (2013) corroborate that the management of climate change is a fragmented affair, making it difficult to realise governance consensus. Acknowledging the role of states in exercising authority within their boundaries, Cadman (2013:2) adds that "climate change governance is simultaneously global, regional and local, state and non-state, and it is characterised by the existence of many forms of authority through which different constellations of actors interact to shape policy outcomes." This implies that the concept is difficult to operationalise. Given these operational challenges, some scholars prefer a fragmentation of climate governance into mitigation and adaptation governance (Meadowcroft 2009; Termeer et al. 2013). Mitigation governance, therefore, is about coordinated interventions to curtail the cause of climate change, while adaptation governance demands ethical and justice reflections in decisions and activities to respond to the impacts of climate change. The climate governance system is also made complex by a fragmentation of efforts and networks (Cadman 2013; Huitema et al. 2016; Ishtiaque 2021). Cadman (2013) argues that this fragmentation might work as an advantage in addressing the governance of climate change.

This chapter adopts a fragmented approach to articulating climate governance and attempts to decipher the climate governance intricacy by narrowing the analysis to adaptation governance involving the use of indigenous knowledge (IK) of African communities affected by climate change impacts. A fragmentation approach acknowledges the separate treatment of mitigation and adaptation while at the same time recognises the various networks and processes that can facilitate governance of climate change (Ross et al. 2013; Termeer 2013; Huitema et al. 2016). Arguably, this preference is considered more appropriate in an African context, which favours an adaptation trajectory. The reasons for this orientation are later explained in this chapter. In this context, the discussion on how to operationalise adaptation governance proceeds by drawing on the applicability of indigenous knowledge systems (IKS), particularly the inclusion of local indigenous communities in Africa who experience climate change impacts in their highly vulnerable environments. The application of IKS is considered more appropriate at local level where communities experience climate change impacts (Sekine et al. 2009; Chanza and de Wit 2016; McNamara and Buggy 2017; Nalau et al. 2018). A critical question that arises in this regard is the role that indigenous people with their IKS can play in the adaptation processes. This is a critical argument given that prescriptions for adaptation that are driven by exogenous science may further disfranchise the already marginalised communities owing to epistemological injustices highlighted by Chanza and de Wit (2013). Against a background of existing heterogeneous terminologies and definitions characterising IKS, this chapter conceptualises it as long-term empirical knowledge held by indigenous communities from their observations of changes happening in their environments and their experiences in using the knowledge to survive under changing climatic conditions. This conceptualisation has an adaptation focus, which is the theme pursued in this chapter. Governance, on the other hand, involves the open engagement of various stakeholders in problem resolution processes (Frohlich and Knieling 2013).

The chapter proceeds as follows. The next section gives an overview of the policy framework that governs the governance of global climate change. This is followed by a narrowed focus to adaptation governance, a level that creates space for meaningful engagement of indigenous populations. The following section then provides a discussion on prospects of strengthening climate change adaptation through the use of IKS, which the chapter argues can be a way to put adaptation governance in practice, and to transform adaptation practices towards resilient indigenous communities in Africa. The chapter concludes by giving pointers on what it means about driving the adaptation governance agenda involving indigenous people at risk of growing climatic threats.

11.2 Policy Framework for Climate Change Governance

The governance of climate change continues to attract scholarly and policy interests at various scales. As a concept, governance is generally understood as involving the shift from the traditional nation-state architecture to a more inclusive concept that recognises the contributions of various structures at global, international, regional and local, and the roles of non-state actors and other stakeholders (Intergovernmental Panel on Climate Change (IPCC) 2018). With reference to climate change governance, Cadman (2013) notes that the concept is made more difficult by the growing numbers of new organisations, mechanisms and proposals relating to policy development and implementation at various levels such as local, regional and global. The intellectual framework that deliberates on climate change is given by the IPCC, while the policy framework governing the same is given by the United Nations Framework Convention on Climate Change (UNFCCC). As the primary international and intergovernmental forum for negotiating the global response to climate change, the UNFCCC is challenged to articulate the climate governance agenda. The UNFCCC apparatus has also developed National Adaptation Programme of Action (NAPAs) and National Adaptation Plans (NAPs). These mechanisms are intended to guide the implementation of adaptation activities at national level (UNFCCC 2012). Sethamo et al. (2020) hint that successful implementation of these mechanisms would require significant levels of participatory engagement with local people affected by climate change at local level. The 2015 Paris Agreement also recognises the critical importance of meaningful community participation and engagement in climate change adaptation processes (UNFCCC 2016). However, these global frameworks have not managed to address the climate governance question since their establishment about three decades ago. In other words, they fall short of addressing the unique climatic challenges that most communities experience at local levels. Since its establishment in 1991, the thrust of the IPCC has been on climate change mitigation, whose focus is largely on stabilising GHGs. This bias has raised ethical, equity and justice issues given the limited contribution by developing countries to the atmospheric destabilisation from GHG emissions (Nyong et al. 2007).

Under the auspices of the UNFCCC, the threat of climate change has been pitched to Sustainable Development Goals (SDGs). In particular, SDG 13 specifies that the global community needs to 'take urgent action to combat climate change and its impacts.' How this action is to be taken is specified in the following list of indicators (UN 2016:17):

- "Strengthen resilience and adaptive capacity to climate-related hazards and natural disasters in all countries.
- Integrate climate change measures into national policies, strategies and planning.
- Improve education, awareness-raising and human and institutional capacity on climate change mitigation, adaptation, impact reduction and early warning.
- Implement the commitment undertaken by developed-country parties to the UNFCCC to a goal of mobilising jointly $100 billion annually by 2020 from all sources to address the needs of developing countries in the context of meaningful mitigation actions and transparency on implementation and fully operationalise the Green Climate Fund through its capitalisation as soon as possible.
- Promote mechanisms for raising capacity for effective climate change-related planning and management in least developed countries and Small Island Developing States, including focusing on women, youth and local and marginalized communities."

These targets are clear about how the climate change management agenda should be conducted, from mainstreaming of climate change into development programmes and disaster management, integration of climate change into national policies, education and awareness raising, resource mobilisation and transparency in resource allocation, to the inclusion of marginalised communities and groups. However, there are already indications that climate change will affect the realisation of the SDGs and that some of the SDG targets will not be met by 2030 (Leal Filho et al. 2020). This means the climate stabilisation agenda and the need to address climate change impacts, which SDG 13 envisage to address, may not be realised.

Van Vossole (2012), Cadman (2013) and Huitema et al. (2016) query the legitimacy of the global climate governance regime. They argue that climate change impacts will unevenly affect the vulnerable, especially the poor and marginalised groups, the majority of which live in developing countries. Climate change remains a concern given the observed and anticipated increase in the frequency and intensity of extreme weather events such as droughts, floods and tropical cyclones and heat waves, including threats on water resources, agriculture and food security systems, health, human security, infrastructure and human settlements, and interruptions in the provision of basic services such as water and sanitation, education, energy and transport (IPCC 2018). A worrying situation is that there is a climate risk divide in which developing countries are affected the most. Accordingly, the stance taken by Africa is about considering adaptation priorities (Codjoe et al. 2014; Chanza 2015; Omuko 2015).

11.3 Focus on Adaptation Governance

The complexity of climate change impacts and the challenges of mitigation require that adaptation be considered as a primary policy objective (Cockfield 2013). Global priorities focus on mitigation while local realities and challenges require adaptation. Proponents of environmental justice have been advocating for vulnerability and adaptation considerations, which are more pertinent in the sub-Saharan Africa (Ross et al. 2013; Omuko 2015; Huitema et al. 2016; McNamara and Buggy 2017). Huitema et al. (2016) refer to adaptation governance as the patterns that emerge from the governing activities of social, political, and administrative actors during the processes and activities of adapting to climate change. This policy paradigm can be considered more appropriate in the context of Africa where climate change impacts tend to be devastating because of the region's high vulnerability as a result of poverty (IPCC 2018). Against the backdrop of the apparent failure of the international mitigation agenda, focus is now shifting the climate governance imperative beyond international and national levels to include adaptation regimes at levels where communities experience climate change impacts. It is at such scales that meaningful engagement of people in climatic responses can be realised, which can be one way of localising SDG 13 implementation. Omuko (2015) indicate that the level appropriate for the participation of local communities is at district or county level, while Chirisa and Chanza (2009) show that local government structures are best positioned to deal with climate change governance issues at finer scales such as census tract level. Raymond et al. (2010) opine that to manage the complexity of environmental problems, particularly climate change, it is critical to embrace participatory research methods that seek to facilitate participatory, multi-level governance processes intended to validate knowledge and to increase involvement of stakeholders in decision-making. The rush by climate change practitioners to address governance issues has seen the mushrooming of approaches such as vulnerability risk assessment (VRA), impact assessment studies, and adaptive capacity assessment (Sethamo et al. 2020). However, these methodologies have been accused of failure to observe the ethical and justice interests of the concerned communities. In Australia, for example, Bardsley and Rogers (2010) reveal that communities engaged in climate adaptation planning through VRA raised concerns about the outcomes of these processes as they were not able to define which systems were vulnerable and the capacities sought to adapt.

With reference to indigenous populations, many scholars raise epistemic injustices which contribute to their vulnerability to climate change (Sekine et al. 2009; Chanza and de Wit 2013; Savo et al. 2016; Basel 2020). Thus, the governance question of climatic responses remains a critical concern. It is from this thinking that IKS can be used to strengthen adaptation governance beyond theory to practice (Chanza and de Wit 2016). Indigenous communities have a long history of adapting to environmental and climatic changes, which many scholars attribute to their reliance on climate-sensitive and natural resources-dependent livelihoods (Codjoe et al. 2014; Weber and Schmidt 2016; Nanfuka et al. 2020). In this regard, the application of local indigenous knowledge can assist in decision-making and promoting adaptive governance (Raymond et al. 2010). It can be argued that the inclusivity demanded by climate governance when developing adaptation strategies should be extended to indigenous communities who are affected by climate change impacts. In the anticipated operation of adaptation governance, therefore, this approach ensures that the knowledge held by people witnessing climate change can be utilised to design adoptable and effective climate change projects that reflect local ownership. Accordingly, there are opportunities for operationalising SDG 13 by involving communities in tackling climate change impacts. The instruments for engagement of local communities in adaptation practice are already in use in the form of community-based adaptation, ecosystem-based adaptation and indigenous-based adaptation. The cases illustrated in the next section show how these instruments can be used to operationalise adaptation governance.

11.4 Articulating the Adaptation Governance Question: Evidence from Cases

11.4.1 Ecosystem-Based Adaptation

According to the Convention on Biological Diversity (CBD) (2009:41), ecosystem-based adaptation (EbA) involves the use of biodiversity and ecosystem services as part of an overall adaptation strategy to help communities adapt to the adverse effects of climate change. This strategy integrates biodiversity and ecosystem services into a portfolio of climate change adaptation governance (Munroe et al. 2012). Borrowed from established disciplines such as community-based natural resources management (CBNRM), the concept has largely been driven by environment or conservation practitioners (Reid 2016). There are also environmental benefits in the form of GHG sequestration and climate stabilisation. Munang et al. (2013) indicate that this approach of using natural capital to adapt to climate change impacts has multiple co-benefits for mitigation, protection of livelihoods and poverty alleviation. In its application, EbA can enhance climate change adaptation and address social and environmental vulnerabilities. Local and indigenous communities can thus devise ecosystem-related solutions for addressing their vulnerability to climate change, thereby building their adaptive capacities. Through it, communities can directly benefit from ecosystem services in the form of food, water, medicines and defence against extreme weather and climatic events as well as other ecosystem services. These benefits can be realised at a local scale, as long as access to ecosystems by communities is not restricted (Doswald et al. 2014; Woroniecki 2019). Chanza and de Wit (2016) affirm that the main activities of EbA, which include vulnerability assessment, capacity building, designing policy measures, and implementation, involve the use of indigenous knowledge held by local people. A typical EbA project involving local participation in South Africa is cited from Mellmann (2015) (Table 11.1).

Table 11.1 Description of EbA project in South Africa

The South African EbA project is implemented in North West region of the country in semi-arid terrestrial ecosystems. The project focuses on wetland and rangeland restoration in order to support resource poor farming communities. Climate change is reinforcing other human related stressors from overstocking of sheep and goats, which lead to land degradation. The project has rehabilitated wetlands to retain freshwater and fodder for the livestock. The link between livelihoods and ecosystems is very evident in the project. The wetland restoration benefits manifest quickly, alien species are being replaced by native vegetation which is beneficial to livestock. There are also promises of improvement in water availability. With their local-based knowledge, the communities have supported the restoration of the wetland areas to reduce climatic risks

Source Mellmann (2015)

11.4.2 Community-Based Adaptation

Community based Adaptation (CbA) can be regarded as a sister concept to EbA. Like its counterpart, CbA is fairly recent and emerges from CBNRM, but propelled by community driven practitioners along the course of learning about the best approach with which to respond to climate change (Reid 2016; McNamara and Buggy 2017). Its emergence has been motivated by the recognition of the human dimensions of climate change impacts, appreciation of local knowledge for strengthening adaptive capacity, and the desire to focus on the spatial scale at which impacts are felt (McNamara and Buggy 2017). Reid et al. (2009:13) define it as "a community-led process, based on communities' priorities, needs, knowledge and capacities, which should empower people to plan for and cope with the impacts of climate change." McNamara and Buggy (2017:445) adopted a more comprehensive scope of the concept from the work of Ayers and Forsyth (2009) as:

- Focus on communities at the local level that are vulnerable to climate change impacts;
- Identify development activities at the local level that bolster the adaptive capacity of communities to live in less predictable climates;
- Develop strategies for adaptation through the active involvement of local stakeholders as well as development and disaster risk-reduction practitioners; and
- Integrate existing cultural norms and address root causes of vulnerability to climate change impacts.

Proponents of this strategy argue about the need to position the community at the centre of designing and implementing coping and adaptive interventions (Forsyth 2013; Wright et al. 2014; Reid 2016; Clarke et al. 2019; Basel et al. 2020). Basel et al. (2020) specifically emphasise the significance of this strategy in addressing complex social-ecological systems in climate adaptation. In its deployment, CbA uses the wealth of knowledge and experience held by communities on dealing with climate change impacts (Reid 2016). Chanza and de Wit (2016) affirm that CbA uses indigenous knowledge and experiences of local communities in responding to climate change. The governance issues observed during the use of CbA relate to a bottom-up approach where communities are given recognition and space by adaptation practitioners to select options and priorities based on their preferences and experiences with weather and climatic incidences (Wright et al. 2014). It is at the community level where the concerns and needs of vulnerable and marginal groups can be incorporated. Special

Table 11.2 Description of the WeValue + VRA hybrid approach in Botswana

Developed from initial work where VDCs were considering climate change issues, adopting VRAs, and drafting local adaptation plans (LAPs). The process involved local and external collaborations in community workshops to identify and prioritize existing and future vulnerabilities, risks, capacities and visions. It also involved elicitation of participant perceptions about ownership, relevance, and participation in adaptation activities. At the end of the assessment, there was evidence that the approach significantly led to deep and meaningful engagement and participation, ownership, leadership responsibility to address local climate change impacts and relevant to the participants of the VRA processes and the anticipated outcomes

Source Sethamo et al. (2020)

groups such as youths and women are also considered at this level. The community structure is central to problem identification and problem solving, planning, local governance and context relevant adaptation (Reid 2016). Ensor (2009) calls this 'governance for community-based adaptation.'

Sethamo et al. (2020) use a hybrid tool that they refer to as WeValue + VRA to demonstrate the practicality of adaptation governance involving local communities in Botswana. They identified Village Development Committees (VDCs), which are the local level governance structures set by the local people to link with national adaptation planning. The methodology combines WeValue, a values-based approach, with VRA method, from which they found that communities were able to engage more deeply and communicate more effectively on the topic of climate change (Table 11.2).

11.4.3 Indigenous-Based Adaptation

Indigenous-based Adaptation (IbA) uses the climate and environment knowledge of indigenous people to adapt to climate change. This concept emerges from the discipline of IKS, which has gained considerable attention owing to decolonial demands to address epistemological and methodological injustices (Chanza and de Wit 2013), and to acknowledge that the climate change challenge cannot be addressed by Western Science alone (Mutambisi et al. 2021). In its application, IbA may not be a stand-alone strategy but can be integrated into EbA and CbA. Scholars of IKS indicate that this knowledge is critical in enabling communities to survive in challenging conditions, which are being worsened by climate change (Nyong et al. 2007; Orlove et al. 2010; Mistry and Berardi 2016; Savo et al. 2016). Aspects that make this strategy critical in governance relate to intra-generational and inter-generational transferability of knowledge and skills; knowledge accessibility and ownership; collective understanding and responsibility to manage natural resources to maintain livelihoods; and the participatory and empowering nature of the knowledge system (Orlove et al. 2010; Chanza and de Wit 2016). This section cites two indigenous practices that can be used to illustrate adaptation governance, which are *nhimbe* (collective work) and *zunde* (citizen mobilisation) in Zimbabwe (Table 11.3).

These three instruments (CbA, EbA and IbA) demonstrate that it is possible to see meaningful operation of adaptive governance at local scales involving indigenous people that are experiencing climate change. Under EbA, local practices and governance processes guiding the exploitation of ecosystem services can be used to provide ecosystem-based solutions in the face of climatic threats. Similarly, CbA is a governance process that provides community-based solutions for

Table 11.3 Examples of indigenous-based practices resembling governance

Practice	Description	Comment
Nhimbe	An indigenous scheme of social capital involving collective pull of resources and labour to enhance household and community food security. It involves mutual support through collective sharing of resources by members of a community. This co-operation ideology fosters responsibility and counteracts dependency syndrome as it challenges everyone to work collectively. Agricultural duties such as planting, weeding and harvesting are done speedily and timeously. In addition, the social capital shared extend to incorporate such other vulnerable members in the villages as widows, child-headed households, people living with disabilities and old people. Largely intended to avoid overreliance on external assistance, this practice is motivated by solidarity, co-operation, trust, honesty, respect, allegiance and reciprocity. When viewed from an adaptation perspective, the concept serves to strengthen the social nets of a community against climatic risks	A similar practice exists in other countries in the form of cooperatives, e.g., *harambee* (Kenya), *ujamaa* (Tanzania), *bulungi bwansi* (Uganda)
Zunde	*Zunde* is a Shona word that means a large gathering of people taking part in a common activity. Its objective is to ensure that food security for a village(s) is always guaranteed. A Chief designates a piece of land for cultivation by his subjects. The yield from this land is stored in granaries (*Zunde raMambo*) at the Chief's place. The main beneficiaries of this strategy are the villagers during times of food shortages such as drought. Its ideology goes beyond crop production activities to include social, economic and political rallying mechanisms. Participation in the *zunde* is an expression of oneness and carried with it social and moral obligations. Food is perceived not only as a means of meeting dietary requirements, but also as a social tool which brings people together to share their successes and/or failures	An old practice in Zimbabwe, now being revitalised through government support to strengthen climatic responses at the grassroots

Source Chanza (2015)

people experiencing climatic risks. The Botswana case has revealed that village level structures can be effective in articulating the adaptation governance agenda. Lastly, IbA experiences of *nhimbe* and *zunde* cited from Zimbabwe showcase the operationalisation of adaptive governance, with similar practices being reported in Kenya, Tanzania and Uganda.

11.5 Discussion

The evidence emerging from this review shows that governance is critical in the success of CbA, EbA and IbA. All these instruments involve bottom-up approaches to the conceptualisation and praxis of climate change adaptation, which operate at finer scales. At such a scale, it is easy for the people to understand climate change issues and the citizens can be challenged to participate in climate initiatives when they get information about the nature and risks they face. Thus, the climate governance agenda articulated through these instruments demonstrate a transition from theory to practice and is underpinned by IKS. Nyong et al. (2007) identify five reasons that undergird the devolution governance agenda involving indigenous people as: culturally rich context; appropriateness and sophistication of the knowledge form in indigenous settings; buy-in from the community; promoting equity, efficiency and environmental integrity; and increased communication and understanding. Like the CBNRM used in natural resources management (Reid 2016), the bottom-up approaches to adaptation presented here are based on the notion that if people are empowered to meaningful participate in decisions about ways to address climate change, through recognising their own knowledge and experiences, they can smoothly collaborate with climate change and development agencies for effective climatic responses.

Notwithstanding the revealing promises for addressing governance issues related to climate change adaptation through CbA, EbA and IbA, there are challenges of up-scaling and mainstreaming these strategies. For example, Reid (2016) warns that some agencies are using the terms 'community-based' or 'bottom-up' as a vehicle for getting funding. On the ground, there may not be clear grassroots-based initiatives for addressing community challenges. The centrality of the communities and their home-grown knowledge and decisions about adaptation choices and methodologies for executing adaptation projects remain questionable. Similarly, Forsyth (2013) questions whether CbA can represent local people fairly and if successful CbA can be assessed. However, the argument provided by this study is that these approaches hold great potential for ensuring governance of climate change at levels where climatic impacts are felt. Similarly, experiences by Practical Action, working with indigenous rural communities in developing countries including Kenya and Zimbabwe reveal that adaptation governance can only be realised if practitioners and policy makers understand it as a process, through which people gain access to resources and information necessary for acquiring knowledge and skills to keep evolving in matching with the continuously changing environment (Ensor, 2009). The findings of this review indicate that IKS create opportunities for demonstrating the praxis of adaptation governance at scales where people affected by climate change can take reasonable action. Wright et al. (2014) indicate that adaptation governance can be enabled through upstream (community-level adaptation) and downstream (national-level adaptation) linkages involving extension services and farmers' organisations. The three approaches of CbA, EbA and IbA can facilitate such upstream and downstream linkages where external agencies would interact with indigenous communities through participatory and iterative learning, tailoring of adaptation technologies to suit local contexts, and working with local institutions. This is part of the action-oriented approach in tackling climate change impacts, which is sought by SDG13.

11.6 Conclusion

Against a background of increased recognition of IKS in climate science, and governance processes in particular, this chapter has argued about the need to make climate governance more realistic by focusing its applicability to the level of adaptation governance. The chapter has argued that the phenomenal growth in scholarship that examines the concept of climate governance, has not resolved the existing contestations in the praxis of climate governance. Accordingly, to make the governance process work, there is need to break the global institutional architecture of climate change that has characterised climate change responses, into mitigation and adaptation processes that are adoptable at national and local levels. The purpose of this chapter was to show the potential for using IKS in strengthening adaptation governance processes. It has demonstrated that the existing global climate change governance architecture cannot properly address the pressing challenges of local communities who experience climate change impacts in their unique environments. The chapter also showed that the existence of the governance frameworks to articulate climate change, although it is still evolving, has not managed to demystify the operational complexities associated with the concept. While this challenge rages on, it is proposed that IKS can be used to address one of the components of climate change governance, which is adaptation governance. This thinking has been supported by identifying appropriate approaches that can be used for meaningful engagement with indigenous communities, which are EbA, CbA and IbA. Although given distinct treatment in the analyses, these approaches are common in their execution in that they all utilise IKS held by local communities that is developed from experiencing changes in their environments.

References

Ayers J, Forsyth T (2009) Community-based adaptation to climate change. Environ: Sci Policy Sustain Dev 51 (4):22–31. https://doi.org/10.3200/ENV.51.4.22-31

Bardsley, D. and Rogers, G. (2010). Prioritizing engagement for sustainable adaptation to climate change: An example from natural resource management in South Australia. *Society and Natural Resources*, 24 (1): 1–17. https://doi:https://doi.org/10.1080/08941920802287163

Basel B, Goby G, Johnson J (2020) Community-based adaptation to climate change in villages of Western Province, Solomon Islands. Mar Pollut Bull 156. https://doi.org/10.1016/j.marpolbul.2020.111266

Cadman, T. (2013). *Introduction: Global Governance and Climate Change*. In, Cadman, T. (eds), Climate Change and Global Policy Regimes. International Political Economy. Palgrave Macmillan, London. https://doi.org/10.1057/9781137006127_1

Chanza N (2015) Indigenous-based adaptation: An imperative for sustainable climate change strategies for Africa. In: Mawere M, Awuah-Nyamekye S (eds) Harnessing cultural capital for sustainability: A pan Africanist perspective. Langaa Publishing House, Bamenda, pp 85–134

Chanza, N. and de Wit, A. (2013). Epistemological and methodological framework for indigenous knowledge in climate science. *Indilinga: African Journal of Indigenous Knowledge Systems*, 12 (2): 203–206.

Chanza, N. and De Wit, A. (2016). Enhancing climate governance through indigenous knowledge: Case in sustainability science. *South African Journal of Science*. 112 (3/4), Art. #2014–0286. https://doi.org/10.17159/sajs.2016/20140286

Chirisa I, Chanza N (2009) How will climate change transform African local governance? – assessing the role of civic engagement. Journal of Public Administration and Policy Research 1(2):035–046

Clarke, A., McNamara, K.E., Clissold, R. and Nunn, P.D. (2019). Community-based adaptation to climate change: lessons from Tanna Island, Vanuatu. *Island Studies Journal*, 14 (1): 59–80. https://doi.org/10.24043/isj.80

Cockfield G (2013) Governing adaptation policies and programmes. In: Cadman T (ed) Climate Change and Global Policy Regimes. Macmillan, Palgrave, pp 61–78

Codjoe SNA, Owusu G, Burkett V (2014) Perception, experience, and indigenous knowledge of climate change and variability: the case of Accra, a sub-Saharan African city. Reg Environ Change 14:369–383. https://doi.org/10.1007/s10113-013-0500-0

Convention on Biological Diversity (CBD). (2009). *Review of the literature on the links between biodiversity and climate change – impacts, adaptation and mitigation.* Technical Series No. 42, Secretariat of the Convention on Biological Diversity (CBD), Montreal, Canada.

Deere-Birkbeck, C. (2009). Global governance in the context of climate change: The challenges of increasingly complex risk parameters. *International Affairs (Royal Institute of International Affairs*, 85 (6): 1173–1194. http://www.jstor.org/stable/40389011

Doswald N, Munroe R, Roe D, Giuliani A, Castelli I, Stephens J, Möller I, Spencer T, Vira B, Reid H (2014) Effectiveness of ecosystem-based approaches for adaptation: review of the evidence-base. Climate Dev 6(2):185–201. https://doi.org/10.1080/17565529.2013.867247

Ensor J (2009) Governance for community-based adaptation. Practical Action Publishing, Rugby, A Practical Action discussion paper

Forsyth T (2013) Community-based adaptation: a review of past and future challenges. Wires Clim Change 4:439–446. https://doi.org/10.1002/wcc.231

Frohlich, J. and Knieling, J. (2013). Conceptualising climate change governance. In, Knieling J. and Leal Filho W. (eds) *Climate Change Governance. Climate Change Management*. Springer, Berlin, Heidelberg. https://doi.org/10.1007/978-3-642-29831-8_2

Gupta J (2016) Climate change governance: History, future, and triple-loop learning? Clim Change 7 (2):192–210

Huitema, D., Adger, W., Berkhout, F., Massey, E., Mazmanian, D., Munaretto, S., Plummer, R. and Termeer, C. (2016). The governance of adaptation: Choices, reasons, and effects. Introduction to the

Special Feature. *Ecology and Society*, 21(3). http://www.jstor.org/stable/26269946

Intergovernmental Panel on Climate Change (IPCC). (2018). *Global Warming of 1.5°C. An IPCC Special Report on the impacts of global warming of 1.5°C above pre-industrial levels and related global greenhouse gas emission pathways, in the context of strengthening the global response to the threat of climate change, sustainable development, and efforts to eradicate poverty.* In, Masson-Delmotte, V., P. Zhai, H.-O. Portner, D. Roberts, J. Skea, P.R. Shukla, A. Pirani, W. Moufouma-Okia, C. Pean, R. Pidcock, S. Connors, J.B.R. Matthews, Y. Chen, X. Zhou, M.I. Gomis, E. Lonnoy, T. Maycock, M. Tignor, and T. Waterfield (eds). In Press

Ishtiaque, A. (2021). Multilevel governance in climate change adaptation: Conceptual clarification and future outlook. In, Fares, A. (eds), *Climate Change and Extreme Events*, Elsevier, https://doi.org/10.1016/B978-0-12-822700-8.00009-3

Jagers, S. and Stripple, J. (2003). Climate governance beyond the State. *Global Governance*, 9 (3): 385–399. http://www.jstor.org/stable/27800489

Knieling, J. and Leal Filho W. (2013). Climate change governance: The challenge for politics and public administration, enterprises and civil Society. In: Knieling J. and Leal Filho W. (eds), *Climate Change Governance. Climate Change Management.* Springer, Berlin, Heidelberg. https://doi.org/10.1007/978-3-642-29831-8_1

Leal Filho WL, Taddese H, Balehegn M, Nzengya D, Debela N, Abayineh A, Mworozi A, Osei S, Ayal DY, Nagy GJ, Yannick N, Kimu S, Balogun AL, Alemu E, Li C, Sidsaph H, Wolf F (2020) Introducing experiences from African pastoralist communities to cope with climate change risks, hazards and extremes: Fostering poverty reduction. International Journal of Disaster Risk Reduction 50. https://doi.org/10.1016/j.ijdrr.2020.101738

Li, A. (2016). Hopes of limiting global warming? China and the Paris Agreement on Climate Change. *China Perspectives*, 1 (105): 49–54. http://www.jstor.org/stable/44091098

McNamara KE, Buggy L (2017) Community-based climate change adaptation: a review of academic literature. Local Environ 22(4):443–460. https://doi.org/10.1080/13549839.2016.1216954

Meadowcroft, J. (2009). *Climate change governance*, Policy Research Working Paper 4941, Background Paper to the World Development Report, World Bank. https://doi.org/10.1596/1813-9450-4941

Mellmann, N. (2015). *Ecosystem-based adaptation – in theory and practice: A case study of projects supported by the International Climate Initiative.* MSc Thesis, Uppsala University, Sweden.

Mistry, J. and Berardi, A. (2016). Bridging indigenous and scientific knowledge. *Science*, 352 (6291): 1274–1275. https://doi.org/10.1126/science.aaf1160

Munang R, Thiaw I, Alverson K, Mumba M, Liu J, Ravington M (2013) Climate change and ecosystem-based adaptation: A new pragmatic approach to buffering climate change impacts. Current Opinion in Environmental Sustainability 5(1):67–71. https://doi.org/10.1016/j.cosust.2012.12.001

Munroe, R., Roe, D., Doswald, N., Spencer, T., Möller, I., Vira, B., Reid, H, Kontoleon, A., Giuliani, A., Castelli, I. and Stephens, J. (2012). Review of the evidence base for ecosystem-based approaches for adaptation to climate change. *Environmental Evidence* 1 (13). https://doi.org/10.1186/2047-2382-1-13

Mutambisi, T., Chanza, N., Matamanda, A.R., Ncube, R., Chirisa, I. (2021). Climate Change Adaptation in Southern Africa: Universalistic science or indigenous knowledge or hybrid. In, Leal Filho, W., Ogugu, N., Adelake, L., Ayal, D., da Silva, I. (eds), *African Handbook of Climate Change Adaptation*. Springer, Cham. https://doi.org/10.1007/978-3-030-42091-8_8-1

Nalau J, Becken S, Schliephack J, Parsons M, Brown C, Mackey B (2018) The role of indigenous and traditional knowledge in ecosystem-based adaptation: a review of the literature and case studies from the Pacific Islands. Weather, Climate, and Society 10 (4):851–865. https://doi.org/10.1175/WCAS-D-18-0032.s1

Nanfuka S, Mfitumukiza D, Egeru A (2020) Characterisation of ecosystem-based adaptations to drought in the central cattle corridor of Uganda. African Journal of Range & Forage Science 37(4):257–267. https://doi.org/10.2989/10220119.2020.1748713

Nyong A, Adesina F, Elasha BO (2007) The value of indigenous knowledge in climate change mitigation and adaptation strategies in the African Sahel. Mitig Adapt Strat Glob Change 12:787–797. https://doi.org/10.1007/s11027-007-9099-0

Omuko, L. (2015). *Climate change governance at sub-national level: key lessons for Kenya's county governments.* SCELG Working Paper 2/2015. Strathclyde Centre for Environmental Lwa and Governance: University of Strathclyde, Glasgow.

Orlove B, Roncoli C, Kabugo M, Majugu A (2010) Indigenous climate knowledge in southern Uganda: the multiple components of a dynamic regional system. Clim Change 100:243–265. https://doi.org/10.1007/s10584-009-9586-2

Raymond CM, Fazey I, Reed MS, Stringer LC, Robinson GM, Evely AC (2010) Integrating local and scientific knowledge for environmental management. J Environ Manage 91(8):1766–1777. https://doi.org/10.1016/j.jenvman.2010.03.023

Reid H (2016) Ecosystem- and community-based adaptation: learning from community-based natural resource management. Climate Dev 8(1):4–9. https://doi.org/10.1080/17565529.2015.1034233

Reid H, Alam M, Berger R, Cannon T, Huq S, Milligan A (2009) Community-based adaptation to climate change: an overview. Participatory Learning and Action 60:11–33

Ross C, Anderson R, Thronson PA (2013) Achieving climate protection: Fostering an essential focus on human rights and human impacts. Notre Dame Journal of Law, Ethics and Public Policy 27:3–42

Savo V, Lepofsky D, Benner JP, Kohfeld KE, Bailey J, Lertzman K (2016) Observations of climate change among subsistence-oriented communities around the world. Nat Clim Chang 6:462–473. https://doi.org/10.1038/NCLIMATE2958

Sekine, H., Fukuhara, K., Uraguchi, A, Tan, C.K., Nagai, M. and Okada, Y. (2009). *The effectiveness of community-based adaptation (CBA) to climate change – from the viewpoint of social capital and indigenous knowledge*. Global Environment Information Centre (GEIC) working paper series 2009–001. GEIC, Tokyo.

Sethamo OA, Masika RJ, Harder MK (2020) Understanding the role of crystallizing local shared values in fostering effective community engagement in adaptation planning in Botswana. Climate Dev 12(5):448–456. https://doi.org/10.1080/17565529.2019.1639488

Termeer, C., Dewulf, A. and Breeman, G. (2013). Governance of wicked climate adaptation problems. In, J. Knieling and W. Leal Filho (eds), *Climate Change Governance, Climate Change Management*. https://doi.org/10.1007/978-3-642-29831-8_3

United Nations (UN) (2016). *Report of the Inter-Agency and Expert Group on Sustainable Development Goal Indicators* (E/CN.3/2016/2/Rev.1) Available online: https://sustainabledevelopment.un.org/content/documents/11803Official-List-of-Proposed-SDG-Indicators.pdf (accessed 9 March 2021)

United Nations Framework Convention on Climate Change (UNFCCC). (2016). *Report of the conference of the parties on its twenty-first session, held in Paris from 30 November to 13 December 2015*. UNFCCC/CP/2015/10/Add.1. https://unfccc.int/resource/docs/2015/cop21/eng/10a01.pdf#page=2 (accessed 9 March 2021)

United Nations Framework Convention on Climate Change (UNFCCC). (2012). *National adaptation plans, technical guidelines for the national adaptation plan process*. https://unfccc.int/files/adaptation/cancun_adaptation_framework/application/pdf/naptechguidelines_eng_high__res.pdf (accessed 9 March 2021)

Van Asselt, H. (2007). *Dealing with the fragmentation of global climate governance: Legal and political approaches in interplay management*. Global Governance Working Paper No 30. Amsterdam et al.: The Global Governance Project. Available at: www.glogov.org (accessed 9 March 2021).

Van Vossole, J. (2012). Global climate governance: A legitimation crisis: capitalism, power, and alienation from Marxist and Polanyian perspectives. *Review (Fernand Braudel Center)*, 35 (1): 1–27. http://www.jstor.org/stable/43233909

Weber A, Schmidt M (2016) Local perceptions, knowledge systems and communication problems around the climate change discourse – examples from the Peruvian Andes. Erdkunde 70(4):355–366. https://doi.org/10.3112/erdkunde.2016.04.05

Woroniecki S (2019) Enabling environments? Examining social co-benefits of ecosystem-based adaptation to climate change in Sri Lanka. Sustainability 11(3):772. https://doi.org/10.3390/su11030772

Wright H, Vermeulen S, Laganda G, Olupot M, Ampaire E, Jat ML (2014) Farmers, food and climate change: Ensuring community-based adaptation is mainstreamed into agricultural programmes. Climate Dev 6(4):318–328. https://doi.org/10.1080/17565529.2014.965654

Nelson Chanza (Ph.D.) is currently a Research Fellow in the Department of Urban & Regional Planning, University of Johannesburg. Before this, he worked as a Senior Lecturer in the Department of Geography at the Bindura University of Science Education, and Lecturer in the Department of Rural & Urban Planning at the University of Zimbabwe. His research contributions are largely in the fields of indigenous knowledge applications in climate science, i.e., local-based climate change indicators, and adaptation and mitigation strategies that are understood by local communities. His research interests are in using the multiple evidence approach in exploring the sustainability science discourse through examining the interface between climate change and society.

Walter Musakwa (Ph.D.) is an Editorial board member for Geospatial Information Science and Geography and Sustainability Journals. His main research focus areas are on the grand global challenges namely; environmental degradation, landscape change, climate change, ecosystem services, sustainable development and livelihoods in Southern Africa. He is also into earth observation and geographic information systems in comprehending the grand global challenges. He is currently an Associate Professor in the Department of Urban and Regional Planning, University of Johannesburg.

Anton de Wit (Ph.D.) is a senior lecturer in the Geosciences Department of the Nelson Mandela University. He lectures environmental management and sustainability related subject matter to diverse audiences in disciplines within the natural sciences, humanities, as well as commerce. His personal research interests primarily revolve around the question of social sustainability.

12 Integration of Indigenous Knowledge Systems in Zimbabwe's Climate Change Policy

Tariro Kamuti

12.1 Introduction

Indigenous knowledge systems (IKS) constitute an integral part of the efforts to effectively tackle the climate crisis in the developing world. IKS has been defined as "the combination of knowledge systems encompassing technology, social, economic and philosophical learning, or educational, legal and governance systems. It is knowledge relating to the technological, social, institutional, scientific and developmental, including those used in the liberation struggles" (Odora Hoppers and Makhale-Mahlangu 1998, pp. 8–9). Indigenous people and their cultures, traditions and customs are intricately linked to a geographical location (Kadir 2019). In their territories, indigenous people have adopted skills that enhance their coping mechanisms against abrupt changes in the weather and climate (Chambers et al. 2017). These indigenous people hold a deep knowledge of their natural environment, making them custodians of areas that are still rich in biodiversity, ecosystems and landscapes (Green and Raygorodetsky 2010; Alexander et al. 2011; FAO and FILAC 2021). The knowledge they possess of their immediate environment has formed the bedrock upon which they adapt to climate variability and change. Despite being among the most vulnerable groups to climatic risks, they contributed the least to it, which is partly due to political and economic marginalisation by their respective governments (Green and Raygorodetsky 2010). It is because IKS played an instrumental role in facilitating adaptation to climate change that several countries in sub-Saharan Africa, including Zimbabwe, have opened up space to ensure climate change is integrated into its climate policy.

In this chapter, I define the integration of IKS as taking into consideration the process of incorporating or including IKS into a country's climate change policy with the aim of bringing a holistic dimension and strengthening efforts to tackle the challenges of climate change. I posit that a climate change policy that excludes IKS will not be effective in helping people to adapt, mitigate or be resilient against climate change-induced impacts on their livelihoods and well-being. Zimbabwe is a signatory to international climate change-related agreements and protocols which embrace IKS as part of its overall strategy to confront climate change (see Sect. 12.3). The country's national climate policy seeks to provide an overall framework to give the country some basic principles and guidance under which the national climate change response strategy will be implemented (Republic of Zimbabwe 2017). To effectively tackle climate change, I

T. Kamuti (✉)
Centre for Gender and Africa Studies, University of the Free State, P.O. Box 339, Bloemfontein 9300, South Africa
e-mail: KamutiT1@ufs.ac.za

E. E. Ebhuoma and L. Leonard (eds.), *Indigenous Knowledge and Climate Governance*, Sustainable Development Goals Series, https://doi.org/10.1007/978-3-030-99411-2_12

argue that the climate change policy of Zimbabwe must integrate IKS in a manner that is contingent to its people's traditions, cultures, value systems and aspirations, while taking care of their needs. This is because IKS is useful in providing a robust view of the environmental changes that occur at a local level (Alexander et al. 2011). In the absence of ways of knowing, underpinned by western scientific knowledge, indigenous people have monitored vegetation, wildlife, and the skies to forecast weather conditions and act accordingly (Chambers et al. 2017; Fitchett and Ebhuoma 2017). However, despite the immense role that IKS plays in adapting to climate change, it is often sidelined when structuring programmes to tackle climate change (Makondo and Thomas 2018).

Zimbabwe's climate change agenda has been based mainly on western science, placing enormous emphasis on national governance and giving less consideration to how people at a local level, such as farmers, perceive and tackle climate change (Bhatasara 2015). Therefore, this chapter seeks to explain the policy processes and governance contexts involved in the integration of IKS into the climate change policy of Zimbabwe. In doing this, it aims to crystallise the role IKS play in Zimbabwe's climate change policy. This chapter is crucial, as findings could illuminate the barriers that undermine IKS from playing a more prominent role in Zimbabwe's climate policy, which in turn could negatively impact on the country's ability to achieve the 13th sustainable development goal (SDG 13).

The chapter adopts an institutional approach to analysing the governance of climate change responses. Governments, with the assistance of external and local actors, are key players in policy formulation and implementation within their jurisdictions as part of their governance systems, which involve institutions. Institutions are "enduring regularities of human action in situations structured by rules, norms, and shared strategies, as well as by the physical world" (Crawford and Ostrom 1995, p. 582), or simply systems of rules, whether formal or informal (Fleetwood 2008). Institutions such as local governments, government agencies and organisations that are defined by rules, norms and shared strategies (Imperial 1999) are critical in mediating the shift in adaptation strategies in the face of climate change to foster sustainability (Berman, Quinn and Paavola 2012). The institutional approach focused on how far the climate change policy in Zimbabwe integrates IKS in combating climate change. Subsequently, stakeholder and institutional mapping and analysis, through analysis of documentary evidence, were used to generate the data that informed this chapter. This involved identifying key institutions that deal with or are connected to climate change and looking at whether they incorporate IKS in their affairs. The steps adopted during the conduct of the institutional approach are explained in the next section.

Following the institutional approach to identify the appropriate policy documents to be analysed, the chapter then proceeds by providing a snapshot of climate change policy processes in Zimbabwe. Next, it identifies the key stakeholders and the roles they play in Zimbabwe's climate change policies. Because the issue of sustainable development is inherently linked to climate change, the next section pays attention to the 2030 Agenda for Sustainable Development (AfSD) with a special focus on SDG 13. Specifically, the section explains the measures and interventions put in place by Zimbabwe to achieve SDG 13 before providing a snapshot of the country's climate change policy. The section that follows shows the extent to which IKS is integrated into Zimbabwe's climate change policy, while the final section contains the concluding remarks and makes recommendations to strengthen the role of IKS in the country's climate policy.

12.2 An Institutional Approach to Selecting Relevant Policy Documents

Tackling climate change requires institutions, hence the need to adopt an institutional approach to analyse how policies relating to climate change are formulated and implemented with the idea of integrating IKS into policy schemes. Stakeholders are regarded as "groups or individuals who can affect or are affected by an issue" (Schiller, Winters, Hanson and Ashe 2013, p. 1) due to their social identity (Crane and Ruebottom 2011) direct involvement, and the possibility of being affected either positively or negatively by a proposed project or intervention in their community (Ozesmi and Ozesmi 2003; Carsten et al. 2005; Chinyio and Olomolaiye 2010). The goal was to identify these stakeholders and their roles (Orts and Strudler 2009) in an effort to deal with climate change, with particular focus on the integral part of IKS in those efforts.

The chapter, therefore, broadly looked at the institutional mechanisms put in place by the Zimbabwean government to tackle climate change. This involved identifying key stakeholders, institutions and state agencies whose mandate directly focuses on, or constitutes part of, overall efforts to tackle climate change at various levels. Interactions among these institutions were also taken into consideration and policy analysis was done around Zimbabwe's approach towards climate change. This involved analysis of policy documents, statutes and related literature on climate change in the Zimbabwean context. For instance, the National Climate Policy of Zimbabwe and related policies were scrutinised as they show the role of all sectors that are linked to climate change issues. In addition, the National Climate Policy, National Climate Change Response Strategy, and the National Adaptation Plan, which are major climate change-related documents that represent the country's climate change policy, were also scrutinised. The major focus of this analysis was on ascertaining the extent to which key stakeholders, institutions and state agencies engaged with and ensure that IKS are integrated into governance structures and processes in order to deal with climate change in a holistic manner from the national to the local level in Zimbabwe. The emphasis was on ascertaining whether IKS play a critical role in Zimbabwe's climate change policy or is an afterthought or tick-box exercise. Overall, this methodological approach encapsulates a theoretical analysis at institutional level by focusing on climate change policy and practical elements of how these institutional processes guide action to combat climate change.

12.3 Stakeholders' Role in Zimbabwe's Climate Change Policy

The institutional analysis highlighted that the Ministry of Environment, Climate, Tourism and Hospitality is the core government agency responsible for issues related to climate change resilience and mitigation in Zimbabwe. Several statutory bodies work with the Ministry of Environment, Climate, Tourism and Hospitality in pursuit of the constitutionally defined mandate of guaranteeing citizens' environmental rights and working towards sustainable development, as enshrined in Sect. 73 of Zimbabwe's constitution on "environmental rights" (Government of Zimnbabwe 2013, p. 42). The Ministry of Environment, Climate, Tourism and Hospitality has a Climate Change Management Department whose mission is to protect the socio-economic wellbeing of the country through prudent management of climate change (Climate Change Management Department 2021). While this is commendable, I argue that the limitation of the Climate Change Management Department is in its failure to provide minimum criteria specifying the exact measures to be put in place by various sectors in order to prepare for climate change. These measures should go beyond the crafting of the Climate Change Response Strategy that has been done by this department (Brown et al. 2012).

The Constitution of Zimbabwe is the overarching law that provides the framework for the governance of natural resources management and sustainable development. Section 73 of the Constitution of Zimbabwe guarantees the citizens' environmental rights and provides for the role of the State to safeguard those rights (Government of Zimbabwe 2013, p. 42). For example, Sect. 73(1)(b) the Constitution places emphasis on having "the environment protected for the benefit of present and future generations through reasonable legislative and other measures that – (ii) promote conservation; and (iii) secure ecologically sustainable development and use of natural resources while promoting economic and social development" (Government of Zimbabwe 2013, p. 42). Concerning the role of the state, Sect. 73(2) states that "the state must take reasonable legislative and other measures, within the limits of the resources available to it, to achieve the progressive realization of the rights set out in this section" (Government of Zimbabwe 2013, p. 43). Climate change issues may fit in within these provisions because they affect the natural environment that people depend on for their existence and livelihoods.

Using the constitution, several statutory bodies have been created with supportive environmental legislation and regulations that concern, for example, issues that directly deal with the management and conservation of forest resources. The Constitution, as a guiding framework, is key in directing ministries and state agencies that deal with natural resources that are mainly affected by climate change. These include the Ministry of Environment, Climate, Tourism and Hospitality, which oversees agencies such as the Environmental Management Agency and the Forestry Commission. For example, the Forestry Commission is critical in the socio-economic advancement of the country by providing governance arrangements in the use of forest resources through implementation of the Forest Act (Chapter 19:05 of 1999) and the Communal Lands Forest Produce Act (Chapter 20 of 1987) (Forestry Commission 2021). The Communal Lands Forest Produce Act (Chapter 20 of 1987) deals with issues related to forest resources in communal areas where IKS could be mainly applied to address climate change. The government later instituted a new statutory body, the Environmental Management Agency (EMA), through the Environmental Management Act (Chapter 20:27) of 2002, which makes the EMA's mandate broader than those of other statutory bodies dealing with the environment (Government of Zimbabwe 2002, p. 8).

Issues to do with climate change in the Constitution also relate to development and the environment besides Sect. 72, which enlists the "rights to agricultural land" and Chapter 16 which elaborates on "agricultural land" (Government of Zimbabwe 2013, p. 41; 136). Section 13 of the Constitution that talks about national development specifies that "the State and all institutions and agencies of government at every level must endeavour to facilitate rapid and equitable development," (Government of Zimbabwe 2013, p. 19). The subsections refer to various sectors of the economy and sections of society in a way that gives room to the incorporation of sustainable development, especially towards achieving the SDGs. The Ministry of Lands, Agriculture, Fisheries, Water and Rural Resettlement deals with issues related to all agricultural land and the inherent natural resources that are affected by climate change. For instance, the management of water is critical as it is one of the natural resources that affect all citizens, yet it is greatly impacted by climate change. The Zimbabwe National Water Authority (ZINWA) that oversees the management of the water resources resorts under this ministry and therefore its policies and action plans should also be done with IKS in mind.

Zimbabwe is a signatory to international climate change-related agreements and protocols which embrace IKS as part of its overall strategy to confront climate change. Globally, the country is a signatory to the United Nations Framework Convention on Climate Change (UNFCCC), AfSD, the Sendai Framework for Disaster Risk Reduction, and the Paris Agreement while, continentally and regionally, the country subscribes to the African Union Agenda 2063 and the SADC Industrialization Strategy respectively (Republic of Zimbabwe 2017). In addition, under the auspices of the AfSD, the country subscribes to the Sustainable Development Goals (SDGs). This chapter is more focused on SDG 13 that speaks to climate action (next section). On the other hand, under the Paris Agreement, the country has its part to play in the Nationally Determined Contribution towards the international effort to keep the global temperature increase under 1.5 °C (Republic of Zimbabwe 2017). The country's National Climate Policy sets the overall institutional arrangements through which the National Climate Change Response Strategy (NCCRS) is going to be put into action; hence, this chapter takes a closer look at these documents (Republic of Zimbabwe 2017). It is partly due to the overwhelming importance of IKS in facilitating adjustment to climate change that the AfSD calls for the utilisation of diverse forms of knowledge to tackle climate change.

12.4 SDG 13 and Indigenous Knowledge Systems in Zimbabwe

The AfSD is a global blueprint that sets commitments ratified by 193 countries in 2015 for transformation towards sustainable development that is rooted in five core areas w– people, planet, prosperity, peace and partnership (Anih 2019; Munsami 2019). The AfSD sets 17 SDGs, 169 targets and 231 indicators to guide countries in implementing national plans of action with the idea of combining their efforts in working towards achieving sustainable development (Anih 2019). It is within this framework – SDG 13 in particular – that issues related to climate change are specified. The AfSD and the 2063 Agenda (which embraces sustainable development for an integrated African continent) that were adopted in 2015, were preceded by the Millennial Development Goals (which was implemented from 2000 to 2015) (Munsami 2019). The AfSD embraces stewardship of nature as part of mechanisms to attain SDGs and accepts that biodiversity is critical for both people and the earth to be in good condition (AFSA 2017).

For Zimbabwe, climate change is seen as a setback in its efforts towards achievement of the SDGs, hence the need to act (Brazier 2017). SDG 13 is the one that speaks directly to climate change by urging nations to take decisive steps to tackle climate change (Campbell et al. 2018). Targets in SDG 13 provide a getaway to integrate IKS in Zimbabwe's climate change policies and practical initiatives. Target 13.1 speaks to bolstering the extent to which countries can be resilient and adapt to climate-induced catastrophes. Target 13.2 aims to incorporate climate change mechanisms into national policies, strategies and planning. This chapter further argues that in cases where Target 13.2 emphasises the integration of climate change efforts into national policies, this integration should include IKS. For instance, in Zimbabwe, which is a less developed country, Target 13.b is crucial because it focuses on uplifting capacity building to facilitate crafting plans on climate change, with special attention to susceptible sections of society. The majority of these communities in less developed countries constitute women, youth and poor people. Target 13.3 is about doing better in education, increasing awareness and the ways in which people, together with their institutions, can forecast, cope, adapt and reduce the impact of climate change (UNEP 2021). The definition of IKS that has been adopted in this chapter constitutes a group of knowledge systems that combine many systems, including for instance philosophical learning and educational systems. Therefore, there is a need to ensure that the country's approaches to increasing awareness about climate change and educating people about it to meet Target 13.3 should be underpinned by IKS. However, this chapter also inherently shows the extent to which Zimbabwe has roped in IKS to tackle climate change to achieve SDG13.

12.5 Zimbabwe's National Climate Policy and Indigenous Knowledge Systems

This section looks in detail at Zimbabwe's National Climate Policy, National Climate Change Response Strategy, and National Adaptation Plan to show the extent to which the integration of IKS has attended to the climate crisis. The idea is to assess the major issues highlighted by the National Climate Policy and relate them to IKS. There is a need to see how IKS is incorporated into the country's climate change responses. In the National Climate Policy, the Government of Zimbabwe has the vision of being a country with a low carbon footprint that is resilient to climate change impacts while making a commitment, in conjunction with the international community, to resolve climate change problems in tandem with attaining the SDGs (Republic of Zimbabwe 2017, p. I). The policy endeavours to forge ahead with economic advancement that fosters its vision where people can cope with climate change challenges while co-existing with nature. To this end, the policy is augmented by the "National Climate Change Response Strategy, National Adaptation Plan, Low Carbon Development Strategy, National Environmental Policy, Renewable Energy Policy, and Forest Policy and other relevant long-term plans towards SDGs such as the Zimbabwe Agenda for Sustainable Socio-Economic Transformation" (Republic of Zimbabwe 2017, p. I; see also Government of Zimbabwe 2014). The policy also acknowledges and seeks to adhere to the UNFCCC which is underpinned by an understanding of belonging to the global commons, but with each country playing its part according to its capacity and contribution.

In its formulation, the National Climate Policy says that it was born out of widespread participation of stakeholders, which was also underpinned by western science outcomes. These evidence-based outcomes are anticipated to be used in the formulation and implementation of the policy with the involvement of all stakeholders. The National Climate Policy identified ten principles, the first of which places emphasis on making decisions on climate issues based on evidence that includes IKS (Republic of Zimbabwe 2017, p. 3). The National Climate Policy asserts that the principles conform to the Constitution of Zimbabwe and they:

> … acknowledge that action on addressing climate change should be country-driven, gender-responsive, participatory and fully transparent approach, taking into consideration vulnerable groups, communities and ecosystems, and should be based on and guided by the best available science and, as appropriate, experiences and knowledge of indigenous people, with a view to integrating adaptation and mitigation into relevant socioeconomic and environmental policies and actions where appropriate (Republic of Zimbabwe 2017, p. 3).

A key observation of this National Climate Policy is that it embeds IKS, although it foregrounds the western-based science ahead of what it considers as suitable experiences and knowledge of indigenous people (Republic of Zimbabwe 2017, p. 3). Therefore, there is a need to take note of how science-based knowledge underpinned by western epistemologies downplays IKS. Nevertheless, it can be argued that all these considerations are made with the idea of taking issues of mitigation and adaptation into consideration in policies (Republic of Zimbabwe 2017, p. 3). Thus, it is crucial to ascertain how this integration of IKS is taking place, especially in the National Climate Change Response Strategy and related supportive provisions that are at the interface of a climate change policy and practical solutions to adapt to and mitigate against climate change.

Chapter Two of the National Climate Policy speaks to ways of coping with climate change with emphasis on drawing sector-specific policies and capacities in "water, agriculture, energy, industry, health, infrastructure, forestry and biodiversity" (Republic of Zimbabwe 2017, p. 5). Policy and strategy documents like the Low Carbon Development Strategy, National Environmental Policy, Renewable Energy Policy, and Forest Policy would fit into this consideration as they incorporate sector-specific approaches to tackle climate change. Chapter Three of the National Climate Policy focuses on reducing the impact of climate change and less carbon emissions on development projects in key sectors such as energy, industrial processes, waste, agriculture, land-use, and forestry. Embracing IKS will go a long way in addressing some of these challenges. In Chapter Four, which focuses on teaching, learning and publicity about climate change, the utilisation of IKS is provided for (Republic of Zimbabwe 2017, p. 14). Here, there is an acknowledgement of how useful IKS are in addition to the widely available knowledge and ways of going about climate change. Indigenous knowledge of climate change is used in a variety of ways, including pinpointing the effect of climate change, making decisions in agriculture, crafting systems in forecasting negative impacts of climate change, and adapting to and reducing the impact of drastic changes in weather conditions by vulnerable communities. This acknowledgement by the National Climate Policy is a positive step in advancing the practical implications and usefulness of IKS in tackling climate change.

Chapter Five deals with weather, climate research and modelling, and contains a section that links institutions to infrastructure development. Institutions are critical in the formulation, implementation, monitoring and evaluation of a climate change policy (Republic of Zimbabwe 2017, p. 16). Institutional processes in infrastructural developments need to consider how those projects are carried out in ways that minimise the impact on the environment. Institutions such as the National Meteorological and Hydrological Services and the Climate Change Management Department should advance research on climate change and generate knowledge that meets the needs of users (Republic of Zimbabwe 2017, p. 16). This kind of research should embrace IKS. Lastly, Chapter Six emphasises the latest techniques and their use in transmitting information that is important for facilitating efforts to tackle climate change. For instance, one can talk about ways in which to facilitate the spread of information concerning climate change, even that which is connected to IKS (Ndlovu 2020).

The National Climate Change Response Strategy (NCCRS) of 2014 preceded the National Climate Policy of 2017 followed by the National Adaptation Plan of 2019 (Government of Zimbabwe 2014, 2019; Republic of Zimbabwe 2017). It would have been ideal to first develop the broad National Climate Policy as an overall guiding framework that incorporates dictates of international and regional obligations and the country's national constitution. This would then inform the National Climate Change Response Strategy, the National Adaptation Plan and other subsidiary national and local strategies and plans across all sectors of the economy and sections of society. Otherwise, the National Climate Change Response Strategy, including other preceding strategies and plans, need to be updated to be in line with the letter and spirit of the National Climate Policy like the National Adaptation Plan that came later. This update will facilitate the vertical integration of the country's climate change policy to sync issues along all tiers of governance, a process that will give room to the integration of IKS and bolster its contribution towards tackling climate change.

The National Climate Change Response Strategy is a comprehensive long-term plan that was crafted to direct the country's response to the impact of climate change and to incorporate these issues into various sectors of the economy from national to the local level (Government of Zimbabwe 2014). This strategy focuses on the long term to help the country withstand climate change while achieving sustainable development by enjoying widespread participation and taking into consideration the country's susceptibility. The executive summary of the National Climate Change Response Strategy refers to IKS and the word 'indigenous' three times. The first instance is under the guiding principles where the response strategy is underpinned by information that integrates IKS, people's ways of living and how they understand their environment (Government of Zimbabwe 2014, p. iv). The second instance is under the part played by the meteorological services in efforts against climate change to increase capacity in building databases on IKS and using that information to foretell trends about the climate trends (Government of Zimbabwe 2014, p. viii, ix). The third instance is under the spread of information where there is a mandate to build awareness, including IKS (Government of Zimbabwe 2014, p. ix).

The National Adaptation Plan directly covers issues related to adaptation through carving out a pathway of particular activities with benchmarks that the country has to follow in line with guidelines from the UNFCCC (Government of Zimbabwe 2019). This involves assessing the capacity to identify shortcomings, scenario planning across all tiers of government, choosing the most pressing adaptation needs and putting in place the relevant structures and resources, and recording progress for short-term assessments and eventual evaluation (Government of Zimbabwe, 2019). A major facet of the National Adaptation Plan is that underscored by the most indigenous knowledge driven by its incorporation into relevant policies and action programmes (Government of Zimbabwe, 2019, p. 7). This is the only part where indigenous knowledge is referred to directly in the National Adaptation Plan. However, another important component of the National Adaptation Plan that I find critical to the integration of IKS, is the inclusion of the basis of administration and planning (that is, the inclusion of institutional framework), as well as how this plan links with the National Climate Change Response Strategy. This synergy is important to enhance the integration of IKS in an effort to tackle climate change.

12.6 Towards the Integration of Indigenous Knowledge Systems in the Climate Change Policy of Zimbabwe

Since most of the IKS are connected to nature, it is easier to include them in broad nature-based solutions to address climate change. Interventions to deal with climate change are physical structures like man-made features, soft approaches like financial instruments and weather-forecasting techniques, and various approaches specifically focusing on nature (USAID 2019, p. 2). Nature-based solutions depend on the use of ecosystems as constituents of how society deals with changes in the environment (Woroniecki 2020, p. 1). One of the nature-based solutions that enhances climate change resilience and mitigation, especially for vulnerable communities, is ecosystem-based adaptation. Ecosystem-based adaptation is a concept that puts people at the centre while recognising that those ecosystems they depend on for their livelihood activities are important as a base to build resilience in the face of climate change (USAID 2019, p. 2). It is on that basis that Zimbabwe's climate change policy should entrench IKS through recognition of the critical role played by ecosystems in providing the natural capital upon which to anchor people's livelihood strategies. The good thing is that people's cultures already have practices based on IKS that enable them to take care of nature; however, more of them need to be formally incorporated into the climate change policy.

These approaches are based on the recognition of the central role of biodiversity in connecting ecosystems to human needs. For example, the United Nations Framework Convention on Climate Change was adopted in conjunction with the Convention on Biological Diversity and the Convention to Combat Desertification at the Rio Earth Summit in 1992 (Lukose 2017). This affirmed the intricate link between nature and climate change, as human action in nature led to its deterioration, thereby causing global warming and climate change. Therefore, addressing these challenges needs acknowledgement of nature as part of the solution, for instance, through limiting greenhouse gas emissions, restoring natural landscapes, combatting desertification, and others. This means that various kinds of ecosystems help enhance the capacity of human beings to withstand the adverse conditions brought about by climate change. These approaches constitute natural climate solutions that have been proven to help mitigate against climate change while contributing to sustainable development (Derbile 2013; Griscom et al. 2021). Zimbabwe, as a signatory to these international agreements that embrace approaches based on the recognition of the central role of biodiversity, should embrace and incorporate IKS in these natural climate solutions.

Since IKS are connected to nature, one would assume that it is logical to involve them in soliciting nature-based solutions to address climate change. For example, studies in Peru, India, Kenya, Ghana and China have shown that indigenous people hold values and practices that uphold nature conservation such that it is recommended that indigenous knowledge be integrated into relevant policies concerning development (Swiderska et al. 2021). This should extend to the integration of indigenous knowledge in approaches towards biodiversity conservation to strengthen capacity to adapt to and mitigating against climate change, thus promoting sustainable development. There is much scope in using indigenous and local knowledge to influence policy that enhances human well-being and nature (Cámara-Leret and Dennehy 2019; Lavorel et al. 2020). Traditional ecological knowledge has helped pastoralists in Ethiopia to cope with adverse changes in vegetation patterns and boost their production of goats (Liao, Ruelle and Kassam 2016). Mechanisms to tackle climate change, including indigenous knowledge, are seen to work well at smallholder level (Mutambisi et al. 2021). This is relevant to Zimbabwe because the majority of its population are located in rural areas where they practice subsistence agriculture, which should tap more from IKS that constitutes their cultures.

In addition, a study in Bangladesh found that formal institutions often do not take into consideration how indigenous and local knowledge is essential in understanding how communities learn about their surroundings in order to cope (Choudhury et al. 2021). These formal institutional processes use the lens of outsiders to define resilience and even disregard the local people as active participants (Choudhury et al. 2021). In addition, the study revealed that local people are even able to perceive the occurrence of cyclones well in advance by using local and indigenous knowledge, for instance, monitoring the warmth of water and characteristics of certain animals (Choudhury et al. 2021). Indigenous knowledge helps people to face environmental crises in their areas proactively (Congretel and Pinton 2020). Integration of indigenous people's institutional arrangements into formal structures of governance helps in combating climate change, nature conservation and managing forest resources (Odora Hoppers 2002; FAO and FILAC 2021). Even in the preparation of future IPCC assessment reports, there is a call to critically integrate indigenous knowledge into the formulation of relevant policies on the adaptation to climate change (Ford et al. 2016; Nakashima Rubis and Krupnik 2018; Potts 2021). We noticed how Zimbabwe's climate change policy foregrounds western-based epistemologies, followed by locally grounded IKS like how, in Bangladesh, formal institutions often lack consideration of how indigenous and local knowledge is used by communities to become accustomed to their local environment. However, this trend in Zimbabwe should change towards formally embracing IKS in climate change action.

It was noted that the National Climate Change Response Strategy of 2014 preceded the National Climate Policy of 2017, followed by the National Adaptation Plan of 2019. There is a possibility that some of the institutional mechanisms included in the National Climate Change Response Strategy that were put in place before 2017 could have been misaligned to the National Climate Policy of 2017 when it was adopted. In that case, the National Climate Change Response Strategy and other preceding strategies and plans have to be updated to be in line with the letter and spirit of the National Climate Policy like the National Adaptation Plan that followed later in 2019. This can be done in tandem with updating the country's NDC goals following the requirements and guidelines of the Paris Agreement, especially that the least developed countries face the challenge of a lack of capacity that hinder implementation (Pauw and Klein 2020; Aragon et al. 2021; Schulz et al. 2021; Steinebach and Limberg 2021). This process of aligning policy processes and governance mechanisms relating to climate change should also be undertaken to strengthen the IKS component in the overall climate change policy of Zimbabwe.

12.6.1 Conclusion

This chapter looked at how the Zimbabwean institutions have gone so far in integrating IKS into their climate change agenda. Zimbabwe is a signatory to international climate change-related agreements and protocols that embrace indigenous knowledge systems as part of its overall strategy to confront climate change. They also include international agreements that deal with sustainable development and disaster risk management. The country's National Climate Policy seeks to provide an overall framework to give the country some basic principles and guidance under which the National Climate Change Response Strategy will be implemented. Examples in the country and elsewhere have shown that IKS are a critical component of measures to stem climate change with practical implications for vulnerable populations who are the custodians of natural resources in their territories. The National Climate Policy proactively speaks to IKS, as there is recognition of indigenous knowledge in attending to climate change. The first of the National Climate Policy's ten principles places climate-related decisions or actions on a pedestal of scientific findings, including IKS. The National Climate Policy has been shown to be the key reference in the country's climate change agenda under the Ministry of Environment, Climate, Tourism and Hospitality, which is supported by sector-specific strategies. Zimbabwe has started to integrate indigenous knowledge systems in its climate policy, but there is a need for further refinement in specifying tangible action in efforts to combat climate change since indigenous knowledge has been sidelined.

The major question that needs further attention is how much does the inclusion of IKS contributes to guiding action on the ground when formulating ways of combatting climate change. Zimbabwe's climate change policy to a large extent foregrounds western-based epistemologies with some acknowledgement of locally grounded IKS. This chapter argues that Zimbabwe's climate change policy and strategies need further refinement in specifying tangible actions in efforts towards combatting climate change, thus helping to achieve targets set under the SDG 13. The updating of these policies will then pave the way to embracing the critical role that IKS play in fighting climate change. Institutions and their governance mechanisms should also be reoriented to formally embrace IKS, which are already grounded in people's traditions and cultures, and how they relate to the natural environment.

References

AFSA (2017) A study of policies, frameworks and mechanisms related to agroecology and sustainable food systems in Africa. Kampala, Uganda: Alliance for Food Sovereignty in Africa

Alexander C, Bynum N, Johnson E, King U, Mustonen T, Neofotis P, Oettlé N, Rosenzweig C, Sakakibara C, Shadrin V, Vicarelli M, Waterhouse J, Weeks B (2011) Linking indigenous and scientific knowledge of climate change. Bioscience 61(6):477–484. https://doi.org/10.1525/bio.2011.61.6.10

Aligica PD (2006) Institutional and stakeholder mapping: frameworks for policy analysis and institutional change. Public Organiz Rev 6(1):79–90. https://doi.org/10.1007/s11115-006-6833-0

Alizart M (2021) The climate coup. Polity Press, Cambridge, UK

Anih S (2019) Passive to active space and the role of space assets in sustainable development. In: Froehlich A (ed) Embedding space in African society: the United Nations sustainable development goals 2030 supported by space applications. Springer, Cham, Switzerland, pp 31–50

Aragon I, Alcobé F, Idriss Y (2021) Updating NDCs: useful guidance to support greater ambition. London, UK: IIED. https://pubs.iied.org/20341IIED

Berman R, Quinn C, Paavola J (2012) The role of institutions in the transformation of coping capacity to sustainable adaptive capacity. Environ Devel 2:86–100

Bhatasara S (2015) Rethinking climate change research in Zimbabwe. J Environ Stud Sci 7(1):39–52. https://doi.org/10.1007/s13412-015-0298-9

Brazier A (2017) Climate change in Zimbabwe: a guide for planners and decision makers. Harare, Zimbabwe: Konrad-Adenauer-Stiftung

Brown D, Chanakira RR, Chatiza K, Dhliwayo M, Dodman D, Masiiwa M, Muchadenyika D, Mugabe P, Zvigadza S (2012) Climate change impacts, vulnerability and adaptation in Zimbabwe. International Institute for Environment and Development (IIED) Climate Change Working Paper No. 3. London, UK: IIED

Cámara-Leret R, Dennehy Z (2019) Information gaps in indigenous and local knowledge for science-policy assessments. Nat Sustain. https://doi.org/10.1038/s41893-019-0324-0

Campbell BM, Hansen J, Rioux J, Stirling CM, Twomlow S, Wollenberg EL (2018) Urgent action to combat climate change and its impacts (SDG 13): transforming agriculture and food systems. Curr Opin Environ Sustain 34:13–20. https://doi.org/10.1016/j.cosust.2018.06.005

Carsten NH, Christensen SM, Tarp P (2005) Rapid stakeholder and conflict assessment for natural resource management using cognitive mapping: the case of Damdoi forest enterprise, vietnam. Agricul Human Values 22:149–167

Chambers LE, Plotz RD, Dossis T, Hiriasia DH, Malsale P, Martin DJ, Mitiepo R, Tahera K, Tofaeono TI (2017) A database for traditional knowledge of weather and climate in the Pacific. Meteorol Appl 24 (3):491–502. https://doi.org/10.1002/met.1648

Chinyio E, Olomolaiye P (2010) Introducing stakeholder management. In Chinyio E, Olomolaiye P (eds) Construction stakeholder management (pp 1–12). West Sussex, United Kingdom: Blackwell Publishing

Choudhury M-U-I, Haque CE, Nishat A, Byrne S (2021) Social learning for building community resilience to cyclones: role of indigenous and local knowledge, power, and institutions in coastal Bangladesh. Ecol Soc 26(1):1–19. https://doi.org/10.5751/ES-12107-260105

Climate Change Management Department (2021) About climate change management department. http://www.climatechange.org.zw/about-climate-change-management-department

Congretel M, Pinton F (2020) Local knowledge, know-how and knowledge mobilized in a globalized world: a new approach of indigenous local ecological knowledge. People Nat 2(3):527–543. https://doi.org/10.1002/pan3.10142

Crane A, Ruebottom T (2011) Stakeholder theory and social identity: rethinking stakeholder identification. J Bus Ethics 102:77–87

Crawford SES, Ostrom E (1995) A grammar of institutions. Am Polit Sci Rev 89(3):582–600

Derbile EK (2013) Reducing vulnerability of rain-fed agriculture to drought through indigenous knowledge systems in north-eastern Ghana. Int J Climate Change Strat Manag 5(1):71–94. https://doi.org/10.1108/17568691311299372

FAO and FILAC (2021) Forest governance by indigenous and tribal people: an opportunity for climate action in Latin America and the Caribbean. Santiago, Chile: FAO

Fitchett JM, Ebhuoma E (2017) Phenological cues intrinsic in indigenous knowledge systems for forecasting seasonal climate in the delta state of Nigeria. Int J Biometeorol 62(6):1115–1119. https://doi.org/10.1007/s00484-017-1495-x

Fleetwood S (2008) Structure, institution, agency, habit, and reflexive deliberation. J Inst Econ 4(2):183–203

Ford JD, Cameron L, Rubis J, Maillet M, Nakashima D, Willox AC, Pearce T (2016) Including indigenous knowledge and experience in IPCC assessment reports. Nat Clim Chang 6(4):349–353. https://doi.org/10.1038/nclimate2954

Forestry Commission (2021) Welcome to the forestry commission. http://www.forestry.co.zw/

Government of Zimbabwe (2002) Environmental Management Act (Chapter 20:27). Government Printer, Harare, Zimbabwe

Government of Zimbabwe (2013) Constitution of Zimbabwe Amendment No. 20 of 2013. Government Printer, Harare. https://freezimbabwe.org/the-constitution

Government of Zimbabwe (2014) National climate change response strategy. Harare, Zimbabwe: Government Printer. http://www.climatechange.org.zw/sites/default/files/National%20Climate%20Change%20Response%20Strategy.pdf

Government of Zimbabwe (2019) National adaptation plan (NAP) roadmap for Zimbabwe. Harare, Zimbabwe: Government Printer

Green D, Raygorodetsky G (2010) Indigenous knowledge of a changing climate. Clim Change 100(2):239–242. https://doi.org/10.1007/s10584-010-9804-y

Griscom BW et al (2020) National mitigation potential from natural climate solutions in the tropics. Philos Trans R Soc B 375:20190126. https://doi.org/10.1098/rstb.2019.0126

Imperial MT (1999) Institutional analysis and ecosystem-based management: the institutional analysis and development framework. Environ Manage 24 (4):449–465

Kadir MYA (2019) Defining 'people' and 'indigenous people' in international human rights law and its application in Indonesia. Int J Minority Group Rights 26(3):373–408. https://doi.org/10.1163/15718115-02602006

Lavorel S, Locatelli B, Colloff MJ, Bruley E (2020) Co-producing ecosystem services for adapting to climate change. Philos Trans Royal Soc b: Biol Sci 375 (1794):20190119. https://doi.org/10.1098/rstb.2019.0119

Liao C, Ruelle ML, Kassam K-AS (2016) Indigenous ecological knowledge as the basis for adaptive environmental management: evidence from pastoralist communities in the Horn of Africa. J Environ Manage 182:70–79. https://doi.org/10.1016/j.jenvman.2016.07.032

Lukose LP (2017) Global warming and climate change: a critique on international law and policy. J Contemp Urban Affairs 1(3):38–42. https://doi.org/10.25034/ijcua.2018.3677

Makondo CC, Thomas DSG (2018) Climate change adaptation: linking indigenous knowledge with western science for effective adaptation. Environ Sci Policy 88:83–91. https://doi.org/10.1016/j.envsci.2018.06.014

Mavhura E, Manyena SB, Collins AE, Manatsa D (2013) Indigenous knowledge, coping strategies and resilience to floods in Muzarabani, Zimbabwe. Int J Disaster Risk Red 5:38–48. https://doi.org/10.1016/j.ijdrr.2013.07.001

Munsami V (2019) Maximising the use of space applications in implementing the sustainable development goals in Africa. In: Froehlich A (ed) Embedding space in African society: the United Nations sustainable development goals 2030 supported by space applications. Springer, Cham, Switzerland, pp 1–29

Mutambisi T, Chanza N, Matamanda AR, Ncube R, Chirisa I (2021) Climate change adaptation in southern Africa: Universalistic science or indigenous knowledge or hybrid. In: African handbook of climate change adaptation, pp 1751–1766

Filho LW, Oguge N, Ayal D, Adeleke L, da Silva I (eds) African handbook of climate change adaptation. (pp. 1751–1766) Cham, Denmark: Springer. https://doi.org/10.1007/978-3-030-45106-6_8

Ofori SA, Cobbina SJ, Obiri S (2021) Climate change, land, water, and food security: perspectives from sub-Saharan Africa. Front Sustain Food Syst. https://doi.org/10.3389/fsufs.2021.680924

Nakashima D, Rubis JT, Krupnik I (2018) Indigenous knowledge for assessment and adaptation: introduction. In: Nakashima D, Krupnik I, Rubis JT (eds) Indigenous knowledge for climate change assessment and adaptation. Cambridge University Press and UNESCO, Cambridge, UK and Paris, France, pp 1–20

Ndlovu D (ed) (2020) The role of the media in environmental discourse in Zimbabwe. Harare, Zimbabwe: Konrad-Adnauer-Stiftung

Odora Hoppers CA (2002) Indigenous knowledge and integration of knowledge systems: towards a conceptual and methodological framework. In: Odora Hoppers CA (ed) Indigenous knowledge and integration of knowledge systems: towards the philosophy of articulation. New Africa Books, Claremont, South Africa, pp 2–22

Odora Hoppers CA, Makhale-Mahlangu P (1998) A comparative study of the development, integration and protection of indigenous systems in the third world: an analytical framework document. Pretoria, South Africa: Human Sciences Research Council

Ozesmi U, Ozesmi S (2003) A participatory approach to ecosystem conservation: fuzzy cognitive maps and stakeholder group analysis in Uluabat Lake, Turkey. Environ Manag 31(4):518–531

Pauw WP, Klein RJT (2020) Beyond ambition: increasing the transparency, coherence and implementability of nationally determined contributions. Climate Policy 20(4):405–414. https://doi.org/10.1080/14693062.2020.1722607

Potts A (2021) The role of culture in climate resilient development. UCLG Committee on Culture Reports, no. 10, and Climate Heritage Network (Working Group 5), Barcelona, 5 November 2021

Raum S (2018) A framework for integrating systematic stakeholder analysis in ecosystem services research: stakeholder mapping for forest ecosystem services in the UK. Ecosyst Serv 29:170–184. https://doi.org/10.1016/j.ecoser.2018.01.001

Republic of Zimbabwe (2017) National climate policy. Harare, Zimbabwe: Government Printer

Schiller C, Winters M, Hanson HM, Ashe MC (2013) A framework for stakeholder identification in concept mapping and health research: a novel process and its application to older adult mobility and the built environment. BMC Public Health 13:428. https://doi.org/10.1186/1471-2458-13-428

Schulz A, Alcobé F, Endalew G (2021) Implementing the Paris agreement: LDC gaps and needs in greenhouse gas inventory reporting. London, UK: IIED. https://pubs.iied.org/20086IIED

Steinebach Y, Limberg J (2021) Implementing market mechanisms in the Paris era: the importance of bureaucratic capacity building for international climate policy. J Eur Publ Policy. https://doi.org/10.1080/13501763.2021.1925330

Swiderska K, Argumedo A, Song Y, Rastogi A, Gurung N, Wekesa C, Li G (2021) Indigenous knowledge and values: key for nature conservation. London, UK: IIED. https://pubs.iied.org/20351IIED

UNEP (2021) Goal 13: climate action. https://www.unep.org/explore-topics/sustainable-development-goals/why-do-sustainable-development-goals-matter/goal-13

USAID (2019) Ecosystem-based adaptation. https://www.climatelinks.org/sites/default/files/asset/document/2019_USAID%20Series%20Synthesis%20Ecosystem-based%20Adaptation.pdf

Woroniecki S, Wendo H, Brink E, Islar M, Krause T, Vargas A-M, Mahmoud Y (2020) Nature unsettled: how knowledge and power shape “nature-based” approaches to societal challenges. Global Environ Change 65(1). doi: https://doi.org/10.1016/j.gloenvcha.2020.102132

Tariro Kamuti (PhD, MPhil, BSc (Hons), BSc, BBA) is a Research Fellow at the Centre for Gender and Africa Studies of the University of the Free State, South Africa. He holds a joint PhD in Geography from Vrije University Amsterdam, The Netherlands and the University of the Free State, South Africa. His research interests include land and environmental policy, governance issues surrounding sustainable utilisation of natural resources, the social dimensions of biodiversity conservation, climate change and sustainable development.

13 Enhancing Climate Change Adaptation Through Indigenous Knowledge Systems and Local Governance in Sub-Saharan Africa: A Systematic Review

Jane Mbolle Chah, Esdras A. R. Obossou, Eromose E. Ebhuoma, Ifeoma Q. Anugwa, and Divine Ewane

13.1 Introduction

Climate change is already affecting every inhabited region across the globe with anthropogenic activities like burning fossil fuels, planting Nitrogen-fixing crops, fertilizer production and greenhouse gas emissions contributing to many observed changes in weather and climate extremes (Amouzou et al. 2019; Ortiz-Bobea et al. 2020, 2021; Zheng et al. 2021). However, Africa is the region most exposed to the adverse impacts of climate change despite its minimal contribution to global warming. Africa's vulnerability to climate change is exacerbated by the interaction of multiple constraints such as endemic poverty, complex governance and institutional dimensions; limited access to capital, including markets, infrastructure and technology; ecosystem degradation; and complex disasters and conflicts at different levels and the continent's low adaptive capacity (Niang et al. 2014). The fifth and sixth assessment report of the International Panel on Climate Change (IPCC) highlighted the observed effects of climate change on the physical, biological and human systems, in particular on water and hence on food production and waterborne diseases in Africa (IPCC 2014). Due to adverse effects of climate change that are anticipated to worsen in the future not just in Africa but the rest of the world, the thirteenth sustainable development goal (SDG 13) urges the world to take urgent action to combat climate change and its effects. SDG 13 offers an opportunity to combine multi-levelled actions towards climate change with development objectives. Some exemplary climate change adaptation initiatives taken by African governments have been reported in the Sixth IPCC report. These are not limited to Ethiopia's programme of adaptation to climate change, which includes sectoral, regional, national, and local community levels (Hunde 2012), Lesotho's coordinated policy framework involving all ministries and stakeholders (Corsi et al. 2012), and Mali's experience with a methodology for integrating climate change adaptation into multiple sectors (Fröde and Manasfi 2013). Nonetheless, other climate change adaptation initiatives are taken by farmers themselves based on their indigenous knowledge systems (IKS).

J. M. Chah (✉) · E. A. R. Obossou · I. Q. Anugwa
Department of Agricultural Extension, University of Nigeria, Nsukka, Nigeria
e-mail: Jane.chah@unn.edu.ng

I. Q. Anugwa
e-mail: ifeoma.irohibe@unn.edu.ng

E. E. Ebhuoma
Department of Environmental Sciences, School of Ecological and Human Sustainability, College of Agriculture and Environmental Sciences, University of South Africa (UNISA), Johannesburg, South Africa
e-mail: ebhuoee@unisa.ac.za

D. Ewane
Department of Animal Science, University of Buea, Buea, Cameroon

E. E. Ebhuoma and L. Leonard (eds.), *Indigenous Knowledge and Climate Governance*, Sustainable Development Goals Series, https://doi.org/10.1007/978-3-030-99411-2_13

IKs refer to the knowledge and skills accumulated over generations that guide human societies in their many interactions with their environment. Berkes (2012) defined IKS as the cumulative body of knowledge, practices, and beliefs, evolving through adaptive processes and transmitted from generation to generation through cultural transmission, about the relationship of living things (including humans) to their environment. In the literature on climate change, common names used to refer to IKS are not limited to traditional knowledge, traditional ecological knowledge, local knowledge, farmers' knowledge, folk knowledge and indigenous science (Kelman et al. 2012; Nalau et al. 2018). This knowledge plays a critical role in climate change adaptation in sub-Saharan Africa (SSA) (Tom et al. 2019). Some rural communities such as Igbide, Uzere and Olomoro in southern Nigeria (Ebhuoma and Simatele 2019) as well as Makoni and Hwedza communities in eastern Zimbabwe (Mapfumo et al. 2016), both in SSA, depend primarily on IKS to predict the weather to make strategic, tactical and operational decisions on crop production, including management of food stocks and social safety nets. This knowledge is often at a much finer spatial and temporal resolution than science underpinned by western epistemologies in most rural communities in SSA (Nakashima et al. 2012). Also, IKS play a critical role in community development planning (Mutasa 2015) by providing communities with ideas hinged on personal and experiential experiences. In addition, through the use of IKS, the cultural identity of local communities could be preserved and strengthened (Sithole 2007). According to Shaw et al. (2008), IKS can empower local communities to adapt to climate change by scaling up community participation aimed at reducing climate vulnerability. Thus, it can be argued that the integration of IKS into local governance could be one of the efficient ways to ensuring more effective and sustainable implementation of climate change adaptation strategies in local communities in SSA.

In this chapter, local governance refers to the processes—political and institutional—through which issues are resolved and the corresponding outcomes implemented at specific sub-national geographic locations (Williams et al. 2020). Argued to bear the brunt of climate action (Pasquini et al. 2015), "local governance is tasked with, amongst other issues, incorporating climate risks in development plans; mobilizing resources for adaptation; adjusting building and land use regulations to consider climate risks; and enhancing disaster preparedness, response and recovery" (Baker et al. 2012; Corfee-Morlot et al. 2011; Reisinger et al. 2011 in Williams et al. 2020, p. 549). According to Iloka (2016), IKS when applied to local governance schemes can address climate change adaptation issues. However, due to the dearth of empirical evidence that has showcased the effectiveness of IKS in assisting governments to upscale climate change adaptation at the community level in SSA, it is expedient for studies to analyse how indigenous peoples could assist local governments in carving out effective climate adaptation policies tailored to local communities. For example, Marango et al. (2016) assert that failure to integrate IKS with scientific knowledge has contributed significantly to the negative effects of climate change being felt aggressively by households in Zimbabwe because interventions drawn from external and more global perspective may be inadequate to effectively tackle it impacts local people and their livelihood activities. Local governments are catalysts for change, often on the frontline of dealing with the effects of climate change at the grassroots level (Williams et al. 2020). In other words, the relevant local leaders have an important role to play in promoting effective climate change adaptation measures. This important role is hinged on the fact that because each localised environment has its nuances that could make climate change be felt differentially. Therefore, the expertise of local people is a necessity because they are most conversant with local realities and how the adverse effects of climate change compromise peoples' livelihood (e.g. see Mavhura et al. 2013; Jiri et al. 2015; Ebhuoma and Simatele 2017). In other words, one-size-fits-all climate change governance policies will not suffice to address the negative consequences in local communities. In this regard, this chapter contributes to the burgeoning discourse that climate change adaptation can be strengthened in SSA through the integration of IKS with local governance, whose climate change adaptation schemes and modus operandi are often underpinned by western scientific knowledge (Maupin 2016).

Although IKS and local governance continue to occupy relevant spaces in the climate science literature partly through deliberations on how these bodies of knowledge—underpinned by their unique epistemologies—can co-exist to improve climate change adaptation (Arruda and Krutkowski 2017; File et al. 2021; Owusu et al. 2018), the relationship between these concepts is poorly documented, especially in SSA. It is against this backdrop that this chapter seeks to establish the relevance of the collaboration between IKS and local governance in facilitating climate change adaptation in SSA. Two main questions are therefore posed to achieve these objectives. They are what are the factors that facilitate effective collaboration between IKS and local governance in climate change adaptation and what barriers stifle this union? Findings from the systematic review resulted in the development of a conceptual framework to show the relationship regarding how IKS can play a vital role in scaling up climate governance at the local level. It is hoped that the findings of this chapter will stimulate conversations that could greatly influence the development of climate change adaptation policies tailored to local communities. This chapter is organised into four parts. Part one highlights the theoretical framework, underlining how IKS and local governance should be integrated to achieve climate change adaptation at the local level. The next part robustly explains how the systematic review was conducted and provides a schematic diagram of the stepwise process. Part three presents the results while the final part provides the conclusion and recommendations, which is hoped, could contribute towards actualising SDG 13 in SSA.

13.2 Methodology

The study employed a desktop research method using secondary data. In particular, a systematic review method was used. A systematic review is an examination of a clearly formulated question that uses systematic and explicit methods to identify, select and critically appraise relevant research and to collect and analyse data from studies that are included in the review. Statistical methods may or may not be used to analyse and summarise the results of the review (Higgins et al. 2011). Through a systematic review, the authors' claims of the rigour of their research can be substantiated, identifying gaps and directions for future research. The findings of this exercise are then synthesised and summarised as the state of knowledge on the topic of interest. In addition, the search terms and criteria for inclusion and exclusion of articles, as well as the papers excluded from the review, are fully reported. Search engines used to conduct the search include Google Scholar, PubMed, Science Direct, Taylor and Francis.

The systematic review process was conducted in four stages. The first stage was to identify the keywords used in the research process. Therefore, *"climate change adaptation" and "indigenous knowledge" and "local governance" and "Africa"; "climate change adaptation" and "local governance" and "Africa"; "climate change adaptation" and "Indigenous Knowledge" and "Africa",* were respectively used as primary search terms and keywords in each of the search engines. 1712 potential articles were identified for review. At this stage, after eliminating duplicates by reading the titles of each article, 1158 articles in total were retrieved. Different criteria for eligibility and exclusion were defined to conduct the search (Table 13.1).

First of all, with regard to the type of literature review we are conducting, only the journal articles with empirical data were considered eligible, which means review articles, book series, book chapters and conference proceedings were all excluded. The selection was limited to articles

Table 13.1 Inclusion and exclusion criteria

Criteria	Eligibility	Exclusion
Literature type	Journal (research article)	Journals (review), book series, book chapter, conference proceedings
Language	English	Non-English
Timeline	Between January 2010–July 2021	Before January 2010
Type of journal	Social science	Hard science
Country	sub-Saharan countries	Any country outside of sub-Saharan Africa

Source Authors' compilation (2021)

published in English. The third stage was the eligibility stage, in which all articles were accessed. The review focused on studies conducted in sub-Saharan countries and published between January 2010 and July 2021. Therefore, studies published before 2010 and studies conducted in countries outside of sub-Saharan Africa were also excluded. Meanwhile, we acknowledge that the grey literature on the subject is extensive, particularly practitioners' reports and evaluations of donor-funded projects, but we have not included them in this study. While this may be a limitation of the current article, the purpose of this study was to provide a focused discussion and analysis as presented in the academic literature. That said, a similar analysis of the grey literature could be examined at a later date, which would give greater robustness to the analysis undertaken in this study.

Based on eligibility and exclusion criteria, 1063 articles were excluded. Next, titles and abstracts were initially scrutinised to determine their relevance. Then, the full texts were downloaded to ascertain if they, directly and indirectly, addressed issues of indigenous knowledge, local governance and climate change adaptation, in sub-Saharan Africa. At the final stage of the review, a total of 26 articles were identified and used for the qualitative analysis (Fig. 13.1). Thereafter, the items considered were evaluated and analysed. Data was extracted by first reading the abstracts and then the full (in-depth) articles to identify appropriate themes and subthemes. A qualitative analysis was carried out using thematic analysis.

13.3 Results and Discussion

13.3.1 Description of the Studies Analysed

A total of 13 sub-Saharan African countries are represented in the articles included in the review. The country with the largest number of studies was South Africa (n = 6) followed by Zimbabwe (n = 5) (Fig. 13.2). Overall, the review revealed that the majority of studies were conducted in South African (n = 13) and East African (n = 9) countries. In contrast, very few studies were conducted in West African countries (n = 4) (Fig. 13.3). Twenty-six papers (1.5% of the initial search) met the inclusion criteria, scoring three or above. An identifier number (#1–26) was assigned to each publication and is used to refer to each one individually. These 26 high-quality papers covered quantitative (n = 2), qualitative (n = 16) and mixed methods (n = 8) approaches.

13.3.2 The Nexus Between IKS and Local Governance in Facilitating Climate Change Adaptation in SSA

Integrating indigenous knowledge systems into other knowledge bases is the best way to ensure more effective and sustainable implementation of climate change adaptation strategies in communities (Makondo and Thomas 2018). Such

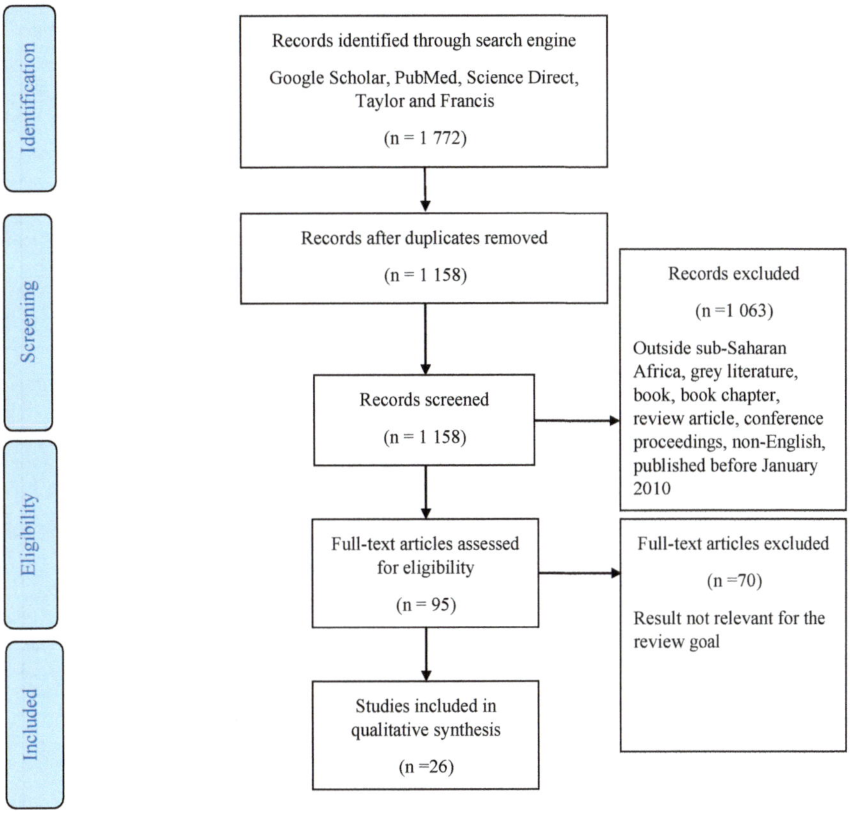

Fig. 13.1 Flow chart of article screening and selection

Fig. 13.2 Distribution of the number of studies per country. *Source* Authors (2021)

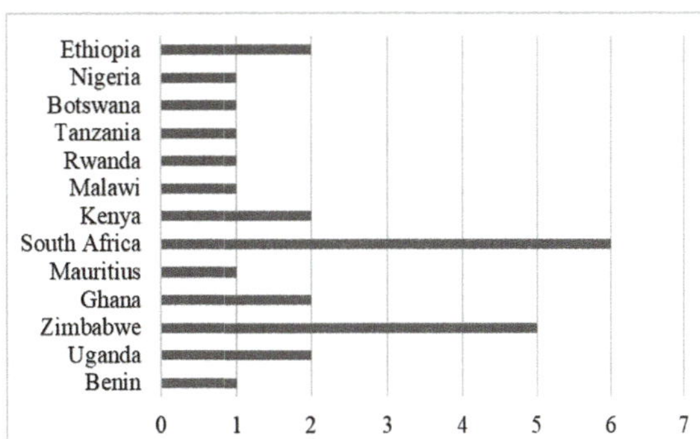

integration will allow for the formation of a system that benefits from the local relevance of indigenous techniques and the increased accuracy, efficiency, cost-effectiveness and sustainability of modern techniques (Mogomotsi et al. 2020), which is desperately needed to facilitate the actualisation of SDG 13. This section highlights the themes that emerged from the review of literature, which have been instrumental in ensuring effective collaborations of IKS into local governance to scale up climate change adaptation in SSA. They include:

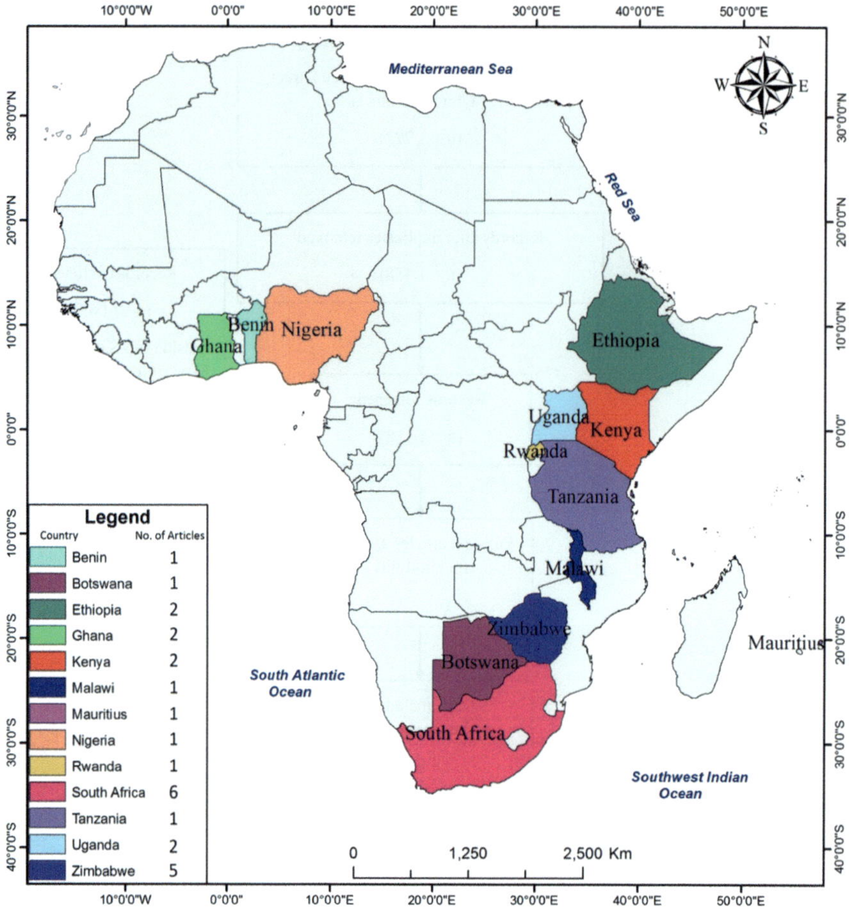

Fig. 13.3 Geographical distribution of the reviewed papers. *Source* Authors (2021)

13.3.2.1 Platforms for Dialogue Between Local Government and Indigenous Stakeholders

The only article from which this theme emerged revealed that a key attribute that has fostered a healthy relationship between IKS and local governance, which in turn facilitated climate change adaptation in local communities, is opening up space for dialogue between local government officials and indigenous stakeholders (Musarandega et al. 2018). To enhance local governance to address climatic risks, the integration of a wide range of stakeholders in participatory processes—with the aim of co-producing knowledge—is overwhelmingly essential (Williams et al. 2018). Co-producing knowledge underpinned by differential epistemology holds enormous value in scaling up adaptation to climate change (Ebhuoma 2020). This includes creating a common platform for dialogue to bridge the differences between formal and informal stakeholders in climate change adaptation planning (Innocenti and Albrito 2011).

Evidence from the literature in Zimbabwe, for example, suggests that the integration of indigenous voices in climate change adaptation processes is attributed to statutory obligation by parliament (e.g. the amended Council of Traditional Leaders Act of 2000) to ensure local traditional authorities are empowered as the Act bestows on traditional authorities presidential appointee status. This provides an avenue for them to inject their expertise in climate change adaptation strategies tailored to specific localised contexts (Musarandega et al. 2018). Due to the statutory backing, local traditional leaders—custodians of IKS—have the latitude to work in liaison with agents such as the Environmental Management Agency (EMA), the Forestry Commission and the Republic of Zimbabwe police to protect the environment from damage. In fact, EMA in its operational framework does recognise the role of LTAs. This platform, therefore, forms a formidable taskforce capable of ensuring that sustainable practices are put in place within rural communities. Thus, it is unsurprising to note that a study conducted in SSA by Pasquini et al. (2013) clamoured for a change in planning and other laws that govern local governments' operation in order to mainstream the climate change adaptation strategies at local level to address the individual-level barriers (such as a lack of understanding of climate change and adaptation options) that impede the expertise of indigenous peoples from featuring in climate change adaptation governance schemes. Nonetheless, it is asserted that Zimbabwe's new constitution—adopted in 2013—strictly regulates the affairs of traditional authorities (Chigwata 2016). This might cause it to be overlooked by some local government officials when coming up with schemes that will affect local people in Zimbabwe.

In light of the above, it can be argued that for IKS to feature prominently in climate change governance, a statutory declaration may be required that allocates equal rights to both traditional authorities and local governments in SSA to contribute equally to climate governance at local level. This may expedite fulfilling the UN SDG 13 agenda which calls for a great deal of expertise to tackle climate change. Also, it will likely truncate the inferiority stigma often associated with IKS as not being able to produce knowledge that is both theoretically and practically relevant, especially when compared with scientific knowledge (Ebhuoma 2020), which results in the knowledge not being taken seriously.

13.3.2.2 Institutional Support for Indigenous People to Implement Local Plans

It is acknowledged that indigenous people often have little influence in policies that shape their lives. Policies developed at national levels are rarely responsive to local needs (Williams et al. 2020) and fail to provide indigenous people with the relevant assets needed for them to be innovative and adapt effectively to climate change (Sharma et al. 2014). For climate change adaptation strategies to build resilience at local level, the knowledge systems of indigenous people must be recognised in formal planning systems and provided with the relevant institutional support by the government at local level. Institutional support entails providing direct access to resources for implementing local plans that are highly inclusive, primarily through collective action (Williams et al. 2020). This is observed in a piloted development planning in Isiolo County, a dryland area in Kenya whose local economy is primarily driven by communal pastoralism and where the authority of customary institutions for governing resources was hampered by the overlapping and often contradictory systems of governance through national laws and regulations (Pattison et al. 2013).

However, the adoption of a new national constitution in August 2010 as well as the modification of local governance systems paved the way to modify decentralised development planning and prioritise local adaptation to climate change. This opened up space for shared learning dialogues that combined indigenous and formal knowledge of the natural environment (Sharma et al. 2014). As documented by Roba (2014, p. 182) in the Kenyan context, "given the focus of devolution of responsibilities for natural resource management to local communities in the constitution, the local government were open and facilitative to processes like development of bylaws based on customary rules and norms". In turn, this supported bottom-up prioritisation of investments—such as transport and communications infrastructure and health and veterinary care—as highlighted by communities through participatory processes needed for scaling up the adaptive capacity and resilience of indigenous people (Sharma et al. 2014). This fostered a greater appreciation on the part of government staff regarding the value of IKS and the factors that underpin crucial pastoral management strategies. Also, the dialogue built the capacity of indigenous people to articulate these issues in a way accessible to other stakeholders. These may be precisely why District Councils' representatives in a study conducted in Mauritius clamoured for mandatory consultations with respective communities when projects are planned and conducted in their respective jurisdictions to identify their infrastructural needs to build their resilience to climate change (Williams et al. 2020). It is worth mentioning that despite the enablers of IKS and local governance collaboration in facilitating adaptation to climate change in SSA, there are barriers that undermine the process. These are underlined in the next section.

13.3.3 Collaboration Challenges Between Local Governance and IKS in SSA

Multiple barriers affect the ability of local governments in SSA to integrate the expertise of indigenous peoples into climate change adaptation. This ranges from the individual, institutional system to the regulatory and socio-cultural systems that make up local communities (Pasquini et al. 2013). Other barriers include the lack of technical know-how and financial and human resources, strict legislations and ineffective monitoring mechanisms, which prevent effective adaptation to climate change (Williams et al. 2020). For example, it is acknowledged that indigenous forecasts that are not limited to plant phenology, animal behaviours and lunar observations inform the decision of farmers in SSA as to when to start cultivating, the types and quantity of food to produce in a particular season (Abate 2016; Apraku et al. 2021; Ebhuoma and Simatele 2019; Egeru 2012; Grey et al. 2020; Kpadonou et al. 2012; Mugambiwa 2018; Napogbong et al. 2020; Ngwenya et al. 2016; Nkomwa et al. 2014; Nkuba et al. 2020; Taremwa et al. 2016; Ubisi et al. 2020). Because indigenous forecasts are not as reliable as they have been in previous decades partly due to the increased occurrences of climate change, there have been increased calls for meteorological forecasts to complement the limitations of IKS (Nkomwa et al. 2014; Ebhuoma 2020; Nkuba et al. 2020). Not only will this help to develop socially acceptable adaptation strategies at local level (Napogbong et al. 2020), but it would also allow the younger generation to connect with the few knowledge keepers for learning and preservation of local heritage (Apraku et al. 2021) and provide more comprehensive and accurate information for local community adaptation (Jiri et al. 2015). Yet, despite various strategies proposed by scholars (Mapfumo et al. 2016; Napogbong et al. 2020; Ubisi et al. 2020) regarding ways to integrate IKS with scientific knowledge, this call has not been heeded in SSA chiefly due to the aforementioned factors. As Grey et al. (2020) argue the lack of recognition and integration of IKS into formal adaptation strategies has resulted in limited success of external interventions in local communities.

To overcome the barriers impeding IKS from featuring in local governance schemes to scale up climate change adaption, File and Derbile (2020) advocate for an endogenous approach to development and adaptation that builds on IKS for effective community education, mobilisation and participatory response to climate change. Also, training and capacity building needs of local government officials must be identified in order to provide targeted training on effective adaptation to climate change (Williams et al. 2020). It is worth mentioning that the capacity of local authorities must be strengthened to improve their knowledge. This will enable them to appreciate the integration of indigenous and modern climate change adaptation strategies (Musarandega et al. 2018). Furthermore, local governments need to be empowered to receive appropriate education and information on climate change, anticipated local impacts and adaptation (Pasquini et al.

2013). Further, appropriate measures to reduce the unintended effects of party politics on the effective functioning of local governance need to be taken seriously (Pasquini et al. 2013), as this may minimise entry points for IKS in climate change adaptation at local level. The organisation of annual local government climate change adaptation summits, as well as the possibility of sharing information and good practices when collaborating with indigenous peoples, identifying challenges and offering specific training workshops should be explored. There is also a need to identify the training and capacity building needs of local governments in order to provide targeted training on climate change adaptation (Williams et al. 2020). The training should be tailored in a way that builds local government's capacity on how to work with indigenous peoples in a way that is respectful to their ways of knowing and does not perceive their knowledge to be inferior to theirs.

13.4 Conceptual Framework: IKS, Local Governance and Climate Change Adaptation

From the findings of the systemic review, a conceptual framework is developed to show the relationship regarding how IKS can play a vital role in scaling up climate governance at local level (Fig. 13.4). The framework illustrates that IKS should be a starting point for adaptation to climate change in developing countries. However, to ensure continuity and sustainability, adaptation to climate change must be based on good governance at local level, which encompasses effective management of indigenous institutions and strongly advocating for local participation. According to Tharakan (2017), many technologies classified as "appropriate" for developing communities to meet basic water, sanitation, and agricultural adaptation have their roots in IKS. In many SSA countries, communities employ IKS as a basis for decision-making to achieve food security, human and animal health, education, and natural resource management (Smit and Masoga 2012). For instance, indigenous people, particularly women, have developed local food security strategies for climate change adaptation over the years, including knowledge of the behaviours of living organisms, wind direction and stars as early indicators of changing weather conditions. These indicators play a pivotal role in the selection of appropriate seeds, is a crucial determinant in the types of agricultural practices to be engaged in the forthcoming planting season and food preservation techniques such as fermentation and sun drying to attain food security (Kamwendo and Kamwendo 2014; Tlhompho 2014). As pointed out by UNESCO (2020), IKS have the potential to contribute significantly to the achievement of SDG 13 because it provides insights that could enable better management of ecological challenges that could require low investments, preventing biodiversity loss and contribute to resilience and recovery in the case of natural disasters. This potential has also been mentioned by the Intergovernmental Panel on Climate Change (IPCC):

> Indigenous knowledge is an invaluable basis for developing adaptation and natural resource management strategies in response to environmental and other forms of change […]indigenous or traditional knowledge may prove useful for understanding the potential of certain adaptation strategies that are cost-effective, participatory and sustainable (IPCC 2007, 2010).

It is important to state that local governments represent a key opportunity to implement local adaptation to climate change impacts. As the level of government closest to the grassroots, local governments have a major role to play in achieving climate change adaptation. Local governments generally hold significant authority in areas relevant to effective governance. Also, they play an instrumental role in allocating social resources, transforming development methods and improving people's livelihoods (United Cities and Local Governments 2016). By considering the intrinsic link between climate change adaptation and local governance, the valorisation of indigenous knowledge could be achieved. In theory, local governance occurs when people live in a community and interact closely enough

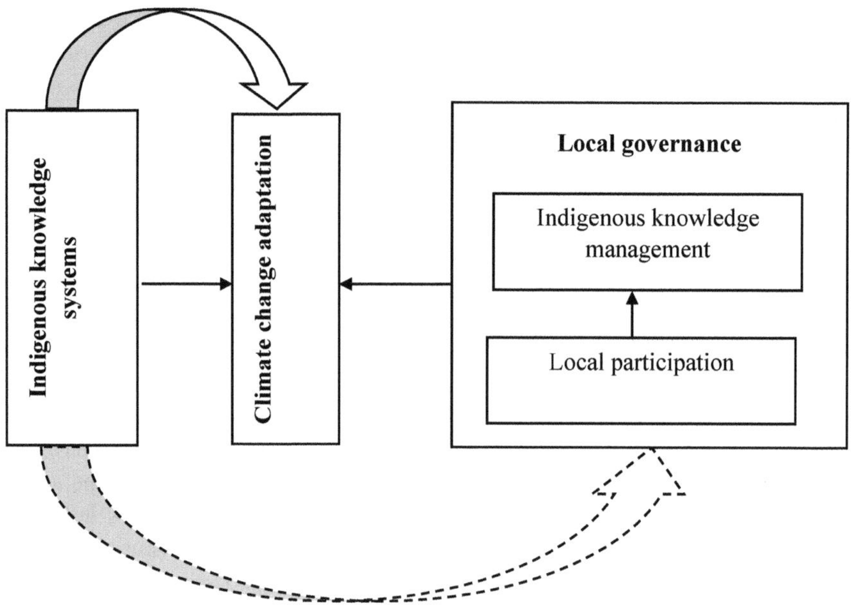

Fig. 13.4 Conceptual framework: Indigenous knowledge systems, climate change adaptation and local governance. The role of local governance is to provide an enabling environment to upscale the management of indigenous knowledge by local people. *Source* Authors' conceptualisation (2021)

(Ndreu 2016) to solve their climate change-related problems. Thus, they must work together as a group to achieve the outcomes they desire. This process involves the use of indigenous knowledge that is consistent with the strategies developed by communities to adapt to the negative effects of climate change. Two fundamental elements to consider regarding local governance include indigenous knowledge management and the way local people participate in this management (Fig. 13.4). Both are not only distinctive elements of local governance but also serve as an indicator of effectiveness. Effective indigenous knowledge management would therefore ensure that indigenous knowledge is created, stored and used to meet the objectives and needs of people and their organisations (Prabha 2021). According to Prabha (2021), indigenous knowledge management involves four stages including: knowledge acquisition; knowledge development through identifying and generating a set of knowledge documents from information that has been acquired from acquisition stage. The third stage is the knowledge access whereby indigenous knowledge is stored and accessed via information retrieval systems. The last stage is management, which mainly deals with knowledge codification which results in making the indigenous traditional knowledge visible, accessible and usable for decision-making by various stakeholders and the end user. Arguably, the framework can be used by development practitioners to analyse the relevance of IKS, local governance in climate change adaptation.

13.5 Conclusion and Recommendations

In sub-Saharan Africa (SSA), climate change has had a deleterious impact on rural communities, which consistently puts the ability of households to maintain their livelihoods in a precarious position. Despite the low capacity of rural households to adapt to climate change, they have been able to consistently engage in their livelihood activities through the skilful application of their IKS. This has fortified the call for locally-driven solutions to be injected into climate governance schemes at local level, as this will contribute directly towards achieving SDG 13. Local governments have a fundamental role to play in promoting IKS (Mogomotsi et al. 2020; Radeny et al. 2019). Thus, it is advocated that IKS should feature prominently in climate change policies, especially at local level. This is because such knowledge provides the necessary baseline information from historical and cultural contexts that helps to understand how things work at local level, which creates better engagement with local people and therefore better policies (Iloka 2016). In this regard, this chapter sought to underline the factors that facilitate effective collaboration between IKS and local governance in scaling up climate change adaptation and highlight the barriers that stifle this union at the local level.

A systematic review of existing literature in SSA was conducted. Findings reveal that providing platforms for dialogue between local government and indigenous stakeholders and providing institutional support for indigenous people to implement local plans were drivers that ensured collaboration between IKS and local government. Barriers to the union ranged from lack of technical know-how and financial and human resources to strict legislations and ineffective monitoring mechanisms of schemes meant to provide IKS with a viable platform to share their expertise with local government officials. In light of these findings, we assert that climate change governance will be holistic and more effective if the IKS of local communities are incorporated to ensure sustainable interventions at grassroots level (e.g. see Diga et al. 2016). Such efforts will improve information gathering and dissemination, resource mobilisation, the establishment of useful networks with relevant stakeholders, capacity building of farmers on different fronts, and provision of leadership in climate adaptation programmes (Makate 2019).

Finally, it is worth highlighting that evidence in the literature showcasing the relevance of IKS and local governance in strengthening adaptation to climate change in SSA has grown at a relatively slow pace. With only four (Roba 2014; Sharma et al. 2014; Musarandega et al. 2018; Williams et al. 2020) out of the 26 selected empirical articles that provided case studies regarding the collaboration of indigenous peoples and local government officials while the remaining literature provided pointers for factors undermining such collaboration, with a concentration of studies in the Southern and Eastern African region, our ability to critically interrogate the literature for emerging patterns is limited. This underscores an important gap in the research, which will need to be urgently addressed if discussions of climate action in relation to IKS and local governance are to inform better decision-making at local level. Nonetheless, the findings in this chapter highlight the drivers that have ensured IKS are integrated with local governance to improve climate governance. Also, through the conceptualisation of a framework carved out from the findings of the result, the chapter provides a valuable roadmap that could guide development practitioners on how to ensure that the voices of indigenous peoples are incorporated in climate governance at grassroots level.

References

Abate T (2016) Contribution of indigenous knowledge to CC and adaptation response in Southern Ethiopia. J Earth Sci Clim Change 7(11):1–9. https://doi.org/10.4172/2157-7617.1000377

Apraku A, Morton JF, Apraku Gyampoh B (2021a) CC and small-scale agriculture in Africa: does indigenous knowledge matter? Insights from Kenya and South Africa. Sci Afr 12:e00821. https://doi.org/10.1016/j.sciaf.2021.e00821

Amouzou KA, Lamers JPA, Naab JB, Borgemeister C, Vlek PLG, Becker M (2019) Climate change impact on water- and nitrogen-use efficiencies and yields of maize and sorghum in the northern Benin dry savanna, West Africa. Field Crops Research 235(4):104–117

Arruda GM, Krutkowski S (2017) Arctic governance, indigenous knowledge, science and technology in times of climate change: Self-realization, recognition, representativeness. Journal of Enterprising Communities, 11(4):514–528. https://doi.org/10.1108/JEC-08-2015-0041

Baker I, Peterson A, Brown G, McAlpine C (2012) Local government response to the impacts of climate change: an evaluation of local climate adaptation plans. Landsc Urban Plan 107(2):127–136. https://doi.org/10.1016/j.landurbplan.2012.05.009

Berkes F (2012) Sacred ecology. Routledge Press

Chigwata T (2016) The role of traditional leaders in Zimbabwe: are they still relevant? Law Democr Dev 20:69–90

Corfee-Morlot J, Cochran I, Hallegatte S, Teasdale PJ (2011) Multilevel risk governance and urban adaptation policy. Clim Change 104(1):169–197. https://doi.org/10.1007/s10584-010-9980-9

Corsi M, Hagemann S, Salgado SC (2012) Africa Adaptation Programme third quarterly report 2012. Prepared by theAAP Inter-Regional Technical Support Component. UNDP Africa Adaptation Programme, pp 54

Diga K, Ngcoya M, Bracking S, Lombo M, Ntombela S (2016) Poverty Reduction Co-benefits through Indigenous Knowledge in Climate Change Adaptation: A Study within the eThekwini Municipality. Journal of Social Sciences, 46(2):87–97. https://doi.org/10.1080/09718923.2016.11893515

Ebhuoma E, Simatele D (2017) Defying the odds: climate variability, asset adaptation and food security nexus in the Delta State of Nigeria. Int J Disaster Risk Reduct 21:231–242

Ebhuoma EE, Simatele DM (2019) 'We know our Terrain': indigenous knowledge preferred to scientific systems of weather forecasting in the Delta State of Nigeria. Clim Dev 11(2):112–123. https://doi.org/10.1080/17565529.2017.1374239

Ebhuoma E (2020) A framework for integrating scientific forecasts with indigenous systems of weather forecasting in southern Nigeria. Dev Pract 30(4):472–484. https://doi.org/10.1080/09614524.2020.1723494

Egeru A (2012) Role of indigenous knowledge in CC adaptation: a case study of the Teso Sub-Region, Eastern Uganda. Indian J Tradit Knowl (IJTK) 11(4)

File DJM, Derbile EK (2020) Sunshine, temperature and wind: community risk assessment of CC, indigenous knowledge and CC adaptation planning in Ghana. Int J CC Strateg Manag 12(1):22–38. https://doi.org/10.1108/IJCCSM-04-2019-0023

Fröde A, Scholze M, Manasfi N (2013) Taking a climate perspective on development: GIZ's climate-proofing for development approach. Climate and Development, 5(2):160–164. https://doi.org/10.1080/17565529.2012.751894

File DJMB, Domapielle MK, Derbile EK (2021) Local Perspectives on the Causes of Climate Change in Rural Ghana: Implications for Development Planning. Ghana J Geography, 13(2)

Fröde A, Scholze M, Manasfi N (2013) Taking a climate perspective on development: GIZ's climate-proofing for development approach. Climate and Development, 5(2):160–164. https://doi.org/10.1080/17565529.2012.751894

Grey MS, Masunungure C, Manyani A (2020) Integrating local indigenous knowledge to enhance risk reduction and adaptation strategies to drought and climate variability: the plight of smallholder farmers in Chirumhanzu district, Zimbabwe. Jamba (potchefstroom, South Africa) 12(1):1–10. https://doi.org/10.4102/JAMBA.V12I1.924

Higgins JPT, Altman DG, Gøtzsche PC, Jüni P, Moher D, Oxman AD, Savović J, Schulz KF, Weeks L, Sterne JAC (2011) The cochrane collaboration's tool for assessing risk of bias in randomised trials. BMJ 343(7829). https://doi.org/10.1136/bmj.d5928

Hunde M (2012) Integrating community-based disaster risk reduction and climate change adaptation into local development planning. Proceedings of Communicating Community Based Adaptation, Sixth International Conference on Community-Based Adaptation, 16–22, April 2012, Hanoi, Vietnam, Session 7, Abstract 3. Organised by the International Institute for Environment and Development in the UK, the Ministry of Agriculture and Rural Development (MARD) of Vietnam, the Ministry of Natural Resources and Environment (MONRE) of Vietnam, and the Bangladesh Centre for Advanced Studies, IIED, London, UK, pp 2021

Iloka NG (2016) Indigenous knowledge for disaster risk reduction: an African perspective. Jàmbá J Disaster Risk Stud 8(1). https://doi.org/10.4102/jamba.v8i1.272

Innocenti D, Albrito P (2011) Reducing the risks posed by natural hazards and CC: the need for a participatory dialogue between the scientific community and policy makers. Environ Sci Policy 14(7):730–733. https://doi.org/10.1016/J.ENVSCI.2010.12.010

IPCC (2007) CC 2007: impacts, adaptation and vulnerability. In: Parry ML, Canziani OF, Palutikof JP, van der Linder PJ, Hanson CE (eds) Contribution of Working Group II to the fourth assessment report of the intergovernmental panel on CC. Cambridge University Press. https://www.ipcc.ch/site/assets/uploads/2018/03/ar4_wg2_full_report.pdf

IPCC (2014) Technical summary. In: CC 2014—impacts, adaptation and vulnerability: Part A: global and sectoral aspects: working group II contribution to the IPCC fifth assessment report. Cambridge University Press pp 34–94. https://doi.org/10.1017/CBO9781107415379.004

IPCC (2010) Review of the IPCC Processes and Procedures, report by the InterAcademy Council (IPCC-XXXII/Doc. 7), 32nd Session, Busan, Seoul, 11–14 October 2010

Jiri O, Mafongoya P, Chivenge P (2015) Indigenous knowledge systems, seasonal 'quality' and CC adaptation in Zimbabwe. Clim Res 66(2):103–111. https://doi.org/10.3354/cr01334

Kamwendo G, Kamwendo J (2014) Indigenous knowledge-systems and food security: some examples from Malawi. J Hum Ecol 48(1):97–101. https://doi.org/10.1080/09709274.2014.11906778

Kelman I, Mercer J, Gaillard J (2012) Indigenous knowledge and disaster risk reduction. Geography 97(1):12–21. https://doi.org/10.1080/00167487.2012.12094332

Kpadonou RAB, Adégbola PY, Tovignan SD (2012) Local knowledge and adaptation to CC in Ouémé valley, Benin. Afr Crop Sci J 20(2):181–192. https://doi.org/10.4314/acsj.v20i2

Makondo CC, Thomas DSG (2018) CC adaptation: linking indigenous knowledge with western science for effective adaptation. Environ Sci Policy 88:83–91. https://doi.org/10.1016/j.envsci.2018.06.014

Makate C, Makate M, Mango N, Siziba S (2019) Increasing resilience of smallholder farmers to climate change through multiple adoption of proven climate-smart agriculture innovations. Lessons from Southern Africa. J Environ Management, 231:858–868. https://doi.org/10.1016/J.JENVMAN.2018.10.069

Mapfumo P, Mtambanengwe F, Chikowo R (2016) Building on indigenous knowledge to strengthen the capacity of smallholder farming communities to adapt to climate change and variability in southern Africa. Clim Dev 8(1):72–82

Marango T, Francis J, Adjibolosso S (2016) Realities of communication and social inclusion in Zimbabwe local government: a case of the Eastern Chimanimani District. Rev Human Factor Stud 22(1):104–128

Maupin A (2016) The SDG13 to combat CC: an opportunity for Africa to become a trailblazer? Afr Geogr Rev 36(2):131–145. https://doi.org/10.1080/19376812.2016.1171156

Mavhura E, Manyena SB, Collins AE, Manatsa D (2013) Indigenous knowledge, coping strategies and resilience to floods in Muzarabani, Zimbabwe. Int J Disaster Risk Reduct 5:38–48

Mogomotsi PK, Sekelemani A, Mogomotsi GEJ (2020) CC adaptation strategies of small-scale farmers in Ngamiland East, Botswana. Clim Change 159(3):441–460. https://doi.org/10.1007/S10584-019-02645-W

Mugambiwa SS (2018) Adaptation measures to sustain indigenous practices and the use of indigenous knowledge systems to adapt to CC in Mutoko rural district of Zimbabwe. Jàmbá J Disaster Risk Stud 10 (1). https://doi.org/10.4102/JAMBA.V10I1.388

Musarandega H, Chingombe W, Pillay R (2018) Harnessing local traditional authorities as a potential strategy to combat the vagaries of climate change in Zimbabwe. Jàmbá: J Disaster Risk Studies, 10(1). https://doi.org/10.4102/JAMBA.V10I1.651

Mutasa M (2015) Knowledge apartheid in disaster risk management discourse: is marrying indigenous and scientific knowledge the missing link? Jàmbá J Disaster Risk Stud 7(1):10. https://doi.org/10.4102/jamba.v7i1.150

Nakashima D, McLean KG, Thulstrup HD, Castillo AR, Rubis JT (2012) Weathering uncertainty: traditional knowledge for CC assessment and adaptation.

Nalau J, Becken S, Schliephack J, Parsons M, Brown C, Mackey B (2018) The role of indigenous and traditional knowledge in ecosystem-based adaptation: a review of the literature and case studies from the Pacific Islands. Weather Clim Soc 10(4):851–865. https://journals.ametsoc.org/view/journals/wcas/10/4/wcas-d-18-0032_1.xml

Napogbong LA, Ahmed A, Derbile EK (2020) Fulani herders and indigenous strategies of CC adaptation in Kpongu community, North-Western Ghana: implications for adaptation planning. 13(3):201–214. https://doi.org/10.1080/17565529.2020.1746231

Ndreu A (2016) The definition and importance of local governance. European University of Tirana

Ngwenya BN, Thakadu OT, Magole L, Chimbari MJ (2016) Memories of environmental change and local adaptations among Molapo farming communities in the Okavango Delta, Botswana-A gender perspective. Acta Tropica. https://doi.org/10.1016/j.actatropica.2016.11.029

Nkomwa EC, Joshua MK, Ngongondo C, Monjerezi M, Chipungu F (2014) Assessing indigenous knowledge systems and CC adaptation strategies in agriculture: a case study of Chagaka Village, Chikhwawa, Southern Malawi. Phys Chem Earth Parts A/B/C 67–69:164–172. https://doi.org/10.1016/J.PCE.2013.10.002

Nkuba MR, Chanda R, Mmopelwa G, Kato E, Mangheni MN, Lesolle D (2020) Influence of indigenous knowledge and scientific climate forecasts on Arable Farmers' climate adaptation methods in the Rwenzori region, Western Uganda. Environ Manag 65 (4):500–516. https://doi.org/10.1007/S00267-020-01264-X

Niang IOC, Ruppel MA, Abdrabo A, Essel C, Lennard JP, Urquhart P (2014): Africa. In: Climate Change 2014: Impacts, Adaptation, and Vulnerability. Part B: Regional Aspects. Contribution of Working Group II to the Fifth Assessment Report of the Intergovernmental Panel on Climate Change. Barros VR, Field CB, Dokken DJ, Mastrandrea MD, Mach KJ, Bilir TE, Chatterjee M, Ebi KL, Estrada YO, Genova RC, Girma B, Kissel ES, Levy AN, MacCracken S, Mastrandrea PR, White LL (eds.) Cambridge University Press, Cambridge, United Kingdom and New York, NY, USA, pp 1199–1265

Owusu K, Yankson PWK, Ayisi EK (2018) Mainstreaming climate change into local governance: financing and budgetary compliance in selected local governments in Ghana. Develop Practice, 28(1):65–80

Pasquini L, Cowling RM, Ziervogel G (2013) Facing the heat: barriers to mainstreaming climate change adaptation in local government in the Western Cape Province, South Africa. Habitat Int 40:225–232. https://doi.org/10.1016/J.HABITATINT.2013.05.003

Pasquini L, Ziervogel G, Cowling RM, Shearing C (2015) What enables local governments to mainstream climate change adaptation? Lessons learned from two municipal case studies in the Western Cape, South Africa. Climate and Development, 7(1):60–70. https://doi.org/10.1080/17565529.2014.886994

Pattison J, Hesse C, Anderson S (2013) Local to national: putting local voices at the heart of national policies. In: Paper presented at the hunger, nutrition and climate justice: a new dialogue: putting people at the heart of global development conference, Dublin, 15–16 April

Prabha D (2021) Indigenous Knowledge Management vs Adam's Knowledge Management. Acta Scientific Agriculture, 5(4):44–45. https://actascientific.com/submission.php

Radeny M, Desalegn A, Mubiru D, Kyazze F, Mahoo H, Recha J, Kimeli P, Solomon D (2019) Indigenous knowledge for seasonal weather and climate forecasting across East Africa. Climatic Change, 156(4):509–526. https://doi.org/10.1007/s10584-019-02476-9

Reisinger A, Wratt D, Allan S, Larsen H (2011) The role of local government in adapting to climate change: lessons from New Zealand. In: Ford JD, Berrang-Ford L (eds) Climate change adaptation in developed nations. Springer Netherlands, pp 303–319. https://doi.org/10.1007/978-94-007-0567-8_22.

Roba G (2014). Strengthening communal Governance of Rangeland in Northern Kenya. In: Herrera P, Davies J, Manzano P (eds) The Governance of Rangelands: collective action for sustainable pastoralism. Routledge, UK

Sharma V, Orindi V, Hesse C, Pattison J, Anderson S (2014) Supporting local climate adaptation planning and implementation through local governance and decentralised finance provision. Dev Pract 24(4):579–590

Shaw R, Uy N, Baumwoll J (2008) Indigenous knowledge for disaster risk reduction. United Nations Office for Disaster Risk Reduction (UNISDR)

Sithole J (2007) The challenges faced by African libraries and information centres in documenting and preserving indigenous knowledge. IFLA J 33(2):117–123. https://doi.org/10.1177/0340035207080304

Smit JA, Masoga M (2012) Indigenous knowledge in Africa: challenges and opportunities. People's Publishers

Taremwa NK, Gashumba D, Butera A, Ranganathan T (2016) CC Adaptation in Rwanda through indigenous knowledge practice. J Soc Sci 46(2):165–175. https://doi.org/10.1080/09718923.2016.11893524

Tharakan J (2017) Indigenous knowledge systems for appropriate technology development. In: Indigenous people. InTech. https://doi.org/10.5772/intechopen.69889

Tlhompho G (2014) African indigenous food security strategies and CC adaptation in South Africa. J Hum Ecol 48(1):83–96. https://doi.org/10.1080/09709274.2014.11906777

Tom MN, Sumida Huaman E, McCarty TL (2019) Indigenous knowledges as vital contributions to sustainability. In: International review of education, vol 65, Issue 1. Springer Netherlands, pp 1–18 https://doi.org/10.1007/s11159-019-09770-9

Ubisi NR, Kolanisi U, Jiri O (2020) The role of indigenous knowledge systems in rural smallholder farmers' response to CC: case study of Nkomazi local municipality, Mpumalanga, South Africa. J Asian Afr Stud 55(2):273–284. https://doi.org/10.1177/0021909619874824

United Cities and Local Governments (2016) The Role of Local Governments in Territorial Economic Development

UNESCO (2020) Indigenous knowledge and CC. https://en.unesco.org/links/climatechange

Williams DS, Rosendo S, Sadasing O, Celliers L (2020) Identifying local governance capacity needs for implementing climate change adaptation in Mauritius. Clim Policy 20(5):548–562. https://doi.org/10.1080/14693062.2020.1745743

Williams DS, Costa MM, Celliers L, Sutherland C (2018) Informal Settlements and Flooding: Identifying Strengths and Weaknesses in Local Governance for Water Management. Water, 10(7):871. https://doi.org/10.3390/W10070871

Ortiz-Bobea A, Ault TR, Carrillo CM, Chambers RG, Lobell DB (2020) The Historical Impact of Anthropogenic Climate Change on Global Agricultural Productivity. Nature Climate Change, 11(4):306–312. https://doi.org/10.1038/s41558-021-01000-1

Ortiz-Bobea A, Ault TR, Carrillo CM, Chambers RG, Lobell DB (2021) Anthropogenic climate change has slowed global agricultural productivity growth. Nature Climate Change,11(4):306–312. https://doi.org/10.1038/s41558-021-01000-1

Zheng K, Tan L, Sun Y, Wu Y, Duan Z, Xu Y, Gao C (2021) Impacts of climate change and anthropogenic activities on vegetation change: Evidence from typical areas in China. Ecological Indicators, 126(7):107648. https://doi.org/10.1016/J.ECOLIND.2021.107648

Jane Mbolle Chah (Ph.D.) is a senior lecturer in the Department of Agricultural Extension, University of Nigeria, Nsukka. Currently, she is a postdoctoral research fellow at Barcelona School of Economics, Spain. She has completed a certificate course on Building Community Resilience at the Coady Institute Canada. Her research interests include climate change, rural livelihoods, rural development, gender studies, agricultural innovation and poverty; evaluation of agricultural projects with particular interest in livestock extension.

Esdras A. R. Obossou (M.Sc.) is a social scientist. He has over 4 years of work and interdisciplinary research experience. He is also a research assistant (part-time) at the University of Parakou, Benin. His research interests include climate change and rural livelihoods, agricultural innovation and innovation systems, rural extension and advisory services, digital agriculture and sustainability transition. Esdras is an alumnus of the prestigious Orange Knowledge Programme (OKP) Scholarship, Netherlands and the German Academic Exchange Service (DAAD) scholarship.

Eromose E. Ebhuoma (Ph.D.) is a postdoctoral research fellow at the Department of Environmental Sciences, College of Agriculture and Environmental Sciences, University of South Africa (UNISA). His research expertise spans climate change vulnerability and adaptation, climate services, climate change policy and governance, local and indigenous knowledge systems, political ecology, rural livelihoods and environmental sustainability in sub-Saharan Africa. He is a steering committee member of the South African Adaptation Network.

Ifeoma Q. Anugwa (Ph.D.) is a lecturer in the Department of Agricultural Extension, University of Nigeria Nsukka. She has over ten years of interdisciplinary and trans-disciplinary research experience. She has conducted several researches in the areas of food and nutrition security, resilience, agricultural policies, climate change, rural livelihoods, information and communication technology, gender studies and poverty. Ifeoma is a member of several global research institutions and is a Fellow of the Norman E. Borlaug Leadership Enhancement in Agricultural Program (LEAP), USA.

Divine Ewane (Ph.D.) is a Lecturer and Acting Head of Department of Animal Science, University of Buea, Cameroon. Ewane has 29 years of experience in merging environment and forest conservation concerns with sustainable development imperatives, including Animal Production at industry and grass roots level. He was deputy director of the Centre for Entrepreneurship Research and Innovation (CERI) at the Catholic University Institute Buea (CUIB) in Cameroon and Lecturer II at Alex Ekweme Federal University Ndufu Alike, Ikwo (AE-FUNAI) in Nigeria.

14 Harnessing Indigenous Knowledge Systems for Enhanced Climate Change Adaptation and Governance: Perspectives from Sub-Saharan Africa

Felix K. Donkor and Kevin Mearns

14.1 Introduction

Sustainable development denotes a form of development that mainstreams socio-economic and environmental imperatives. Several countries in sub-Saharan Africa have aligned the sustainable development goals (SDGs) in their development agenda (Ebhuoma 2021). The 17 SDGs underpinning Agenda 2030, are complementary as progress in one aspect has positive spin-offs on other facets. Moreover, a hallmark of the SDGs is the imperative to Leave No One Behind, as United Nations member states commit to enhance the wellbeing and development of those socially marginalised, and hence left farthest behind (UN 2021). A further critical feature of sustainable development is the concept of intra and inter-generational equity. This indicates that present development should be inclusive and not destroy the capacity of future generations to thrive. Issues of inter-generationally equity are very common in discussions bordering management of the environment and its biological integrity (Summers and Smith 2014). Moreover, the SDGs are commended for attempting to address urgent problems plaguing the global community, in a holistic manner (UNWTO 2020).

Climate change represents one of the gravest challenges to sustainable development (Donkor et al. 2019) and the attainment of the SDGs. The United Nations indicates that the cost of climate-induced disasters to the global economy run into hundreds of billions of dollars yearly. Furthermore, geo-physical disasters, which are largely attributed to climate change, have cost the lives of over 1.3 million individuals, whilst injuring circa 4.4 billion people between 1998 and 2017 (UN 2021). In addition, the earth's natural resources, which underpin the environmentally sensitive livelihoods of the most vulnerable and are acknowledged as crucial in the framework of the SDGs, are equally threatened by climate change and compromising their attainment. Thus, climate change is a serious contemporary sustainability challenge that requires robust management (Apraku et al. 2021). This makes it necessary to address the issue of climate change governance and processes of knowledge co-creation that informs its effective management. This issue has become a key policy topic as the timeline for attaining the SDGs draws closer, coupled with calls to build back better sustainably in the post COVID-19 era (Donkor 2020).

Reviews by the United Nations (UN) show that despite evidence of progress across the

F. K. Donkor (✉) · K. Mearns
College of Agriculture and Environmental Sciences (CAES), University of South Africa (UNISA), 28 Pioneer Ave, Florida Park, Roodepoort 1709, South Africa
e-mail: donkofk@unisa.ac.za

K. Mearns
e-mail: mearnkf@unisa.ac.za

E. E. Ebhuoma and L. Leonard (eds.), *Indigenous Knowledge and Climate Governance*, Sustainable Development Goals Series, https://doi.org/10.1007/978-3-030-99411-2_14

SDGs, there is still a lot of work to be done (Donkor 2020). This is because the pace of action required is not commensurate with the scale of challenges faced. The so-called Decade of Action (2020–2030) refers to a call on the international community to catalyse sustainable measures and relevant interventions addressing urgent challenges. This involves issues ranging from poverty to gender, climate change, inequality, and closing the finance gap (UNDESA 2020; Sarrasanti et al. 2020). Consequently, the United Nations encourages social actors to galvanise action towards three core dimensions: (1) global action to achieve effective leadership, (2) improved resources, and (3) smarter solutions for realising the SDGs. Climate change intersects with all dimensions of this call. Moreover, it has been observed that some of the geographic regions that have little contribution to climate change are the worst affected, such as Africa (Mavhura and Mushure 2019). With climate impacts projected to increase in severity and frequency, it is important to reinforce adaption measures to build climate resilience (Forsyth 2018). In acknowledgement of the urgent need to address the threat of climate change, Goal 13 of the SDGs is centred on addressing the threat of climate change (UN 2021). Reinforcing the resilience of vulnerable communities to climate change comes with direct benefits for Goal 13 as well as the other SDGs such as no poverty (SDG 1), zero hunger (SDG 2), reduced inequality (SDG 10) and sustainable cities and communities (SDG 11).

The SDG 13 has, as part of its targets, a desire to enhance resilience and adaptive capacity to climate-related hazards (UN 2021). It also promotes approaches to reinforcing climate change-associated planning and management. However, adaptation is knowledge intensive. This is because no meaningful natural resources conservation or for that matter, sustainable development, can be attained in an absence of sound management (Ionela et al. 2015; Pereira et al. 2017). Knowledge is therefore an important resource for climate change adaption. It is hence important to harness the diverse forms of knowledge to effectively address climate change, which includes indigenous knowledge systems (IKS) and the involvement of relevant actors in climate change governance (Thakur et al. 2020).

Indigenous groups have close and unique connections to nature, mainly because their livelihoods and overall existence is closely tied to it, and hence, are particularly prone to the negative effects of climate change. They therefore often bear the brunt of climate change impacts (Spear et al. 2019). This further highlights the urgency of exploring climate action strategies to protect the environment and facilitate sustainability. Different communities have also developed unique ways of understanding and caring for the ecosystem, which is reflected in the indigenous epistemologies and knowledge (Liao et al. 2016). It is, hence, common to find indigenous groups living sustainably or in harmony with nature. Although some argue that their contributions towards environmental management have been accidental, others contend that the efforts of indigenous groups make conscious and deliberate efforts conserving the environment (High 2015). Etchart (2017) is of the view that indigenous people have, until recently, been viewed as victims of the impacts of climate change, instead of agents of environmental management. This situation resulted in the poor consideration given to IKS. This is reflected in the omission of indigenous groups in the framework of the Millennium Development Goals (MDGs) (Etchart 2017). This was later corrected in the SDG goals of 2015. This highlights the importance of IKS and the need to integrate them in climate adaptation measures (Ebhuoma et al. 2020a; Tantoh et al. 2019).

Indigenous groups have a vested interest in preventing environmental degradation and deforestation as their lives often revolve directly around nature (Donkor et al. 2020). It is therefore critical to acknowledge the capacity of their knowledge systems and decision-making processes to help safeguard the environment (Ebhuoma et al. 2020b; Mafongoya and Ajayi 2017). Ultimately, the complex nature of climate change necessitates inclusive approaches that enhance knowledge co-creation. However, this is often not the case (Jetoo 2019). One of the core groups whose voices are often side-lined in climate policy discourse are indigenous communities. The UN climate summit in 2019 acknowledged that conserving the land rights of indigenous peoples and local communities can be a scalable climate solution. Moreover, indigenous peoples and local communities are vital to stewarding and conserving the lands and forests of the world that are crucial to climate change management (UNESCO 2017; Etchart 2017). Against this background, this chapter highlights the various dimensions in which IKS has been utilised to facilitate climate change adaptation among rural households in sub-Saharan Africa. In so doing, it reinforces the burgeoning call for this vital knowledge base to be provided a viable platform for it to feature in issues pertaining to climate change governance and environmental sustainability. Deploying the requisite suite of measures that addresses the effects of climate change effectively and holistically, is an important element in the dynamics of climate governance (Coen et al. 2020). Thus, this chapter also discusses the implication of IKS for effective climate change governance.

14.2 Methodology

A systematic literature review of existing scholarship on IKS and climate change management/governance was conducted. Several systematic, comprehensive bibliographic searches were performed in scientific databases. This includes the Science Direct data base, with a focus on the geographic scope being Africa. A search revealed 275 journal articles. Further scanning of these journals led to the shortlisting of 15 case studies, of which nine were selected. Search terms were organised around the core themes of this study: "Indigenous knowledge systems"; "Climate change governance"; "Sustainability"; and "Environmental management", which were coupled with Boolean operators like AND as well as OR. Examples of such searches include indigenous knowledge and climate governance. The reference section of some of the articles were also perused given the related themes addressed therein. Principal inclusion criteria included peer-reviewed journals, articles written in English, articles produced within a ten-year framework (giving priority to more recent literature), indigenous knowledge systems that were related, and articles demonstrating the role of indigenous knowledge systems in climate governance related to Africa (see Table 14.1). Literature screening, which involves a three-pronged method of examining the topic, abstract and then the entire article, was followed with extracting relevant information. The study of literature ensures a critical appraisal of the current state of scholarship on a thematic area of interest within a specific timeframe (Ralph et al. 2014).

The thematic content analysis approach was adopted, whereby the major themes arising from the literature review were stringed into similar paragraphs through *meaning condensation*. Meaning condensation refers to summaries of repetitive themes into succinct formulations (Torraco 2016). This analysis technique revolves around five principal processes: the entire text is initially selected and evaluated to afford a comprehensive overview; the *meaning units* are identified by the researcher; the core thematic areas regarding the meaning units are paraphrased tersely; the meaning units are then cross-examined; and finally, the core themes of the entire text are combined to form a descriptive statement (Torraco 2016). The several themes emerging were condensed into four core themes to facilitate discussion (Table 14.1).

14.3 Results and Discussions

14.3.1 The Role of IKS in Enhancing Land-Based Livelihoods

A theme that emerged from the review of literature is the enormous role played by IKS in facilitating climate change adaptation to ensure continuity of land-based livelihoods (Liao et al. 2016; Guodaar et al. 2021). McLean (2012) is of the view that IKS are locally fine-tuned, which is vital for climate change adaptation and long-term community resilience. Indigenous groups

Table 14.1 Outline of literature case studies and relation to study theme

References	Study area	Main findings
Guodaar et al. (2021)	Ghana	Blended approaches that integrate IKS with scientific data are vital for producing effective and sustainable adaptation policies
Ogar et al. (2020)	Nigeria	IKS has successfully conserved biodiversity across the globe but is not accorded the same respect as Western Science as an approach to knowing. The two knowledge systems need to be coupled in a restitutive and rights-based manner
Mekonnen et al. (2021)	Ethiopia	IKS is the basis for successful community-based climate adaptation and natural resources management. Despite the changing dynamics of communities, IKS will serve as the basis of weather forecasting, agriculture and natural resources management. This will safeguard vulnerable livelihoods and natural resources
Apraku et al. (2021)	Kenya and South Africa	Local specificity is the main ingredient missing in climate change adaptation strategies in Africa. This limits our understanding of how IKS is deployed by specific communities to mitigate climate impacts. Greater integration of IKS will help foster syncretic agroecological knowledge amongst farmers and resource managers
Ramborun et al. (2021)	Mauritius	Access to scientific knowledge is often a challenge due to poor extension services and illiteracy. However, IKS is readily available and proven effective in enhancing the agricultural productivity through measures like mulch, no-tillage and tillage in Mauritius
Mavhura and Mushure (2019)	Zimbabwe	IKS involves spiritual and cultural values in many rural environments such as the Nharira community. These resources have proven invaluable in safeguarding protected areas[,] which are under threat together with forest and wildlife resources
Thakur et al. (2020)	South Africa	IKS reflects the adaptive skills of indigenous groups and cost-effective tool for conserving natural resources like water The usage of indigenous technical knowledge in natural resource management helps mitigate climate impacts and facilitates sustainable development Given that it is transmitted orally, it is important to document it, hence the South African government's introduction of the Indigenous Knowledge Bill in 2016 is a welcome development
Spear et al. (2019)	Africa	Farmers in semi-arid regions have historically coped using long established practices such as the place-based climate forecasts based on observations which help farmers in arid areas to cope have become affected by climate change such that they are less accurate due to uncertainties. Whilst some farmers rely on only national meteorological information, some employ place-based forecasting and others combining the two. Promoting the merger of the two will enable farmers enhance adaptation to climate variability and change [sic]
Liao et al. (2016)	Horn of Africa	Pastoralists in the horn of Africa, successfully adapt to the spread of woody plants by increasing their goat holdings, and wealthy ones are invest[ed] in camels. Indigenous knowledge guides the management practices in livestock holdings helping cope with climatic effects and enhance adaptive environmental management in the pastoral systems [sic]

Source Authors' compilation (2021)

interpret and respond to the effects of climate change in creative ways (Ojeda and Covarrubias 2011; UNDESA 2015). This significantly helps to adapt to climate change. Basdew et al. (2017) in their assessment of rural Kwa-Zulu Natal in South Africa, explain that IKS have since time immemorial, informed the management measures of peasant farmers and helped them adapt to ecological challenges such as climate change (Liao et al. 2016). For example, IKS enables local farmers to forecast seasonal changes and precipitation patterns.

Food security and sustainable agricultural practices are core issues threatened by climate change (Mekonnen et al. 2021). Yet, the access to improved agricultural approaches can be a challenge for rural-based farmers who need it most (Basdew et al. 2017). This is further aggravated by the poor penetration of agricultural extension services to remote areas. Whilst access is an issue, it is worth noting that the manner of access, or medium of transmitting climate adaptation information, is equally important. For example, it has been observed that scientific weather information (which is important to resource poor farmers) is represented in a stereotypical format with technical explanations and diagrams or figures. Thus, it is unsurprising that some local farmers fail to make sense or understand such information. Therefore, they are unable to make the requisite decisions and implement much needed response measures (Basdew et al. 2017). Basdew et al. (2017) observe that in the Swayimane community of rural Kwa-Zulu-Natal, more than 73% of the farmers are reliant on IKS to guide their farming activities such as shifting cultivation and crop rotation, among others. The IKS helped forecast the coming season, determining planting periods, schedules for irrigation, methods of controlling pests, and the prediction of rainfall, amongst others (Elia 2014; Basdew et al. 2017). The use of IKS in the prediction or guidance of agricultural practices is informed by the use of some indicators. For example, Basdew et al. (2017) note that in the Swayimane community of rural KwaZulu-Natal, the blooming of the peach plant (*Prunus persica*) is interpreted as the requisite time for planting or the onset of the summer season. Also, the blooming of flowers such as the pear plant (*Pryus communis*), and Roses (*Rosa damascena*), means it is the right time for ploughing. In this regard, the use of unique application of IKS to predict future weather outcomes, can help enhance and support the adaptive capacity of vulnerable groups (Donkor and Mazumder 2020). Consequently, it can also help to facilitate effective climate governance at the local level (Apraku et al. 2021), which will fast-track the attainment of SDG 13.

14.3.2 IKS as an Instrument for Localising Climate Action Through Tailored Measures

Findings from the systematic review of the literature on climate change and IKS reveal that effective and inclusive climate adaptation policies can be achieved when IKS are incorporated, instead of relying solely on scientific knowledge (Ogar et al. 2020). Given the variable effects of climate change on different geographic regions, tailored measures are needed (For an example, see Donkor et al. 2019). This should be developed in close collaboration with target communities (Ogar et al. 2020). This will help to ensure maximum impact of climate adaptation measures in the development process (UNESCO 2020). Moreover, resilience is enhanced as measures are tailored to the peculiar challenges faced by target areas and their specific values are put into consideration. Improved coordination and cooperation measures with proponents of IKS will enhance climate change adaptation measures and facilitate the exchange of best practices, whilst limiting the duplication of efforts (Beckford et al. 2010; Etchart 2017). This is partly because IKS are built on a profound connection with the environment, and intense spiritual, cultural, socio-economic relationship with the ecosystem (Kaya and Selati 2013; Mavhura et al. 2019). This makes indigenous groups also uniquely placed in anticipating, preparing for, and responding to the effects of climate change (Ziervogel and Opere 2010; Nick and Duwe 2021).

IKS is often transferred to the wider community through traditional modes such as small gatherings. There is very little technological usage (Donkor et al. 2017). Some rural areas in SSA are without internet connectivity (in particular) or technology (in general). It is hence often a difficult task to access such resources, unlike the position of people in urbanised areas. Some contend that because of the absence of outside channels such as mass media, the insights and measures of IKS are regarded as trustworthy by community members. There are, however, some pointers as to why indigenous knowledge is often looked down upon, resulting in some scientists doubting its efficacy for providing effective tailored interventions to tackle climate change at the grassroots level. For example, Spear et al. (2019) argues that indigenous knowledge is premised on culture and differs from one culture to another. It is also argued that data from IKS fail to project quantities in terms of precipitation but rather project the various planting and harvesting periods (Basdew et al. 2017; Thakur et al. 2020). Another issue that is often cited is the difficulties in attempting to validate IKS (Ogar et al. 2020). Scientific knowledge and information largely follow formalised processes that are authenticated using empirical data. Such results are often published and corroborated with a number of theories and guidelines. Therefore, the needs exist to consider ways of validating and documenting IKS, an issue that has received scant attention in the literature. This will help showcase and communicate its value as a tool, which can play a phenomenal role in scaling up climate governance and addressing the urgent impacts of climate change. Increased focus on IKS approaches is therefore critical for successful and relevant environmental policies, especially in developing countries found in sub-Saharan Africa. There are a number of arguments on how IKS can be effectively harnessed for successful climate adaptation. Ajibade and Eche (2017) are of the view that it is vital to emphasise policy measures that promote the maximum representation of indigenous adaptive capacity. This includes measures that safeguard the integrity of and access to indigenous communities, enhance traditional approaches for the sustenance of plant or livestock diversity, and facilitate the transfer of IKS values, attitudes and worldviews.

14.3.3 IKS as a Vehicle for Inclusive Knowledge Co-Production to Facilitate Sustainability

Another thematic area brought to the fore from the systematic review of literature indicates that profound insight of IKS will enable policy planners and climate specialists to effectively combine IKS with scientific knowledge (Apraku et al. 2021; Ramborun et al. 2021). Such integration of the two knowledge systems will spur development measures that are ecologically and socio-economically suitable and thus, more beneficial for sustainability (Mavhura et al. 2019). In order for development interventions to meaningfully attain the improvement and enrichment of agricultural livelihoods (especially farmers' livelihoods) in sub-Saharan Africa, IKS needs to be integrated with scientific knowledge (Ramborum et al. 2021), and engage the participation of grassroots communities across the value chain (Zuma-Netshiukhwi et al. 2013; Liwenga 2017).

Mpandeli et al. (2017) are of the view that farmers across the Southern Africa region have employed a number of information sources to deal with climate risks. They posit that farmers in the use of IKS harness environmental information through the observation of cloud formations, wind formation, the behaviours of insects as well as the nature of plants flowering. In the Limpopo province of South Africa, these measures they observe acts as climate observation indicators as well as a means of off-setting climate risks. Moreover, in Swaziland, the structure of the Emahlokohloko birds' nests (*Ploceus spp.*) on plants, is considered an indicator for predicting floods (Mpandeli et al. 2017). It has also come to light, that the sound of some specific bird species is listened to in predicting rainfall, whilst the produce of specific wild fruit plants is observed to monitor famine. Other indigenous methods used by the Swazis to predict natural hazards include wind direction, the shape of the crescent moon and the behaviour of certain animals for example, snakes and bees (Mpandeli et al. 2017).

These observations are not isolated incidences but are also observed beyond Southern Africa. Pastoralists in East Africa, for example, as part of their IKS repertoire, harness seasonality, by lunar observation, plants and animal behaviour and cloud formation. These observations have been crucial to informing the timing, types and quantity of food to produce in a growing season. SSA Farmers' reliance on indigenous forecast has led to an advocacy within the scientific community for it to be merged with scientific forecast, complementing the other in a manner that they are unable to do individually (Guodaar et al. 2021; Raymond et al. 2010). Some (Ziervogel and Opere 2010; Ebhuoma 2020) argue that a useful starting point to facilitate co-production of forecast will be for scientific forecast to compensate for the limitation of IKS. Doing this will likely build indigenous farmers' trust in scientific forecast, especially for those who have been adversely affected by an inaccurate scientific forecast (Ebhuoma 2020). This will enable farmers to access a more holistic and rich resource to improve their farming initiatives and other agricultural endeavours. Such integration will enhance agricultural output, as the integrated solutions will significantly make a difference amongst rural farmers (Liao et al. 2016). Some also argue that the failure to combine IKS and scientific knowledge will robustly cause core stakeholders to doubt the credibility and authenticity of IKS premised outputs (Reyes-Garcia et al. 2016).

Despite studies underlying strategies for integrating scientific forecasts with indigenous forecasts (Tengö et al. 2014; Ebhuoma 2020), the processes, approaches and frameworks of integrating these two knowledge systems need to be given more consideration, especially since most recommended strategies have not been conducted in real-life scenarios to ascertain the most plausible way of ensuring that the voices of indigenous stakeholders are not marginalised in the process.

14.3.4 IKS as an Instrument for Gender Mainstreaming in Climate Adaptation

The use of IKS in coping and adapting to climate change amongst rural farmers manifests in unique traditional methods of conserving water and irrigating farms (Donkor and Mearns 2018). Women, often due to their domestic duties of feeding and providing water for the household, are frequently engaged in activities that facilitate direct interaction with the environment (Seleti and Nthunbo 2014). They therefore develop intimate knowledge of the environment and are critical resource stakeholders in IKS. This is crucial because access to capital, land, credit and extension services amongst others are limited for women compared to men, even though they are in the majority as small-scale farmers in most rural areas of SSA (Seleti and Nthunbo 2014). They represent circa 70% of the overall farming populace. Estimates from the United Nations, indicates that if women have equal access to productive resources as men, it can translate into improved yields on their farms by 20–30 per cent, uplifting circa 100 to 150 million individuals out of hunger. Thus, helping women farmers to enhance their climate adaptation coping strategies will benefit a large section of the vulnerable populace and boost community and household resilience.

Lukhele-Olorunju and Gwandure (2018) argue that women develop indigenous technologies to address the impacts of climate change on crop, livestock and poultry production. In rural Zimbabwe, for example, seeds that weigh very little and seen floating in a winnowing basket, are considered unhealthy and are not planted. Deformed seeds, *kuputana*, are regarded as of low quality amongst the Shonas in Zimbabwe. The Swati people of Swaziland consider cracked seeds as unsuitable, given their belief that a germinating plant is nourished by the seed prior to the onset of roots (Lukhele-Olorunju and Gwandure 2018). Also, among the Swati people, a bull with undesirable features traditionally referred to as *chisiri,* is removed from the herd or not allowed to breed, so as not to pass on the defective genes, a practice adopted by indigenous women in both Congo (Moula et al. 2013) and Cameroon (Kah 2013), respectively. Similar actions are observed in the rearing of other livestock and poultry. Hens that are prolific layers and hatch many chicks, are kept and not butchered (Lukhele-Olorunju and Gwandure 2018). Underweight roosters referred to as *zvi-totorito*, are sent to the market and not kept for breeding. Women have introduced grain preservation techniques, keeping grains in large clay pots that retain low temperatures to avert the decaying of seeds. It is argued that this technique is able to prevent decay for up to five years, and can be traced to eighteenth and nineteenth century Zimbabwe. The foliage of the neem tree *Azarrdirachta indica*, can serve as a substance for fertilising, pesticide control, manuring, soil conditioning and fumigation in agriculture (Eifediyi et al. 2013).

Also, indigenous women have been acknowledged to use their wealth of knowledge to safeguard their crops from insect infestation. In many parts of Nigeria, for example, the desiccated leaves are ground into a powdery form to save the grains from insect attacks (Lukhele-Olorunju and Gwandure 2018). Since climate governance is multi-tiered, ranging from local actors to the international level, IKS provides an avenue for rural women to be proactively engaged at the local level, and not be left behind in combating climate change and its impacts (SDG 13).

14.4 Discussion

Climate change represents one of the biggest challenges confronting the 21st Century (Coen et al. 2020), posing new threats and opportunities for livelihoods and communities (Donkor and Mearns 2020). Climate change therefore features prominently in both the erstwhile Millennium Development Goals (MDGs) and the prevailing SDGs. Adopting integrated measures will help produce effective and timely responses to the climate emergency (Meadowcroft 2010). Climate change governance provides an avenue for knowledge co-production to inform policy and cooperation amongst stakeholders to address this complex global challenge (Ebhuoma et al. 2020b). IKS is hence, an essential element in this dynamic of climate governance that needs and deserves more focus. This is more so as the SDG 13 highlights the need for urgent actions to address climate change. Such urgent actions have to be felt at the grassroots to safeguard the critical climate-sensitive livelihoods that are threatened by climate change. IKS, due to it being locally fine-tuned, enhances climate change adaptation and long-term community resilience (Thakur et al. 2020). Despite the importance of IKS, it is often looked down upon. This compromises it capacity to complement scientific knowledge in addressing climate change and enhancing climate governance. This points to the need for IKS practitioners to enhance documentation of IKS and its validation processes. Moreover, showcasing and communicating its value can help to scale up climate governance and address the palpable effects of climate change (Apraku et al. 2021). This is more so in an era of misinformation and climate conspiracies, where credible information is needed to guide governance measures. As calls for more inclusivity in climate governance processes gain traction, the significant use of IKS by women makes a vehicle for facilitating their engagement in climate governance. Furthermore, IKS serves as an instrument for inclusive knowledge co-production that enhances climate governance, and therefore, facilitate sustainability.

14.5 Conclusion

Climate induced extreme events are occurring at unprecedented rates in history, interrupting both ecosystems and livelihoods for individuals all over the world. With the effects of climate change becoming ever palpable, robust and innovative response measures are being explored to enhance adaptation in unique ways. One resource that is increasingly gaining currency is indigenous knowledge systems, as highlighted by the Conference of Parties to the UN Framework Convention on Climate Change, the International Indigenous People's Forum on Climate Change (IIPFCC). IKS characterises the adaptive skills of indigenous communities and is a cost-effective tool for mitigating the effects of climate change as well as conserving natural resources. Thus, the usage of indigenous technical knowledge in natural resource management helps mitigate climate impacts and facilitates sustainable development. Moreover, access to scientific knowledge is often a challenge in rural areas, due to poor extension services. IKS can be integrated into extension services to complement and enhance extension services. However, given that it is largely transmitted orally, it is important to document it and safeguard it for posterity. The involvement of indigenous actors in research and climate decision-making processes, is vital for enhanced knowledge co-creation, for the benefit of sustainability. As the world commences the final decade of action to deliver the SDGs, there is the acknowledgement that the indigenous knowledge systems that have been sustainably employed and practiced since prehistoric times, need to be integrated across climate change adaptation processes.

References

Ajibade LT, Eche JO (2017) Indigenous knowledge for climate change adaptation in Nigeria. In: Mafongoya PL, Ajayi OC (eds) 2017, Indigenous knowledge systems and climate change management in Africa, CTA, Wageningen, The Netherlands, p 316

Apraku A, Morten JF, Gyampoh BA (2021) Climate change and small-scale agriculture in Africa: does indigenous knowledge matter? Insights from Kenya and South Africa. Sci Afr 12:1–13

Basdew M, Jiri O, Mafongoya PL (2017) Integration of indigenous and scientific knowledge in climate adaptation in KwaZulu- Natal, South Africa. Change Adapt Socio-Ecol Syst 3(1): 56–67. https://doi.org/10.1515/cass-2017-0006

Beckford CL, Jacobs C, Williams N, Nahdee R (2010) Aboriginal environmental wisdom, stewardship, and sustainability: lessons from the Walpole island first nations, Ontario, Canada. The J Environ Educ 41 (4):239–248

Coen D, Kreienkamp J, Pegram T (2020) Global climate governance (Elements in Public and Nonprofit Administration). Cambridge University Press, Cambridge. https://doi.org/10.1017/9781108973250

Donkor FK (2020) Recovery following economic and social chaos: challenges and opportunities for building resilient communities. Encyclopaedia of the SDGs. Major reference works (or MRWs). Springer Nature, Cham-Switzerland

Donkor FK, Tantoh H, Ebhuoma E (2017) Social learning as a vehicle for catalysing youth involvement in sustainable environmental management CODESRIA Bulletin Number 3, 2017 [ISSN 0850 - 8712]: Ecologies, economies and societies in Africa

Donkor FK, Howarth C, Ebhuoma E, Daly M, Vaughan C, Pretorius L, Mambo J, MacLeod D, Kythreotis A, Jones L, Grainger S, Golding N, Anderson JA (2019) Climate services for development: the role of early career researchers in advancing the debate. Environ Commun 13(5):561–566. https://doi.org/10.1080/17524032.2019.1596145

Donkor FK, Mazumder RK (2020) Women and the environment: southern perspectives and global implications. Encyclopaedia of the SDGs. Major reference works (orMRWs). Springer Nature, Cham-Switzerland

Donkor FK, Mearns K, Ojong E, Tantoh H, Ebhuoma E, Hadisu A, Mavuso S, Mbewe P, Mabeza C, Leclerc A (2020) Attitudinal changes towards agriculture through the generational lens and impact on engagement in related activities: case study from a mountainous area. In Squires V, Gaur M (eds) Food security and land use change under conditions of climate variability, Springer nature, Cham, Switzerland

Donkor FK, Mearns K (2018) Household head-related social capital: the trump card for facilitating actual uptake of innovation in rural smallholder systems. In: Leal Filho W (ed) Handbook of climate change resilience. Springer, Cham. https://doi.org/10.1007/978-3-319-71025-9_88-1

Donkor FK, Mearns K (2020) The nexus of climate change impacts on urban tourism industry: a case study on managing drought. Exploratory Environ Sci Res 1(1):78–86

Ebhuoma EE (2020) A framework for integrating scientific forecasts with indigenous systems of weather forecasting in southern Nigeria. Dev Pract 30(4):472–484

Ebhuoma EE (2021) COVID-19 hard lockdown in south Africa: lessons for climate stakeholders pursuing the thirteenth sustainable development goal. J Asian Afr Stud. https://doi.org/10.1177/00219096211043922

Ebhuoma EE, Donkor FK, Ebhuoma OO, Leonard L, Tantoh HB (2020a) Subsistence farmers' differential vulnerability to drought in Mpumalanga province, South Africa: under the political ecology spotlight. Cogent Soc Sci 6(1):1792155. https://doi.org/10.1080/23311886.2020.1792155

Ebhuoma EE, Simatele MD, Leonard L, Osadolor OE, Donkor FK, Tantoh BH (2020b) Theorising indigenous farmers' utilisation of climate services: lessons from the oil-rich Niger delta. Sustainability 12 (18):7349; https://doi.org/10.3390/su12187349

Eifediyi EK, Ahamefule HE, Remison SU (2013) Effects of neem seed cake on the growth and yield of okra (*Abelmoschus esculentus* (L.) Moench) in Ilorin, north central Nigeria. J Trop Agric, Food, Environ Extension 12(2):20–27. https://doi.org/10.4314/as.v12i2.3

Elia E (2014) Indigenous knowledge use in seasonal weather forecasting in Tanzania: a case of central semi-arid Tanzania. South Afr J Libr Inform Sci 80:18–27

Etchart L (2017) The role of indigenous peoples in combating climate change. Palgrave Commun 3:17085. https://doi.org/10.1057/palcomms.2017.85

Forsyth T (2018) Is resilience to climate change socially inclusive? Investigating theories of change processes in Myanmar. World Dev 111:13–26. https://doi.org/10.1016/j.worlddev.2018.06.023

Guodaar L, Bardsley DK, Suh J (2021) Integrating local perceptions with scientific evidence to understand climate change variability in northern Ghana: a mixed-methods approach. Appl Geogr 130:102440. https://doi.org/10.1016/j.apgeog.2021.102440

High C (2015) Victims and warriors: violence, history and memory in Amazonia University of Edinburgh. University of Illinois Press, Urbana, IL

Ionela G, Constantin BM, Dogaru L (2015) Advantages and limits for tourism development in rural area (Case Study Ampoi and Mureş Valleys). Procedia Econ Finance 32:1050–1059. https://doi.org/10.1016/S2212-5671(15)01567-1

Jetoo S (2019) Stakeholder engagement for inclusive climate governance: the case of the city of Turku. Sustainability 11:6080. https://doi.org/10.3390/su11216080

Kah HK (2013) Gender and livestock farming in Laimbweland, Cameroon, 1980S–2011. J Sustain Dev Afr 15(1):1–17

Kaya HO, Seleti NY (2013) African indigenous knowledge systems and relevance of higher education in South Africa. The Int Educ J: Comp Perspect 12 (1):30–44

Liao C, Ruelle ML, Kassam KS (2016) Indigenous ecological knowledge as the basis for adaptive environmental management: evidence from pastoralist communities in the Horn of Africa. J Environ Manage 182:70–79. https://doi.org/10.1016/j.jenvman.2016.07.032

Liwenga ET (2017) Indigenous knowledge practices for climate change adaptation among agro-pastoral communities in the semi-arid areas of East Africa. In Mafongoya PL, Ajayi OC (eds) 2017, Indigenous knowledge systems and climate change management in Africa, CTA, Wageningen, The Netherlands, p 316

Lukhele-Olorunju P, Gwandure C (2018) Women and indigenous knowledge systems in rural subsistence farming the case of climate change in Africa. Afr Insight 47(4):1–13

Mafongoya PL, Ajayi OC (eds) (2017) Indigenous knowledge systems and climate change management in Africa. CTA, Wageningen, The Netherlands, p 316

Mavhura E, Mushure S (2019) Forest and wildlife resource-conservation efforts based on indigenous knowledge: the case of Nharira community in Chikomba district, Zimbabwe. Forest Policy Econ 105:83–90. https://doi.org/10.1016/j.forpol.2019.05.019

McLean KG (2012) Land use, climate change adaptation and indigenous peoples. https://unu.edu/publications/articles/land-use-climate-change-adaptation-and-indigenous-peoples.html

Meadowcroft J (2010) Climate change governance. World Bank, Washington, DC

Mekonnen Z, Kidemu M, Abebe H, Semere M, Gebreyesus M, Worku A, Tesfaye M, Chernet A (2021) Traditional knowledge and institutions for successful climate adaptation in Ethiopia. Curr Res Environ Sustain 3:1–11

Moula N, Hornick J, Ruppol P, Antoine-Moussiaux N, Leroy P (2013) Production of animal protein in the Congo Basin, a challenge for the future of people and wildlife. http://www.kaowarsom.be/documents/Conferences/MOULA.pdf. Accessed 2 Nov 2021

Mpandeli S, Maponya P, Liphadzi S, Backeberg G (2017) Indigenous knowledge systems for managing climate change in South Africa. In: Mafongoya PL, Ajayi OC (eds) 2017, indigenous knowledge systems and climate change management in Africa, CTA, Wageningen, The Netherlands, p 316

Nick E, Duwe M (2021) Climate governance systems in Europe: the role of national advisory bodies. Ecologic Institute, Berlin; IDDRI, Paris

Ogar E, Pecl G, Mustonene T (2020) Science must embrace traditional and indigenous knowledge to solve our biodiversity crisis. Once Earth 162:1–4

Ojeda SA, Covarrubias VF (2011) The epistemological status of the sustainability discourse. Eidos 15:142–183

Pereira T, Shackleton S, Donkor FK (2017) Integrating climate change adaptation (CCA) and disaster risk reduction (DRR) for greater local level resilience: lessons from a multi-stakeholder think-tank. POLICY BRIEF Number 16. Grahamstown: Rhodes University Press

Ralph N, Birks M, Chapman Y (2014) Contextual positioning: using documents as extant data in grounded theory research. SAGE Open 4(3):17. https://doi.org/10.1177/2158244014552425

Ramborun V, Facknath S, Lalljee B (2021) Indigenous/traditional climate-smart practices effects on soil fertility and maize yield in Mauritius agricultural system. J Agric Food Res 3:1–10. https://doi.org/10.1016/j.jafr.2020.100096

Raymond MC, Fazey I, Reed MS, Stringer LC, Robinson GM, Evely AC (2010) Integrating local and scientific knowledge for environmental management. J Environ Manage 2010:91

Reyes-García V, Fernández-Llamazares A, Guèze M, Garcés A, Mallo M, Vila-Gómez M, Vilaseca M (2016) Local indicators of climate change: the potential contribution of local knowledge to climate research. Wiley Interdisc Rev Clim Change 7 (1):109–124

Sarrasanti N, Donkor FK, Santos C, Tsagkari M, Wannous C (2020) It's about time we care about an equitable world: Women's unpaid care work and COVID-19. IEEE Eng Manage Rev. https://doi.org/10.1109/EMR.2020.3031313

Seleti YN, Tlhompho G (2014) Rural women subsistence farmers, indigenous knowledge systems and agricultural research in South Africa. J Hum Ecol 48(1):33–41

Spear D, Selato JC, Mosime B, Nyamwanza AA (2019) Harnessing diverse knowledge and belief systems to adapt to climate change in semi-arid rural Africa. Climate Services 14:31–36. https://doi.org/10.1016/j.cliser.2019.05.001

Summers JK, Smith LM (2014) The role of social and intergenerational equity in making changes in human well-being sustainable. Ambio 43(6):718–728. https://doi.org/10.1007/s13280-013-0483-6

Tantoh HB, Simatele DM, Ebhuoma E, Donkor FK, McKay TJM (2019) Towards a pro-community-based water resource management system in Northwest Cameroon: practical evidence and lessons of best practices. GeoJ. https://doi.org/10.1007/s10708-019-10085-3

Tengö M, Brondizio ES, Elmqvist T, Malmer P, Spierenburg M (2014) Connecting diverse knowledge systems for enhanced ecosystem governance: the multiple evidence base approach. Ambio 43(5):579–591

Thakur R, Rane AV, Harris G, Thakur S (2020) Future prospective and possible management of water resources in respect to indigenous technical knowledge in South Africa. In: Singh P, Milshina Y, Tian K, Gusain D, Bassin JP (eds) Water conservation and wastewater treatment in BRICS Nations, pp 353–363. https://doi.org/10.1016/B978-0-12-818339-7.00018-7

Torraco RJ (2016) Writing integrative literature reviews: using the past and present to explore the future. Hum Resour Dev Rev 15(4):404–428. https://doi.org/10.1177/1534484316671606

UNDESA (2015) Climate change. https://www.un.org/development/desa/indigenouspeoples/climate-change.html

UNDESA (2020) SDGs final decade of action. https://www.un.org/en/desa

UN (2021) Sustainable development goals. https://sdgs.un.org/goals/goal13

UNESCO (2017) What is local and indigenous knowledge? http://www.unesco.org/new/en/natural-sciences/priority-areas/links/related-information/what-is-local-and-indigenous-knowledge/

UNESCO (2020) Indigenous knowledge and climate change. https://en.unesco.org/links/climatechange

UNWTO (2020) Tourism and rural development. https://www.unwto.org/world-tourism-day-2020/tourism-and-rural-development-technical-note

Ziervogel G, Opere A (eds) (2010) Integrating meteorological and indigenous knowledge-based seasonal climate forecasts for the agricultural sector. Lessons from participatory action research in sub-Saharan Africa. CCAA learning paper, 1. 2010. https://idl-bnc-idrc.dspacedirect.org/bitstream/handle/10625/46185/132676.pdf?sequence=1&isAllowed=y. Accessed 22 Oct 2021

Zuma-Netshiukhwi G, Stigter K, Walker S (2013) Use of traditional weather/climate knowledge by farmers in the south-western free state of south Africa: agrometeorological learning by scientists. Atmosphere 4:383–410

Felix Kwabena Donkor (Ph.D.) is a member of Future Earth's Knowledge Action Network on the Food Water Energy Nexus and the Early Careers Network. His research interests include sustainable rural livelihoods, indigenous knowledge systems, environmental governance and sustainable development. He serves as deputy of the South African Adaptation Network. He is a founding member of Green Cheque Eco-social and the Inala Food Sovereignty and Climate Justice Forum at the University of the Witwaterstrand where he completed his doctoral studies and co-manages a community food garden.

Kevin Mearns (Ph.D.) is a full Professor in the College of Agriculture and Environmental Sciences (UNISA), South Africa. His research interests include Sustainable tourism, Environmental management, Ecology and Ecotourism. He is involved in a number of community projects such as Sustainability benchmarking for the South African Tourism Industry, Social ecology of conservation areas and Climate change and tourism amongst others.

Factors Fuelling the Underrepresentation of Indigenous Knowledge in Climate Change Governance in Kenya and Ethiopia: A Systematic Review

15

Eromose E. Ebhuoma

15.1 Introduction

Climate variability and change have negatively impacted all sectors of the global economy. However, agriculture, particularly in Sub-Saharan Africa (SSA), has been one of the hardest hit sectors partly because the slightest variation in weather conditions have triggered huge losses in crop productivity (Simatele and Simatele 2015; Ebhuoma et al. 2019; Haile et al. 2019). Yet, indigenous farmers in SSA are not docile victims to climate variability and change. By drawing on their Indigenous Knowledge Systems (IKS), farmers have been able to make informed decisions and adapt their agricultural farming practices in order to consistently grow their food and obtain their livelihoods (Mavhura et al. 2013; Fitchett and Ebhuoma 2018; Nyadzi et al. 2020), albeit not without challenges. While there are nuanced differences in the definition of IKS (e.g. see Maurial 1999; Castellano 2000), this chapter adopts Warren's (1991) and Orlove et al (2010) conceptualisation. They argue that IKS refers to institutionalised local knowledge rooted in cultural identity, experiential with real-life scenarios and passed on from one generation to the next by word of mouth and repetitive engagements. The role of IKS in facilitating climate change adaptation and sustainable resource management is well recognized by international agencies. For example, Chap. 26 of Agenda 21—a non-binding action plan of the United Nations with regards to sustainable development—underscores the need to include indigenous people and their communities in both natural resource management and conservation schemes in order to scale-up sustainable development. Also, the United Nations Scientific Conference Organisation (UNESCO) and the International Council for Science Union (ICSU) applauded the immense contribution by IKS and made a passionate plea for its inclusion in all forms of human engagements (Shisanya 2017). Other globally recognised bodies such as the Intergovernmental Panel on Climate Change (IPCC) (IPCC 2014) and the International Labour Office (ILO) (ILO 2017) have echoed somewhat similar rhetoric. Thus, it is no surprise that scholars (Chanza and de Wit 2016; Magni 2017) have reiterated that IKS should feature prominently in climate change governance, a call that can facilitate the actualisation of 13th Sustainable Development Goal (SDG)—climate action, if heeded.

Climate governance refers to actions adopted by states and non-state actors to mitigate and adapt to the impacts of climate change (Scobie 2016). Climate change governance is increasingly becoming inevitable owing to unprecedented nature in which climate change has, and is

E. E. Ebhuoma (✉)
Department of Environmental Sciences, School of Ecological and Human Sustainability, College of Agriculture and Environmental Sciences, University of South Africa (UNISA),

E. E. Ebhuoma and L. Leonard (eds.), *Indigenous Knowledge and Climate Governance*, Sustainable Development Goals Series, https://doi.org/10.1007/978-3-030-99411-2_15

expected to occur in SSA where the most vulnerable are often poor. As Meadowcroft (2009) notes, part of what will constitute good governance would entail showcasing the responses adopted by governments to tackle climate change. Governments that wish to be considered as 'legitimate' by its citizens and the international community will 'have to display their credentials on this front' to reinforce their commitment towards achieving SDG 13. Thus, climate change governance is undeniably mandatory. While it is recognised that technological innovations (Meadowcroft 2009) and diverse knowledge systems (Ebhuoma 2020) are needed to accelerate climate change governance in key economic sectors including agriculture, IKS have failed to feature prominently in climate change adaptation schemes including its integration with scientific knowledge in SSA (Ebhuoma et al. 2019) including East Africa (Iticha and Husen 2019), a region with one of the highest populations of indigenous farmers on the African continent.

An estimated 70% of the labour force in East Africa rely on food production for their primary source of livelihood (NEPAD 2013). In the last two decades, however, farmers have been regularly besieged by reoccurring drought conditions (Haile et al. 2019) and erratic rainfall patterns (Nicholson 2017). Notwithstanding, they have been able to engage in food production through the utilisation of their IKS, despite being resource-constrained. A key facilitator to ensuring effective application of IKS in food production is the role played by indigenous institutions. Indigenous institutions have been instrumental to producing place-based weather forecasts, often spearheaded by elders knowledgeable in the use of biological and animal indicators. Other ways in which IKS has facilitated food production in the face of climate variability include soil conservation through zero tilling practices in cultivation and fallow system of cultivation, which facilitates forest development necessary to sequestrate and store carbon (Nyong et al. 2007). Thus, it can be asserted that IKS have become the lifeblood for numerous East African farmers as food production, void of its application, may have been abysmal for these cohorts who are resource-constrained.

In light of the above, this chapter analyses drivers undermining IKS from featuring in climate governance schemes in East Africa with a particular focus on Kenya and Ethiopia, countries where extreme weather events have become the norm in the last decade (Berhanu and Beyene 2015; Ouma et al. 2018) and IKS have played an instrumental role in facilitating food production (Garedew et al. 2017; Shisanya 2017; Iticha and Husen 2019). By pursuing this objective, it is hoped that this chapter could advance the discourse on how to ensure inclusivity of indigenous people in schemes aimed at building their resilience to climatic risks, because whatever is done for any targeted group of individuals, without their input, is not theirs. This chapter proceeds in four parts. First, it provides a snapshot of climate governance perspectives in Kenya and Ethiopia. Next, it addresses the study area description and the systematic approach adopted to select relevant literature. The chapter then discusses the common themes identified from the literature pertaining to Kenya and Ethiopia as well as the nuanced differences. Finally, the chapter offers some policy implications, especially its implication for SDG 13.

15.2 Climate Change Governance: Perspectives from Kenya and Ethiopia

According to the IPCC, climate change governance entails a more inclusive approach that recognises all spheres of government (global, international, regional, national, local) and the responsibilities of the private sector, nongovernmental actors and the civil society' (Chanza and de Wit 2016, p. 3). At national level, a crucial aspect of climate governance is to demystify the institutional conditions and dynamics that influences how organisations at sub-national levels respond to climate change. This includes institutions like local governments and decentralised state agencies that operate at the 'meso-level' between the central state and communities, in the administrative spaces encompassing districts, municipalities and provinces (Field et al. 2012; Aggarwal et al. 2014). Such organisations are pivotal to facilitating climate change adaptation at the household level. As Renner (2020, p. 2) acknowledges, 'they are often responsible for implementing national climate change policies and interventions in practice, while at the same time being accountable to the local population'.

In Africa, Kenya is one of the 'first countries on the continent to enact a comprehensive law and policy to guide national and sub-national responses to climate change' (Bellali et al. 2018, p. 25) in its bid to reduce vulnerability at communal and household levels. For example, Sect. 49, Subsection 1 of the Environmental Management and Coordination Act (EMCA) of 1999 stipulates that the relevant authority shall in consultation with other lead agencies advocate for the use of renewable energy. Also, the Constitution of Kenya (CoK) (2010) has become the bedrock of major institutional and legal framework for actions geared towards tackling climate change. Article 42 of the CoK underscores every citizen's right to a healthy environment as well as the protection of the environment for present and future generations by all means necessary. The Constitution was instrumental to the formulation of the Climate Change Act (2016). Specifically, 'Article 10 sets out national values and principles of governance, such as sustainable development, devolution of government and public participation, that were considered mandatory when making or implementing the climate change law' (Naeku 2020, p. 175). Following the ratification of the legally binding climate change Paris Agreement that came into force in December 2015 during the Paris Climate Conference of Parties (COP 21), 'the Act primarily seeks to ensure development, management, implementation enhances climate change resilience and low carbon development for sustainable development' (p. 175). Consequently, Kenya domesticated the Agreement by incorporating it into its law through its Nationally Determined Contribution (NDC). The country sets its NDC under two key components, mitigation and adaptation. 'The adaptation component seeks to ensure enhanced resilience to climate change towards the attainment of Vision 2030 by mainstreaming climate change adaptation concerns into the Medium Term Plans. The mitigation contributions are geared towards reducing Green House Gas (GHG) emissions by 30% by 2030 relative to the 'Business As Usual' scenario of 143 Metric tons of carbon dioxide equivalent (MtCO2 Eq) (Naeku 2020).

Unlike Kenya, Ethiopia's climate change governance schemes are not comprehensive. Notwithstanding, the nation has not been docile especially considering the fact that it is one of the countries in the world that is extremely vulnerable to climate change. At the Company of Parties (COP 17) meeting held in Durban, in 2011, the Ethiopian government launched its Climate Resilient Green Economy (CRGE). The aim was to ensure that Ethiopia grow and develop its economy in a way that is climate-resilient (Colvin and Mukute 2018). Indeed, since agriculture is a mainstay of the livelihood activities in the country, building households resilience to climate change is a top priority of the CRGE. Other programmes introduced to systematically grow its economy while combating climate change include the introduction of Growth and Transformation Plan (GTP), whose focus is to improve natural resources conservation and restoration of degraded lands, and Ethiopian Programme of Adaptation to Climate Change (EPA-CC). The EPA-CC was conceptualized to replace the National Adaptation Programme of Action (NAPA) owing to its failure to channel enormous attention to carving out ways for the country to become water secure, a major issue confronting Ethiopia (Adepeju 2018). Also, because the nation is consistently besieged by reoccurring droughts and flooding, Ethiopia developed its Intended Nationally Determined Contribution (INDC) to facilitate households' adaptation to droughts and floods (Adepeju 2018), and as a response to proposed actions it will undertake towards pledging its commitment to the ratification of the Paris Agreement.

Despite these feats achieved within a short timeframe in Kenya and Ethiopia, having such developed institutions at the national level does

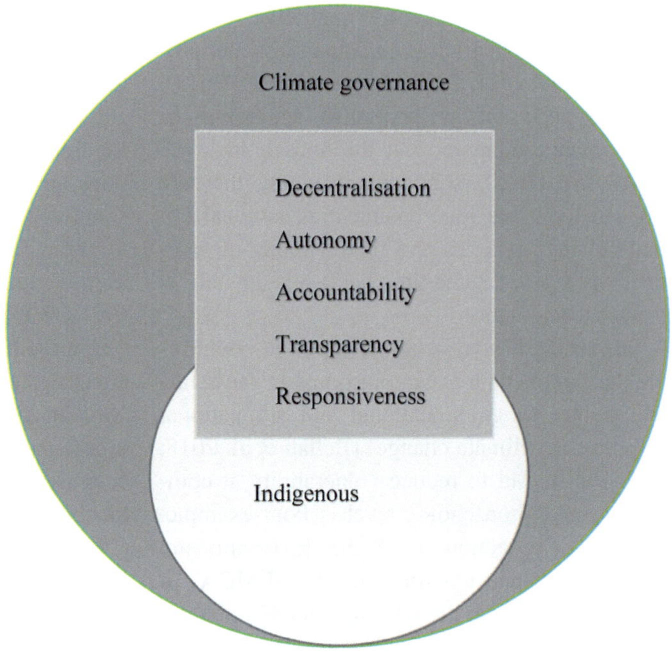

Fig. 15.1 Nexus between climate governance and indigenous knowledge. Adapted from Chanza and de Wit (2016)

not guarantee effective climate change governance at local government or county levels (Ojwang et al. 2017; Smucker et al. 2020). One of the major criticisms that continue to plague these countries is the weak incorporation of IKS into such formal schemes (Naeku 2020). Thus, pursuing the objective of this chapter is critical because a close scrutiny of climate governance, as Chanza and de Wit (2016) acknowledge, would point towards good climate governance-indigenous knowledge connectedness (Fig. 15.1).

15.3 Materials and Methods

15.3.1 Study Area Description

Kenya and Ethiopia, classified as semi-arid countries, are situated in East Africa (Fig. 15.2). 'Kenya is highly vulnerable to climate change and climate variability. The temperature is projected to increase by up to 3 °C, rainfall will become erratic and more intense, droughts will be prolonged and more frequent, and the coasts will see up to one meter of sea-level rise. The country's vulnerability is aggravated by the fact that an estimated 42% of its GDP and 70% of its overall employment are derived from natural resource sectors. The Kenyan people are living with the consequences: decreased agricultural production, loss of livestock, property losses, and famine' (Bellali et al. 2018, p. 25).

Like Kenya, Ethiopia is heavily impacted by climate change. It is classified as one of the countries most vulnerable to climate change, with climate change constituting a major threat to the livelihood of over a quarter of its citizens that live in extreme poverty and an estimated 84% of its citizens that rely on rain-fed agriculture for their livelihood. Ethiopia is home to a great diversity of peoples speaking more than 80 languages. Still, the country has no legislation that protects or address the rights of indigenous peoples (IWGIA 2011).

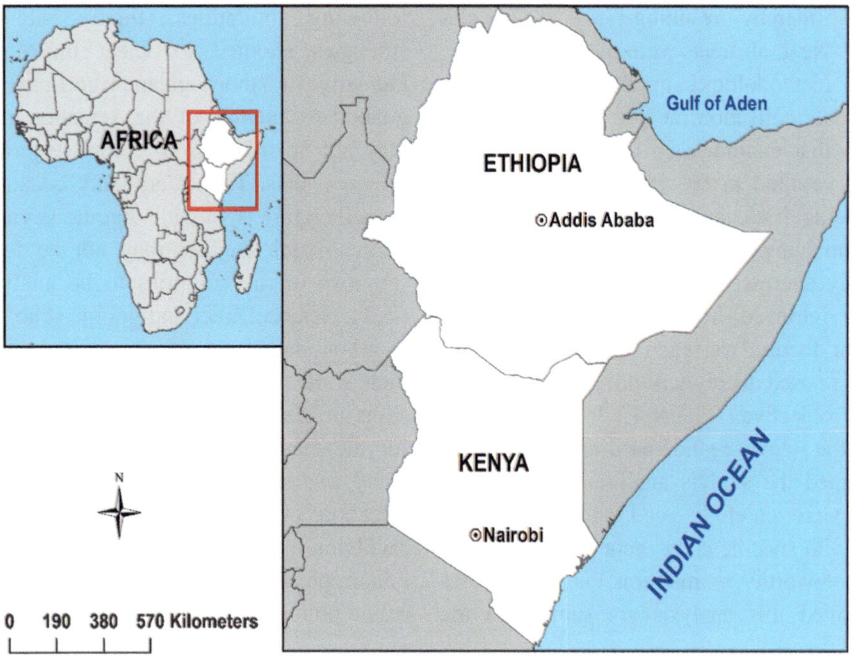

Fig. 15.2 May of Ethiopia and Kenya. *Source* Author (2021)

15.4 Methodology

To accomplish the objective of this chapter, a systematic review was conducted to retrieve relevant literature from both google scholar and ScienceDirect, among the list of the most effective academic search engines (Tober 2011). According to Clarke (2011, p. 64), 'the purpose of a systematic review is to deliver a meticulous summary of all the available primary research in response to a research question. The methodology applied is explicit and precise and aims to minimise bias, thus enhancing the reliability of the conclusions drawn'. Thus, keywords entered in both search engines were consistent to ensure uniformity and they were formulated based on the author's expertise on IKS that such keywords would aid in identifying relevant literature from which the chapter's objective can be achieved.

Specific keywords entered in both search engines include *'climate change governance in Kenya'; 'climate change governance in Ethiopia'; 'multi-level climate governance in Kenya'; 'multi-level climate governance in Ethiopia'; 'co-production of indigenous and scientific knowledge in Africa'; 'indigenous knowledge and climate policy in Africa'; 'indigenous knowledge and climate policy in East Africa'; 'indigenous knowledge and climate policy in Kenya'* and *'indigenous knowledge and climate policy in Ethiopia'*. In both search engines, advanced searches were used to modify the searches to literature published from 2016 to 2021 due to the ratification of the SDGs by global leaders including Kenyan and Ethiopian representatives on 25–27 September 2015 at the United Nations 70th General Assembly. Thus, this chapter thought it befitting to analyse the extent to which indigenous knowledge has been accommodated in climate governance policies in selected countries in East Africa post SDG ratification. In ScienceDirect, however, subject areas were restricted to social sciences, environmental sciences, agricultural and biological sciences and earth and planetary sciences.

In total, 4541 articles were retrieved. After screening based on titles, a total of 4397 were eradicated, and thus, 144 articles were remaining. Thereafter, duplicate articles, 27 in total, were eradicated, thereby resulting in 117 articles remaining. Next, abstracts were read and, in some cases the methodologies applied were read to ascertain data collection method and study areas in studies that examined to ascertain the study area. This resulted in 86 articles being screened out for being irrelevant to the scope of study, thereby remaining 31 articles. Finally, two articles that closely addressed the issue under investigation were retrieved from the references of an article that focused on each country. Thus, 33 articles were used deeply scrutinized to answer the chapter's objective. Figure 15.3 schematically maps out the stepwise guide used to select articles that informed the study's analysis. The selected literature were carefully read and thematically analysed with specific codes generated.

It is noteworthy to mention that the studies that informed this analysis are subject to the following limitations. Firstly, although the strategies adopted to select the literature is indicative of a thorough and rigorous process that guarantees quality of the selection process, it may not be sufficient to completely rule out all selection bias. This is arguably because indigenous literature is so wide that the keywords used in the search engines may not be fully representative of the countries to be analysed. Secondly, ScienceDirect and google scholar, chosen as a boundary for the analysis, even considering their wide scope, cannot prevent possible omissions in identifying relevant pieces of literature for the analysis. Despite these drawbacks, the study was extensive enough to permit the identification of important trends and key issues revolving around the factors undermining the underrepresentation of IKS in climate governance policies.

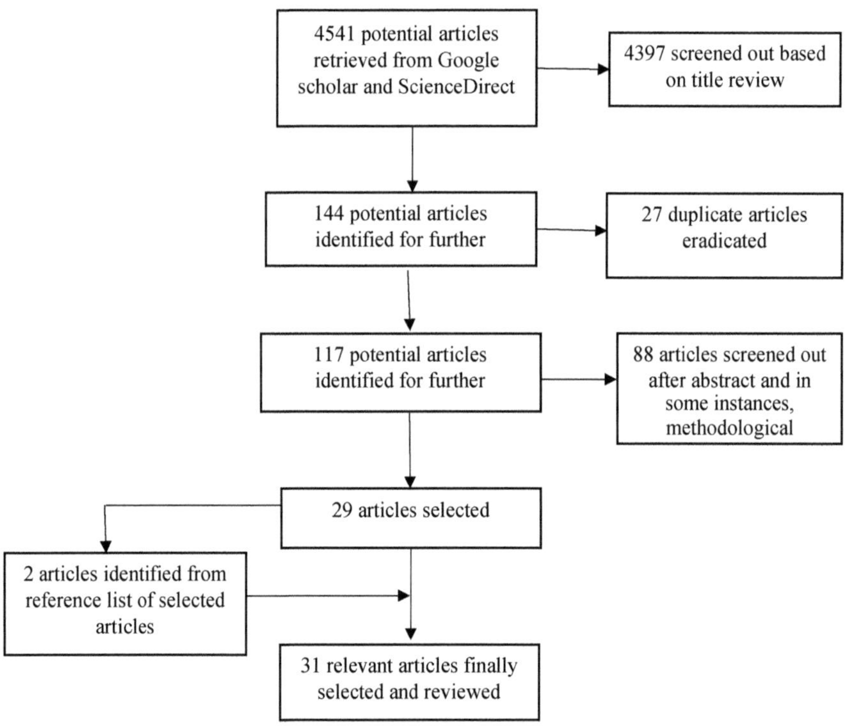

Fig. 15.3 Flow chart of article screening and selection process

15.5 Results and Discussion

15.5.1 Commonalities from Kenya and Ethiopia

Following the systematic review of the literature on Kenya and Ethiopia, four identical themes emerged. These are adaptation constraints, knowledge production, political and financial constraints (Tables 15.1 and 15.2). Under adaptation constraints, analysis from both countries underlined policies not tailored to local context (Iticha and Husen 2019; Adeyeye et al. 2019; Matewos 2020; Ndalilo et al. 2020) as a major bottleneck that stampedes indigenous peoples' participation in climate governance schemes as it evokes the notion that their local realities with regards to the impact of climate change counts for little or nothing. Also, top-down approach of generating adaptation strategies (Ogega 2017; Adeyeye et al. 2019) without listening to them to find out what their needs are were also found as a factor that undermines indigenous peoples participation in climate governance schemes. As Ebhuoma (2020) highlights, this issue partly stems from development practitioners downplaying the value of indigenous people as custodians of valuable knowledge, which may spiral into minimal efforts to engage them in stakeholders forum.

In the context of knowledge production, the identical codes that emerged are hegemony of Western scientific knowledge; incomprehensibility of indigenous epistemology and indigenous knowledge complementary to scientific knowledge. It is argued that hegemony of scientific knowledge, which often trumps indigenous ways of knowing (Adeyeye et al. 2019), may be fuelled by those knowledgeable on indigenous knowledge (e.g. traditional meteorological forecasters) may be unwilling to disclose their knowledge source (Musembi and Cheruiyot 2016), and scientists not making concerted efforts to unpack the epistemologies underlying IKS (Balehegn et al. 2019; Tilahun et al. 2019). For example, Nkuba et al. (2020), in their analysis, traced the 'science' behind indigenous peoples use of biotic indicators to predict future weather outcomes, dispelling the notion of scientific knowledge as the only credible means of producing practically relevant knowledge (see McGloin et al. 2009). Thus, to fortify value of IKS in the wider society, documentation of IKS to garner appreciation in wider society (Desalegn et al. 2017) and integration into East African curriculum may be necessary. In Kenya, for example, despite countless efforts aimed at reforming the postcolonial administration, Kenya's curriculum has largely continued to be influenced by the British—its colonial masters—

Table 15.1 Factors impeding indigenous knowledge from climate governance schemes in Kenya

Themes	Codes
Adaptation constraints	• Misconception of maladaptation • Adaptation policies not tailored to local context • Culture disintegration, resulting in a lot of young people not engaging IKS • Weak feedback mechanisms • Top-down adaptation strategies
Knowledge production	• Hegemony of scientific knowledge • Indigenous knowledge complementary to scientific knowledge • Trivialization of local institutions as custodians of knowledge (often results in minimal engagement of local institution)
Political constraints	• Disregard for indigenous rights (e.g. land grabbing) • Inadequate policing mechanisms of projects on indigenous lands
Financial constraints	• Lack of financial support • Failure to monitor environmental projects on indigenous lands

Source Author's compilation (2021)

Table 15.2 Factors impeding indigenous knowledge from climate governance schemes in Ethiopia

Themes	Codes
Adaptation constraints	• Adaptation measures not tailored to local context • No defined role in climate governance schemes • Top-down adaptation strategies
Knowledge production	• Hegemony of scientific knowledge • LIKS complementary to scientific knowledge
Political constraints	• Interference with local institutions or forms of governance • Disregard for indigenous rights (e.g. land grabbing) • Lack of trust in government • Lack of understanding the complex dynamics of local institutions
Financial constraint	• Lack of financial support for local institutions

Source Author's compilation (2021)

system of education (Lumonya 2020). Similar observation has been made in SSA countries (Munyaradzi 2015). Arguably, the uptake of a curriculum heavily embedded in Western ideology in SSA national schools has created a ripple effect whereby its IKS are being marginalised. However, few institutions in SSA have begun the process of decolonizing their curricula to include IKS. For example, several tertiary institutions in South Africa offer both Bachelor's and Master's degree programmes on IKS.

In the context of indigenous knowledge being complementary to scientific knowledge, some find the narrative disrespectful not least because indigenous sovereignty is not respected (Latulippe and Klenk 2020) and the notion that IKS is static (Adeyeye et al. 2019). According to Adeyeye et al (2019), IKS is a 'living thing', which will continue to evolve. By implication, new knowledge will always be generated based on adaptation and innovation because IKS is livelihood knowledge. It is unsurprising, therefore, to observe that there is growing recognition that science should complement IKS (Ziervogel and Opere 2010; Ebhuoma 2020), with others arguing that IKS is deserving of its space without any form of scientific interference or meddling (Latulippe and Klenk 2020).

With regards to political constraints, disregard for indigenous rights through schemes such as land grabbing (Tarfasa et al. 2018) is an issue that could hamper indigenous peoples participation in climate governance schemes. Access to indigenous lands by investors, usually approved by the respective country's government and, paradoxically traditional authorities in some cases, is well documented as a catalyst for land grabbing (Gilbert 2016; Obeng-Odoom and Van Gyampo 2017). Such acts may cast doubts in the minds of indigenous people regarding whose interest the government is trying to protect in climate governance policies, thereby limiting their participation when stakeholder meetings are scheduled. This is underpinned by natural resources depletion that plays a pivotal role in catering to the welfare of indigenous people. In a study by Shisanya (2017) in Kenya, for example, a participant argued:

> Long ago, we had different varieties of indigenous plant resources in our river basin that served many purposes. But with the introduction of factories in this area, the waste is dumped in rivers leading to extinction of most indigenous plant resources along river valleys (p. 215).

For financial constraints, lack of financial support (Matewos 2020) is a stumbling block for indigenous people as it paints a picture that issues pertaining to indigenous people are not a priority in government schemes. While dearth of finances to support indigenous institutions may be a reflection of the state of both the Ethiopian and Kenyan economies, the inability to empower local institutions financially could, arguably, implicitly indicate that the respective state governments are not in full support of local norms, cultures and values that inform climate adaptation.

15.5.2 Nuanced Differences

In Kenya, additional codes that emerged adaptation constraints include culture disintegration and misconception of maladaptation. Cultural disintegration stems from the fact that elders who are custodians of IKS sometimes fail to pass it down to the next generation to make it widespread (Shisanya 2017), sometimes for fear of losing the social esteem associated with being custodians of knowledge that is valid and useful (Musembi and Cheruiyot 2016). Indeed, failure to effectively communicate IKS across generations may be contributing to its extinction. With the treat of IKS going extinct globally (Battiste 2002), stringent measures such as documentation to foster its appreciation (Desalegn et al. 2017) and including it in the African curriculum are necessary to preserve this vital knowledge strand.

Misconception of maladaptation, albeit implicitly highlighted in the literature, stemmed from the erroneous viewpoint that some adaptation strategies adopted by indigenous people are counter intuitive (e.g. Wetende et al. 2018). For example, it is argued that famers in Kenya failed to manage livestock waste (slurry) which could lead to creation of suitable breeding grounds for debilitating vectors that contribute to livestock disease flare-ups, which have contributed to heavy livestock disease burden (Wetende et al. 2018). Yet, it is the only feasible alternative to ensure their cultivated crops grow optimally under the prevailing circumstances of being resource-constrained (Ebhuoma 2020). Thus, the implication of such erroneous theorisation of indigenous people may be twofold. First, it may result in climate policy developers thinking that a top-down approach of adaptation strategies may be the ideal thing to do. Second, it may cascade into climate policy developers into thinking that indigenous people have nothing valuable to offer in terms of climate governance, thereby not making concerted efforts to get their input in the formation of climate governance policy. Consequently, this may catalyse the lack of defined role given to IKS in climate governance schemes, a nuanced finding in Ethiopia (Godana 2016).

Uneven access to governance stems from assigning roles that undermines the participation of indigenous people (den Besten et al. 2014; Ruwhiu and Carter 2016). This have, in turn, translated into lack of trust in government (Adeyeye et al. 2019). In an indigenous community, Sidama, in Ethiopia, seventeen of the eighty-six civic organizations registered in the community have engaged in various projects specific to agriculture, water and sanitation, education, livelihood and food security, and cooperation and marketing. It was observed that these philanthropic organisations contributed to climate change adaptation through life-saving schemes, creating awareness and reducing vulnerabilities. However, no project-specific civic organization was observed to have engaged directly in climate change adaptation at a local level. Most of them had no office base in their primary mission area. Most of them were pooled in the Bureau of Finance of the region, where they signed a memorandum of understanding and were coordinated. Except at this level, there were no rooms where they were coordinated below the regional level. On the other hand, a cross-sectional household survey analysis indicated that 51% (N = 400) of the households confirmed the importance of institutions for food security and risk-management service delivery. In comparison, 47% and 52% of households agreed to the establishment of capacity-building (farmer training centers (FTCs)) and cooperative enhancement offices, respectively. Similarly, 73% and 53% of the respondents confirmed the importance of environmental protection, educational, and credit services institutions, respectively (Bekele et al. 2020). These sort of negligence will likely cause a withdrawal of indigenous people from climate governance schemes. Yet, the same community had well structured and delineated indigenous institutions with a pool of valuable knowledge that can enable the government can draw from and collaborate to scale up climate adaptation.

Even when schemes have been availed in indigenous community, they are often not adequately financed, which cascades into inadequate policing mechanisms of projects engineered in indigenous communities, a finding specific to Kenya (Muhati et al. 2018). The study highlighted lapses in government agencies ability to handle land-use changes to ensure licensed loggers only harvest dead wood not live trees. Failure to enforce sustainable logging facilitated forest cover loss through harvesting of live trees such as Olea europeae ssp. africana, Strychnos henningsii, and Diospyros abyssinica used for building traditional houses, thanks to their termite resistance qualities, may fuel the notion that their plight is not relevant as logging is done by urban dwellers (Muhati et al. 2018). In contrast with Kenya, political constraint in Ethiopia is inadequate understanding of the complex dynamics of indigenous institutions (Balehegn et al. 2019), usually underpinned by 'informalisation' from outsiders lenses and interference with indigenous institution's form of governance (Behailu et al. 2016). As Homann (2004) argues, indigenous institutions has been interfered with by modern governmental institutions, and its management is dwindling without meaningful attitude changes toward the modern management. Such interference will cripple indigenous people's participation in climate governance schemes.

15.6 Conclusion and Policy Implications

Climate change and variability has severely affected food production in East Africa. Yet, through the skillful application of IKS, smallholder farmers have been able to outwit the odds and continue to grow their food. The acknowledgement of the valuable role played by IKS in scaling up adaptation to climate change among smallholder farmers have galvanized calls for it to be integrated when formulating climate governance policies. Yet, this call has hardly been heeded to. This study, therefore, sought to unpack the factors fueling the underrepresentation of IKS in climate governance in both Kenya and Ethiopia, two East African countries that food production have been adversely compromised partly due to climatic risk. The study finds that adaptation, financial and political constraints and the notion that indigenous people are not custodians of vital knowledge, albeit nuanced differences exist between both countries in some instances, as factors undermining the effective representation of indigenous knowledge in climate governance schemes.

Without delicately addressing these issues, the actualisation of SDG 13 may be hampered in both Kenya and Ethiopia. Because the effect of climate change are experienced locally, the engagement of local actors to needed to reflect initiatives that would reflect their lived realities in climate governance schemes. In other words, place-based polices are the best ways to minimise the exclusion of IKS in climate governance. The question that should always be at the back of climate professionals minds when engaging with indigenous peoples is what sort of assistance would be meaningful to them? This can only be achieved by actively engaging indigenous people in stakeholder forums. It is worthwhile to run with this mindset when engaging with indigenous people so that the notion that whatever is done for us, without us, is not ours, may not be brewed in the minds of indigenous people as this would likely stifle climate governance schemes instituted in their respective communities.

References

Adepeju IZ (2018) Managing climate risks in Africa: insights from South Africa and Ethiopia. Unpublished Master's thesis, University of Newfoundland, Canada. https://research.library.mun.ca/13509/1/Ibrahim_ZainabAdepeju_master.pdf. Accessed 25 Mar 2021

Adeyeye Y, Hagerman S, Pelai R (2019) Seeking procedural equity in global environmental governance: Indigenous participation and knowledge politics in forest and landscape restoration debates at the 2016 World Conservation Congress. Forest Policy Econ 102006:1–14

Aggarwal A, Chhetri N, Cull T, Gustavo FJ, Haggar J, Hutchinson G et al (eds) (2014) Rural areas. In: Contribution of working group II to the Fifth assessment report of the intergovernmental panel on climate change. Cambridge University Press, Cambridge. https://doi.org/10.1017/CBO9781107415416.024

Balehegn M, Balehey S, Fu C, Liang W (2019) Indigenous weather and climate forecasting knowledge among afar pastoralists of North Eastern Ethiopia: role in adaptation to weather and climate variability. Pastoralism Res Policy Practices 9(8):1–14

Battiste M (2002) Indigenous knowledge and pedagogy in First Nations education: A literature review with recommendations. Indian and Northern Affairs, Ottawa, Ontario, Canada

Behailu BM, Pietilä PE, Katko TS (2016) Indigenous practices of water management for sustainable services: case of Borana and Konso, Ethiopia. Sage Open 6(4).https://doi.org/10.1177/2158244016682292

Bekele F, Tolossa D, Woldeamanuel T (2020) Local institutions and climate change adaptation: appraising dysfunctional and functional roles of local institutions from the bilate basin agropastoral livelihood zone of Sidama, Southern Ethiopia. Climate 8:149. https://doi.org/10.3390/cli8120149.

Bellali J, Strauch L, Oremo F, Ochieng B (2018) Multilevel climate governance in Kenya: activating mechanisms for climate action. Adelphi/ILEG, Berlin

Berhanu W, Beyene F (2015) Climate variability and household adaptation strategies in southern Ethiopia. Sustainability 7:6353–6375

Castellano MB (2000) Updating Aboriginal traditions of knowledge. In: Dei GJS, Hall BL, Rosenburg DG (eds) Indigenous knowledges in global contexts. University of Toronto Press, Toronto, Ontario, Canada

Chanza N, De Wit A (2016). Enhancing climate governance through indigenous knowledge: case in sustainability science. S Afr J Sci 112(3/4):7 (Art. #2014-0286). https://doi.org/10.17159/sajs.2016/20140286

Clarke J (2011) What is a systematic review? Evid Based Nurs 14(3):64. https://doi.org/10.1136/ebn.2011.0049

Colvin J, Mukute M (2018) Governance in Ethiopia: impact evaluation of the African climate change and resilience alliance (ACCRA) project. Oxfam publications. https://oxfamilibrary.openrepository.com/bitstream/handle/10546/620412/er-governance-ethiopia-effectiveness-review-090218-en.pdf?sequence=7. Accessed 10 Mar 2021

den Besten JW, Arts B, Verkooijen P (2014) The evolution of REDD+: an analysis of discursive-institutional dynamics. Environ Sci Policy 35:40–48

Desalegn A, Radeny M, Mungai C (2017) Indigenous knowledge in weather forecasting: lessons to build climate resilience in East Africa. https://ccafs.cgiar.org/news/indigenous-knowledge-weather-forecasting-lessons-build-climate-resilience-east-africa-0. Accessed 23 Mar 2021

Ebhuoma E, Simatele M, Tantoh H, Donkor F (2019) Asset vulnerability and systems thinking as a twin methodology for highlighting factors that undermine efficient food production. Jamba J Disaster Risk Stud 11(1):a597. https://doi.org/10.4102/jamba.v11i1.597

Ebhuoma EE (2020) A framework for integrating scientific forecasts with indigenous systems of weather forecasting in southern Nigeria. Dev Pract 30(4):1–13

Field CB, Barros V, Stocker TF, Qin D, Dokken DJ, Ebi KL et al (eds) (2012) Managing the risks of extreme events and disasters to advance climate change adaptation. A special report of working groups I and II of the Intergovernmental Panel on Climate Change. Cambridge University Press, Cambridge, 582p https://doi.org/10.1017/CBO9781139177245.

Fitchett J, Ebhuoma E (2018) Phenological cues intrinsic in indigenous knowledge systems for forecasting seasonal climate in the Delta State of Nigeria. Int J Biometeorol 62(6):1115–1119

Garedew B, Ayiza A, Haile B, Kasaye H (2017) Indigenous knowledge of Enset (Ensete ventricosum (Welw.) Cheesman) cultivation and management practice by Shekicho people, southwest Ethiopia. J Plant Sci 5(1):6–18

Gilbert J (2016) Land grabbing, investors, and indigenous peoples: New legal strategies for an old practice? Community Dev 51(3):350–366

Godana DG (2016) The role of indigenous knowledge in rangeland management in Yebello Woreda Southern Oromia Ethiopia. Arts Soc Sci J 7(2):1–8

Haile GG, Tang Q, Sun S, Huang Z, Zhang X, Liu X (2019) Droughts in East Africa: causes, impacts and resilience. Earth Sci Rev 193:146–161

Homann S (2004) Indigenous knowledge of Borana pastoralitsts in natural resource management: a case Study from Southern Ethiopia. Dissertation, Cuvillier Verlag, Goettingen

International Labour Office (ILO) (2017). Indigenous peoples and climate change: from victims to change agents through decent work. https://www.ilo.org/wcmsp5/groups/public/—dgreports/—gender/documents/publication/wcms_551189.pdf. Accessed 8 Mar 2021

IPCC (2014). Climate change 2014: impacts, adaptation, and vulnerability, part B: regional aspects. In: Barros VR, Field CB, Dokken DJ, Mastrandrea MD, Mach KJ, Bilir TE, Chatterjee M, Ebi KL, Estrada YO, Genova RC, Girma B, Kissel ES, Levy AN, MacCracken S, Mastrandrea PR, White LL (eds) Contribution of working group II to the fifth assessment report of the intergovernmental panel on climate change. Cambridge University Press, Cambridge, United Kingdom and New York, NY, USA, p. 688. https://www.ipcc.ch/site/assets/uploads/2018/02/WGIIAR5-PartB_FINAL.pdf. Accessed 23 Apr 2021

International Work Group for Indigenous Affairs (IWGIA) (2011) Indigenous peoples in Ethiopia. https://www.iwgia.org/en/ethiopia/714-indigenous-peoples-in-ethiopia.html. Accessed 27 July 2021

Iticha B, Husen A (2019) Adaptation to climate change using indigenous weather forecasting systems in Borana pastoralists of southern Ethiopia. Climate Dev 11(7):564–573

Latulippe N, Klenk N (2020) Making room and moving over: Knowledge co-production, Indigenous knowledge sovereignty and the politics of global environmental change decision-making. Curr Opin Environ Sustain 42:7–14

Lumonya JS (2020) A critique of competency based curriculum: towards integration of indigenous knowledge systems in Africa. Unpublished Master's thesis, University of Nairobi, Kenya

Magni G (2017) Indigenous knowledge and implications for the sustainable development agenda. Eur J Educ 52:437–447. https://doi.org/10.1111/ejed.12238

Matewos T (2020) The state of local adaptive capacity to climate change in drought prone districts of rural Sidama, southern Ethiopia. Clim Risk Manag 27:100209. https://doi.org/10.1016/j.crm.2019.100209

Maurial M (1999) Indigenous knowledge and schooling: A continuum between conflict and dialogue. In: Semali LM, Kincheloe JL (eds) What is Indigenous knowledge: voices from the academy. Falmer Press, New York, pp 59–77

Mavhura E, Manyena SB, Collins AE, Manatsa D (2013) Indigenous knowledge, coping strategies and resilience to floods in Muzarabani, Zimbabwe. Int J Disaster Risk Reduct 5:38–48

McGloin C, Marshall A, Adams M (2009) Leading the way: indigenous knowledge and collaboration at the Woolyungah indigenous centre. J Univ Teach Learn Pract 6:1–17

Meadowcroft J (2009) Climate change governance. In: Background paper to the 2010 world development report, Policy research working paper 4941, The World Bank. https://openknowledge.worldbank.org/bitstream/handle/10986/4135/WPS4941.pdf?sequence=1. Accessed 23 Apr 2021

Muhati GL, Olago D, Olaka L (2018) Land use and land cover changes in a sub-humid Montane forest in an arid setting: A case study of the Marsabit forest reserve in northern Kenya. Global Ecol Conser 16:1–17

Munyaradzi M (2015) Indigenous knowledge and public education in Sub-Saharan Africa. Afr Spectr 50(2):57–71

Musembi DK, Cheruiyot HK (2016) A case for validation of indigenous knowledge in forecasting rainfall among the Kamba community of Makueni county, lower Eastern Kenya. J Meteorol Related Sci 9(2):26–34. https://doi.org/10.20987/jmrs.4.08.2016

Naeku MJ (2020) Climate change governance: An analysis of the climate change legal regime in Kenya. Environ Law Rev 22(3):170–183

Ndalilo L, Wekesa C, Mbuvi MTE (2020) Indigenous and local knowledge practices and innovations for enhancing food security under climate change: examples from mijikenda communities in coastal Kenya. In: Gasparatos A et al (eds) Sustainability challenges in sub-saharan Africa II. Science for Sustainable Societies. Springer, Singapore. https://doi.org/10.1007/978-981-15-5358-5_3

NEPAD (2013) Agriculture in Africa: transformation and outlook. https://www.un.org/en/africa/osaa/pdf/pubs/2013africanagricultures.pdf. Accessed 19 Apr 2021

Nicholson SE (2017) Climate and climatic variability of rainfall over Eastern Africa. Rev Geophys. https://doi.org/10.1002/2016RG000544

Nkuba RM, Chanda R, Mmopelwa G (2020) Indigenous knowledge systems and indicators of rain: evidence from Rwenzori Region, Western Uganda. Weather Clim Soc 12:213–234

Nyadzi E, Werners SE, Biesbroek R, Ludwig F (2020) Techniques and skills of indigenous weather and seasonal climate forecast in Northern Ghana. Climate Dev. https://doi.org/10.1080/17565529.2020.1831429

Nyong A, Adesina F, Osman EB (2007) The value of indigenous knowledge in climate change mitigation and adaptation strategies in the African Sahel. Mitig Adapt Strat Glob Change 12:787–797

Obeng-Odoom F, Van Gyampo RE (2017) Land grabbing, land rights and the role of the courts. Geogr Res Forum 37:127–143

Ogega OB (2017) Use of scientific and indigenous knowledge in adapting to climate change and variability at the Kenyan coast. Unpublished Masters thesis, University of Nairobi, Kenya

Ojwang L, Rosendo S, Celliers L, Obura D, Muiti A, Kamula J, Mwangi M (2017) Assessment of coastal governance for climate change adaptation in Kenya. Earth's Future 5(11):1119–1132

Orlove B, Roncoli C, Kabugo M, Majugu A (2010) Indigenous climate knowledge in southern Uganda: the multiple components of a dynamic regional system. Clim Change 100:243–265

Ouma J, Olang L, Ouma G, Oludhe C, Ogallo L, Artan G (2018) Magnitudes of climate variability and changes over the arid and semi-arid lands of Kenya between 1961 and 2013 Period 7:27–39

Renner J (2020) New power structures and shifted governance agendas disrupting climate change adaptation developments in Kenya and Uganda. Sustainability 12:2799

Ruwhiu D, Carter L (2016) Negotiating "meaningful participation" for indigenous peoples in the context of mining. Corp Gov 16(4):641–654

Scobie M (2016) Regional climate governance. In Farazmand A (ed) Global encyclopedia of public administration, public policy, and governance. https://doi.org/10.1007/978-3-319-31816-5_2845-1

Shisanya CA (2017) Role of traditional ethnobotanical knowledge and indigenous institutions in sustainable land management in Western Highlands of Kenya. In: Indigenous people, 159p. InTech. https://doi.org/10.5772/intechopen.69890

Simatele D, Simatele M (2015) Climate variability and urban food security in sub-Saharan Africa: lessons from Zambia using an asset-based adaptation framework. S Afr Geogr J 97:243–263

Smucker TA, Oulu M, Nijbroek R (2020) Foundations for convergence: Sub-national collaboration at the nexus of disaster risk reduction, climate change adaptation, and land restoration under multilevel governance in Kenya. Int J Disaster Risk Reduct:101834

Tarfasa S, Balana BB, Tefera T, Woldeamanuel T, Moges A, Dinato M, Black H (2018) Modelling smallholder farmers' preferences for soil management measures: a case study from South Ethiopia. Ecol Econ 145:410–419

Tilahun M, Taye G, Tesfaye G (2019) Indigenous knowledge and scientific weather forecast: implication for climate change adaptation strategies in Dessa'a Tigray, Northern Ethiopia. Indilinga–Afr J Indigenous Knowl Syst 18(2):145–159

Tober M (2011) PubMed, ScienceDirect, Scopus or Google Scholar—Which is the best search engine for an effective literature research in laser medicine? Med Laser Appl 26(3):139–144

Warren DM (1991) Using indigenous knowledge in agricultural development. World Bank Discussion Paper 127, World Bank, Washington, DC

Wetende E, Olago D, Ogara W (2018) Perceptions of climate change variability and adaptation strategies on smallholder dairy farming systems: insights from Siaya Sub-County of Western Kenya. Environ Develop 27(1):14–25

Ziervogel G, Opere A (2010) Integrating meteorological and indigenous knowledge-based seasonal climate forecasts in the agricultural sector. https://idl-bnc-idrc.dspacedirect.org/bitstream/handle/10625/46185/132676.pdf?sequence=1&isAllowed=y. Accessed 10 Sept 2016

Eromose E. Ebhuoma (Ph.D.) is a postdoctoral research fellow at the Department of Environmental Sciences, College of Agriculture and Environmental Sciences, University of South Africa (UNISA). His research expertise spans climate change vulnerability and adaptation, climate services, climate change policy and governance, local and indigenous knowledge systems, political ecology, rural livelihoods and environmental sustainability in sub-Saharan Africa. He is a steering committee member of the South African Adaptation Network.

16 The Role of Indigenous Knowledge Systems in Climate Change Governance and Action: Conclusions, Recommendations and Future Research Directions

Llewellyn Leonard and Eromose E. Ebhuoma

16.1 Snapshot

The role of indigenous knowledge systems (IKS) in facilitating climate change adaptation and its potential for the co-production of knowledge and enhancing climate governance in sub-Saharan Africa (SSA) has received significant scholarly attention (Nkomwa et al. 2014; Chanza 2015; Ebhuoma and Simatele 2017; Nyadzi et al. 2021). Paradoxically, IKS has received scant attention for its role in influencing climate change governance on the African continent, although articles by Chanza and de Wit (2016) and Adeyeye et al. (2019) are notable exceptions. The minimal inclusion of IKS in climate governance in SSA can have significant ramifications for the achievement of the thirteenth Sustainable Development Goal (SDG 13)—climate action—owing to the viewpoints of notable international organisations. As highlighted in the introductory chapter, some internationally reputable organisations like the United Nations Educational, Scientific and Cultural Organisation (UNESCO), the Intergovernmental Panel on Climate Change (IPCC) and United Nations Framework Convention on Climate Change (UNFCCC), recognise the involvement of indigenous peoples as fundamental in achieving the ambitious targets of the SDGs (UNESCO 2017). Thus, to address the gap in literature on indigenous knowledge, a dialogue that illuminates the extent to which IKS features in climate change governance and policy schemes in SSA, including the factors that undermine such processes, have been provided in this volume. To make this volume more holistic and comprehensive, a platform was first created to showcase the various ways in which IKS upscale climate change in various countries in SSA and how this unique knowledge has been integrated with scientific knowledge. This was necessary to emphasise the overwhelming importance of critically exploring the need for IKS to feature in climate governance and policy schemes, including the factors that undermine such processes.

The various case studies on the role of IKS in facilitating adaptation to climate change—theme one—analysed how IKS facilitate the conservation of natural resources and builds the resilience of agrarian and pastoral communities in the face of climate variability and change, thereby enabling indigenous peoples to obtain their livelihood. For example, Moyo (Chap. 2) demonstrated that tenure systems restricted the movement of minority Mbororo pastoralists in

L. Leonard (✉) · E. E. Ebhuoma
Department of Environmental Sciences, School of Ecological and Human Sustainability, College of Agriculture and Environmental Sciences, University of South Africa (UNISA), Johannesburg, South Africa
e-mail: llewel@unisa.ac.za

E. E. Ebhuoma
e-mail: ebhuoee@unisa.ac.za

E. E. Ebhuoma and L. Leonard (eds.), *Indigenous Knowledge and Climate Governance*, Sustainable Development Goals Series, https://doi.org/10.1007/978-3-030-99411-2_16

the Ndop plains of Cameroon and increased conflict over rangeland management due to pasture shortages triggered by climate change. However, the skilful utilisation of indigenous indicators has proven instrumental in reducing pressure on natural pasture, providing enough time for pasture regeneration, and reducing degradation through the effective practice of transhumant pastoralism. According to Gentle and Thwaites (2016, p. 173), 'transhumant pastoralism refers to the seasonal migration of livestock and humans between various agro-ecological zones'. It is an old practice in numerous mountain regions. Also, in Moyo's chapter, IKS facilitated conflict resolution through the intervention of MBOSCUDA, a traditional institution created in 1992, in Yaoundé, from the fusion of the Fulani Social Development (FUSUDA), Fulani Foundation of Cameroon (FUFOUCAM) (FUFOUCAM) and Mbororo Development Association (MDA). In a similar vein, Moyo-Nyoni (Chap. 3) aptly illustrated how traditional leaders in Lupane district, in the Matabeleland North Province of Zimbabwe, guided by customary laws, resolve conflicts among individuals that emanate from their quest to access scarce resources (e.g., nutritious pastures) triggered by climate change.

A noteworthy feature is the bottom-up approach adopted to resolve conflicts. For example, whenever a conflict arises, it is first handled at an interpersonal/family level. If the conflict remains unresolved, the matter is reported to the village head's court. If still unresolved, the matter is taken up to the Kraal head's court. In the event the aggrieved parties cannot come to a compromise, the issue is taken to the headman's court and, if still unresolved, it is escalated to the traditional court, guided by the traditional leaders and those designated to help them, according to Zimbabwe's Traditional Leader's Act (TLA). In Zimbabwe, studies show that the TLA has been instrumental in empowering traditional leaders with the authority to ensure effective governance at the local level (Chigwata 2016). Also, Moyo-Nyoni's chapter highlights that indigenes of Lupane district practice forest conservation by prohibiting the felling of certain species of trees for firewood to preserve the trees for the future, although some contravene this rule. In addition, traditional healers who make medicine from trees are trained not to destroy the trees when sourcing for materials such as leaves, bulks or parts of tree roots, hinged on the belief that their ancestors and other spirits reside in the forests. Thus, it can be asserted that indigenous peoples are not only interested in obtaining a livelihood from natural resources but attach enormous importance to natural resources conversation to ensure the sustainability of livelihood activities (see also Mafongoya and Ajayi 2017; Magni 2017). This viewpoint aptly crystallised in the contribution by Lawal, Ajayi and Akinyemi (Chap. 4).

The chapter revealed that the close relationship the indigenous Yorùbá people of Nigeria have with the land has sustained forest conservation for centuries through effective management practices deeply rooted in cultural and spiritual beliefs and value systems. Specifically, taboos prohibiting people from harvesting fuelwoods from forests believed to house spiritual deities have checkmated unsustainable harvesting practices (see also Daniel et al. 2015; Daye and Healey 2015). By preserving forests, it helps to mitigate climate change through carbon sequestration (Kihonge 2017; Maru et al. 2020). It is partly the cultural values and belief systems held by indigenous peoples that provide an avenue for indigenous women in Central Africa to employ social capital and practice community-based adaptation to adapt to climate change; a viewpoint captured by Henry, Ebhuoma and Leonard (Chap. 5).

Thus, the aforementioned chapters provided positive perspectives on the potential for IKS to contribute significantly towards actualising both SDG 13—climate action—and Agenda 2063—the Africa we want—through climate change adaptation, natural resources conservation, and its ability to resolve conflicts. With the enormous role IKS play in facilitating climate change adaptation and natural resource conservation, it is only befitting to analyse climate change experts' knowledge of IKS. Such findings are crucial as they would crystallise how easy it would be to bring the 'best of both worlds'—scientific and indigenous knowledge—together to facilitate the actualisation of SDG 13 and Agenda 2063. This is precisely what Kavu, Kajongwe and Ebhuoma (Chap. 6) set out to achieve. The authors analysed the knowledge of climate change experts in Zambia, Malawi and Mozambique regarding the importance of IKS in facilitating climate change adaptation among rural households. Findings suggest that climate change experts are acutely aware of the various techniques employed by indigenous people to adapt to climate change. This includes using indigenous plants to act as a buffer against floods, constructing climate-resilient buildings, and forecasting droughts using indigenous indicators, among others. Building on climate change experts' IKS knowledge-base could, perhaps, be a useful entry point for incorporating IKS with scientific knowledge in the pursuance of SDG 13, which in turn, would reinforce indigenous peoples as holders of valuable knowledge (Mercer et al. 2010; Ebhuoma 2018).

Findings similar to the chapters mentioned above showcased the importance of IKS for their capacity to offer credible solutions to addressing climate change. These stimulated debates around issues of power given to researchers to 'invade' indigenous, often rural, communities; collect and record indigenous knowledge; and represent their views (e.g., see Chambers et al. 2017; Kwame 2017). The positionality of both the researcher and researched (indigenous people in this case) affects the intellectual access to information and shapes how the information is presented (Qin 2016). Therefore, Magaya and Fitchett (Chap. 7) documented a researcher's experiences as an indigenous insider and outsider in conducting research in two provinces in South Africa, exploring how IKS helps build farmers' resilience to climatic risks. A crucial finding of the chapter is that the researcher's deep-rooted understanding of the social norms and cultural nuances, including being able to speak the local dialect of the communities, aided the acquisition of rich data and assisted in affirming the validity of the results. Also, it shaped the ontology of the researcher in a way that revered the knowledge and expertise of indigenous people. This finding is crucial as it lays the foundation for ways in which people whose ontologies are shaped by western science should engage and integrate IKS to scientific knowledge when attempting to scale up climate resilience without being disrespectful to the knowledge of indigenous people, ensuring they do not feel disenfranchised in the process.

In the second theme—IKS and scientific knowledge collaboration to scale up climate resilience—Nyadzi, Nyamekye and Ludwig (Chap. 8) showed that the co-production and knowledge integration of both indigenous ways of weather forecasting and meteorological forecasts improved the design, information generation and uptake of climate information services (CIS) in the Kumbungu district of Northern Ghana. The ability of indigenous knowledge to contribute meaningfully to improving weather forecast in CIS is a testament to its unique epistemology, which cannot be downplayed as development and climate practitioners strive to actualise SDG 13 in the decade of action (2020–2030) (Sarrasanti et al. 2020). The recognition of unique epistemology that informs indigenous ways of knowing may partly have been the catalyst for Rabumbulu and Komane (Chap. 9) highlighting avenues to reconcile indigenous and scientific Lesotho by showcasing how residents of the Maseru district draw on their indigenous knowledge to address droughts and land degradation. An important finding was the adoption of keyhole gardening, implemented by farmers to prevent soil loss and conserve water. Such gardens are typically enclosed by a stone wall and consist of a raised, circular bed of soil. Since these systems are currently being built on a very small scale, scientific experience in the form of modern, inexpensive engineering techniques could perhaps be applied to develop them on a larger scale.

Despite Rabumbulu and Komane's intriguing finding, Chap. 10, by Llewellyn Leonard, is a stark reminder of the ills that could truncate indigenous epistemology, and consequently, its integration with scientific knowledge to facilitate the actualisation of SDG 13. While various explanations have been captured in the literature why indigenous people may feel aggrieved to work with people whose philosophy is underpinned by western scientific knowledge, Leonard meticulously unravelled how the preference of genetically modified seeds may be undermining indigenous expertise in SSA. These sorts of preferences for scientific seeds and anything hinged on western ideologies may be a major contributing factor undermining IKS from featuring effectively in climate change governance, or it may have had a ripple effect in preventing IKS from featuring effectively in climate change governance, even in Africa. This last chapter sets the scene for the final theme—the integration of IKS in climate change governance and planning—in this volume.

The last theme sets off by considering the prospects of strengthening adaptation governance through IKS mentioned by Chanza, Musakwa and de Wit (Chap. 11). The chapter argues that global climate governance does not effectively articulate how indigenous people are affected by climate change in SSA. Also, the chapter highlights that the prospects for IKS integration in adaptation regimes go beyond emancipatory and empowering agendas to guarantee the effectiveness of adaptation projects in indigenous communities. In a somewhat related vein, in Chap. 12, Tariro Kamiti mentions that although Zimbabwe has made provision for IKS inclusion in its climate change policy, he calls for further refinement in specifying tangible actions where indigenous people can play a more effective role in tackling climate change in schemes; such as soliciting nature-based solutions to address climate change. In Chap. 13, Chah, Esdras, Ebhuoma, Aungwa and Ewane partly underlined the factors that have facilitated the incorporation of IKS into local governance to strengthen climate change adaptation in SSA. The chapter shows that creating viable platforms for dialogue between indigenous stakeholders at different levels and providing institutional support for indigenous people to implement local plans are enablers to IKS and local governance collaboration at local levels. In Chap. 14, Donkor and Mearns explored tangible avenues to harness IKS for enhanced climate change adaptation and governance in SSA. A crucial finding of the chapter is that it argues rural women are a key instrument to scale up climate change adaptation due to their intimate knowledge of the environment. Rural women are critical resource stakeholders in IKS. Ultimately, the final chapter by Eromose Ebhuoma highlights the factors that undermine IKS from featuring effectively in climate change governance schemes in Ethiopia and Kenya. Key issues excavated by the chapter include adaptation constraints, knowledge production, political and financial constraints, albeit nuanced differences emerged from the analyses of the two countries.

16.2 The Way Forward for Indigenous Knowledge System, Climate Adaptation, Resilience and Governance

The importance of IKS on the African continent for climate change adaptation and resilience can robustly support the African Agenda 2063: The Africa we want (see African Union Commission 2015). How IKS can be integrated into governance must be a cornerstone to achieving the Agenda 2063 strategic framework that aims to deliver inclusive and sustainable development. It is a tangible demonstration of the Pan-African drive for unity, progress and collective prosperity pursued under Pan-Africanism. All chapters in this volume, to varying degrees, highlighted the importance of IKS to enhance climate adaptation and resilience, and promote good governance, aligning with the priority areas of the seven aspirations of Agenda 2063 (see Table 16.1). It will be important for future scholars to robustly explore the linkages between Agenda 2063, the role of IKS, and the SDGs, specifically SDG 13 calling for climate action.

As highlighted in Chap. 1, SDG 13 is a decisive call for the world to take crucial steps to tackle climate change and its impact. It requires integrating climate change into national policies, improving climate education, and building institutional capacity to scale up climate change mitigation and adaptation strategies. This

Table 16.1 Goals & priority areas of agenda 2063

Aspiration	Goals	Priority areas
(1) A prosperous Africa, based on inclusive growth and sustainable development	(1) A high standard of living, quality of life and well-being for all citizens	• Incomes, jobs and decent work • Poverty, inequality and hunger • Social security and protection, including persons with disabilities • Modern and liveable habitats and basic quality services
	(2) Well-educated citizens and skills revolution underpinned by science, technology and innovation	• Education and Science Technology & Innovation (STI) skills-driven revolution
	(3) Healthy and well-nourished citizens	• Health and nutrition
	(4) Transformed economies	• Sustainable and inclusive economic growth • STI-driven manufacturing/industrialisation and value addition • Economic diversification and resilience • Hospitality/tourism
	(5) Modern agriculture for increased productivity and production	• Agricultural productivity and production
	(6) Blue/ocean economy for accelerated economic growth	• Marine resources and energy • Ports' operations and marine transport
	(7) Environmentally sustainable and climate-resilient economies and communities	• Sustainable natural resource management and biodiversity conservation • Sustainable consumption and production patterns • Water security • Climate resilience and natural disaster preparedness and prevention • Renewable energy
(2) An integrated continent politically united based on the ideals of Pan-Africanism and the vision of African Renaissance	(8) United Africa (Federal or Confederate)	• Framework and institutions for a united Africa
	(9) Continental financial and monetary institutions are established and functional	• Financial and monetary institutions
	(10) World-class infrastructure crisscrossing Africa	• Communications and infrastructure connectivity

(continued)

Table 16.1 (continued)

Aspiration	Goals	Priority areas
(3) An Africa of good governance, democracy, respect for human rights, justice, and the rule of law	(11) Democratic values, practices, universal principles of human rights, justice and the rule of law entrenched	• Democracy and good governance • Human rights, justice, and the rule of law
	(12) Capable institutions and transformative leadership in place	• Institutions and leadership • Participatory development and local governance
(4) A peaceful and secure Africa	(13) Peace, security and stability are preserved	• Maintenance and preservation of peace and security
	(14) A stable and peaceful Africa	• Institutional structure for AU instruments on peace and security
	(15) A fully functional and operational African Peace and Security Architecture (APSA)	• Fully operational and functional APSA pillars
(5) An Africa with a strong cultural identity, common heritage, values, and ethics	(16) African cultural Renaissance is pre-eminent	• Values and ideals of Pan-Africanism • Cultural values and African Renaissance • Cultural heritage, creative arts and businesses
(6) An Africa whose development is people-driven, relying on the potential offered by African people, especially its women and youth, and caring for children	(17) Full gender equality in all spheres of life	• Women and girls' empowerment • Violence and discrimination against women and girls
	(18) Engaged and empowered youth and children	• Youth empowerment and children
(7) An Africa that is a strong, united, resilient and influential global player and partner	(19) Africa as a major partner in global affairs and peaceful co-existence	• Africa's place in global affairs • Partnership
	(20) Africa takes full responsibility for financing her development	• African capital market • Fiscal system and public sector revenues • Development assistance

Source African Union Commission (2015)

includes a need to strengthen resilience and adaptive capacity to climate-related hazards. Africa's Agenda 2063, specifically goal 7, highlights the need for environmentally sustainable and climate-resilient economies and communities, with some priority areas including sustainable consumption and production patterns, climate resilience, and natural disaster preparedness and prevention. Thus, the African Agenda recognises that climate change is a major challenge for the continent's development, negatively affecting African communities' ability to achieve the SDGs and Agenda 2063 aspirations. IKS can be central to realising Agenda 2063 for an Africa that prospers based on inclusive growth and development, founded on the ideals of Pan-Africanism, tied to a strong cultural identity and common heritage. An Africa whose development is people-driven, relying on the potential offered by African people is envisaged.

This volume was important as it emphasised the urgent need to integrate IKS in climate change governance and planning, which will be essential for realising the African Agenda aspiration linked to an Africa of good governance, democracy, respect for human rights, justice, and the rule of law. As many chapters highlighted, local communities and their indigenous systems have been excluded from decision-making and development processes by certain groups deemed more powerful. Thus, there is a need for an environmental justice approach that emphasises good procedural governance inclusive of communities and their IKS in climate change policies, adaptation and resilience development planning, and any decision-making to enable more resilient communities. Such good governance approaches must emphasise State accountability, transparency, and the development of strategies with communities, not externally imposed interventions that ignore IKS and indigenous ways of life. As part of SDG 13 for climate action, this will also require African citizens to hold government and local leaders accountable for lack of action and appropriate interventions to incorporate and recognise IKS to secure climate adaptation and resilience in communities. Essentially, all stakeholders must engage in dialogue to define relationships specifying their respective rights and responsibilities. Good governance approaches can ensure environmentally sustainable and climate-resilient economies and communities, and would assist in enshrining the Agenda 2063 goals.

16.3 Further Recommendations and Areas for Future Research

In light of the above, a key recommendation is the advocacy for place-based policies, with traditional institutions being given statutory powers to lead key discussions on issues pertaining to climate change adaptation. The engagement of local actors is needed to reflect initiatives that would highlight their lived realities in climate governance schemes. Place-based policies are the best ways to minimise the exclusion of IKS in climate governance. As Ebhuoma (Chap. 13) puts forward, the question that should always be at the back of climate professionals' minds when engaging with indigenous peoples is, "what sort of assistance would be meaningful to them?" Also, based on development practitioners' tendency to draw on issues to which people are emotionally attached, another key recommendation is to highlight how climate change could affect those things people attach importance to in order to trigger the adoption of pro-environmental behaviour. This recommendation is hinged on finding from both Chaps. 4 and 5. The chapters reveal that since community members are acutely aware of the possible consequences that could adversely affect them due to harm inflicted on their forests, they jealously guard against unsustainable harvesting.

It may be worthwhile for future research to comprehensively explore whether physical scientists will be willing to learn from indigenous peoples or see their science as based on a trial-and-error method. Western science 'supporting' indigenous practices will be important to enable better adaptation and resilience to combat climate change impacts. Another recommendation is to determine precisely the amount of carbon sequestrated in indigenous communities through forest conservation. While Chaps. 4 and 5 provide evidence of forest conversation coupled with the numerous tree-planting campaigns globally (Daniel et al. 2015; Daye and Healey 2015), not determining the amount of carbon sequestered could undermine the process for individuals, especially those in urban areas, to see the need to adhere strictly to nature conservation. The window of opportunity to inject this vital piece of knowledge—IKS facilitating climate change adaptation and natural resource conservation—in climate governance and policy schemes is agonisingly small.

While various African countries are starting to draw on local and IKS, examples of how this plays out on the ground, including the hiccups that undermine the success of such processes, have not been adequately documented in Africa. As such, there is a need for researchers to explore how power relations, specifically between governments, the private sector and local communities unfold (as highlighted in Chap. 10) to enable good governance approaches, promote food security, and enable appropriate climate change interventions in communities. This also includes how indigenous knowledge is recognised and incorporated into policy, development, and planning processes. It is important to examine how social dimensions such as gender, spiritual beliefs and ethnicity may influence and determine how IKS are incorporated into climate change development strategies. Such analyses must be linked to an Africa whose development is all-inclusive and people-driven, relying on the potential offered by all African people. As the 2015 report by the Department of International Relations and Cooperation, South Africa, entitled 'African Union Agenda 2063: Voices of the African People: A Call to Action' notes, African development should be strengthened by tapping into indigenous and inclusive knowledge systems.

Finally, we advocate that there must be research collaboration among scholars from different sub-regions on the continent to engage in comparisons of IKS, ways of combating climate change, and securing livelihoods for a more informed Pan-African perspective. Africans need to learn from one another, as they largely share similar cultures and backgrounds. This is important if Pan-Africanism is to be realised towards a future where the African people have common interests and should be unified in solving the continent's challenges. Owing to the short timeframe countries have to achieve the SDGs, we are arguably running a race against time. The time to act is now.

References

Adeyeye Y, Hagerman S, Pelai R (2019) Seeking procedural equity in global environmental governance: Indigenous participation and knowledge politics in forest and landscape restoration debates at the 2016 world conservation congress. Forest Policy Econ 102006:1–14

African Union Commission (2015) Agenda 2063: the Africa we want. https://au.int/en/agenda2063/overview . Accessed 8 Oct 2021

Chambers LE, Plotz RD, Dossis T, Hiriasia DH, Malsale P, Martin DJ, Mitiepo R, Tahera K, Tofaeono TI (2017) A database for traditional knowledge of weather and climate in the Pacific. Meteorol Appl 24 (3):491–502. doi:https://doi.org/10.1002/met.1648

Chanza N (2015) Indigenous-based adaptation: an imperative for sustainable climate change strategies for Africa. In: Mawere M, Awuah-Nyamekye S (eds) Harnessing cultural capital for sustainability: a pan Africanist perspective. Langaa Publishing House, Bamenda, pp 85–134

Chanza N, De Wit A (2016) Enhancing climate governance through indigenous knowledge: case in sustainability science. South African J Sci 112(3/4), Art. #2014–0286. doi:https://doi.org/10.17159/sajs.2016/20140286

Chigwata T (2016) The role of traditional leaders in Zimbabwe: are they still relevant? Law, Democ Devel 20:69–90

Daniel SK, Jacob DE, Udeagha AU (2015) Tree species composition in selected sacred forests in Nigeria. Int J Molec Ecol Conserv 5(7):1–10. doi:https://doi.org/10.5376/ijmec.2015.05.0007

Daye DD, Healey JR (2015) Impacts of land-use change on sacred forests at the landscape scale. Global Ecol Conserv 3:349–358

Department of International Relations and Cooperation (2015) African Union Agenda 2063: Voices of the African People: A Call to Action. http://www.dirco.gov.za/department/agenda2063/au_agenda_2063_report.pdf

Ebhuoma E, Simatele D (2017) Defying the odds: climate variability, asset adaptation and food security nexus in the delta state of Nigeria. Int J Disaster Risk Red 21:231–242

Ebhuoma E (2018) A systems thinking approach to underline ethical principle that drives indigenous people to pursue environmental sustainability when adapting to climate change. Proceedings of the international conference on sustainable development, Columbia University, New York. http://ic-sd.org/wp-content/uploads/sites/4/2018/10/Eromose-E-Ebhuoma.pdf. Accessed 25 July 2019

Gentle P, Thwaites R (2016) Transhumant pastoralism in the context of socioeconomic and climate change in the Mountains of Nepal. Mt Res Dev 36(2):173–182

Nkomwa EC, Joshua MK, Ngongondo C, Monjerezi M, Chipungu F (2014) Assessing indigenous knowledge systems and climate change adaptation strategies in agriculture: A case study of Chagaka Village, Chikhwawa, Southern Malawi. Phys Chem Earth 67–69:164–172

Kihonge EW (2017) Assessing contribution of sacred natural sites to climate change effects on dryland ecosystem. Case study: the Gabbra community, Marsabit County. M.Sc. thesis, University of Nairobi. http://erepository.uonbi.ac.ke/bitstream/handle/11295/101333/Kihonge_Assessing%20Contribution%20of%20Sacred%20Natural%20Sites%20to%20Climate%20Change%20Effects%20on%20Dryland%20Ecosystemcase%20Study-%20the%20Gabbra%20Community%2C%20Marsabit%20County.pdf?sequence=1&isAllowed=y

Kwame A (2017) Reflexivity and the insider/outsider discourse in indigenous research: my personal experiences. AlterNative Int J Ind Peoples 13(4):218–225. doi:https://doi.org/10.1177%2F1177180117729851

Mafongoya PL, Ajayi OC (eds) (2017) Indigenous knowledge systems and climate change management in Africa, CTA, Wageningen, The Netherlands, p 316

Magni G (2017) Indigenous knowledge and implications for the sustainable development agenda. Eur J Educ 52:437–447

Maru Y, Gebrekirstos A, Haile G (2020) Indigenous ways of environmental protection in Gedeo community, Southern Ethiopia: a socio-ecological perspective. Cogent Food Agricul 6:1. doi:https://doi.org/10.1080/23311932.2020.1766732

Mercer J, Kelman I, Taranis L, Suchet-Pearson S (2010) Framework for integrating indigenous and scientific knowledge for disaster risk reduction. Disasters 34 (1):214–239

Nyadzi E, Ajayi OC, Ludwig F (2021) Indigenous knowledge and climate change adaptation in Africa: a systematic review. CAB Rev Perspect Agricul Veter Sci Nutr Nat Res (16):029. https://www.cabi.org/cabreviews/review/20210181994

Qin D (2016) Positionality. In Naples N, Hoogland RC, Wickramasinghe M, Wong WC (eds) The Wiley Blackwell Encyclopedia of gender and sexuality studies. doi:https://doi.org/10.1002/9781118663219.wbegss619

Sarrasanti N, Donkor FK, Santos C, Tsagkari M, Wannous C (2020) It's about time we care about an equitable world: Women's unpaid care work and COVID-19. IEEE Eng Manage Rev. doi:https://doi.org/10.1109/EMR.2020.3031313

United Nations Educational, Scientific and Cultural Organisation (UNESCO) (2017) Culture: at the heart of the SDGs. https://en.unesco.org/courier/april-june-2017/culture-heart-sdgs. Accessed 21 Aug 2021

Llewellyn Leonard (Ph.D.) is Professor at the Department of Environmental Science, School of Ecological and Human Sustainability, University of South Africa. Before joining academia, he worked for a human rights environmental organisation working to support vulnerable communities exposed to environmental risks to shape policy and has worked to link communities in Africa, Asia, Latin America and Europe to engage in solidarity and networking. Previously he served as Vice Dean: Research at the University of Johannesburg, College of Business and Economics. His research interest include environmental justice, governance, democracy and human rights, civil-society-state-industry relations, urban risks, sustainable development; risk society and political economy/ecology to name a few. He sits on the editorial board of the Southern African Geographic Journal.

Eromose E. Ebhuoma (Ph.D.) is a postdoctoral research fellow at the Department of Environmental Sciences, College of Agriculture and Environmental Sciences, University of South Africa (UNISA). His research expertise spans climate change vulnerability and adaptation, climate services, climate change policy and governance, local and indigenous knowledge systems, political ecology, rural livelihoods and environmental sustainability in sub-Saharan Africa. He is a steering committee member of the South African Adaptation Network.

GPSR Compliance

The European Union's (EU) General Product Safety Regulation (GPSR) is a set of rules that requires consumer products to be safe and our obligations to ensure this.

If you have any concerns about our products, you can contact us on ProductSafety@springernature.com

In case Publisher is established outside the EU, the EU authorized representative is:

Springer Nature Customer Service Center GmbH
Europaplatz 3
69115 Heidelberg, Germany

Batch number: 10166133

Printed by Printforce, the Netherlands